W9-CFQ-447

Wireless Communication Technology

Wireless Communication Technology

Roy Blake

Niagara College of Applied Arts and Technology

Delmar
Thomson Learning™

Africa • Australia • Canada • Denmark • Japan • Mexico
New Zealand • Philippines • Puerto Rico • Singapore
Spain • United Kingdom • United States

Delmar Staff:

Business Unit Director: Alar Elken
Executive Editor: Sandy Clark
Acquisitions Editor: Gregory L. Clayton
Developmental Editor: Michelle Ruelos Cannistraci
Editorial Assistant: Jennifer Thompson
Executive Marketing Manager: Maura Theriault
Channel Manager: Mona Caron

Marketing Coordinator: Paula Collins
Executive Production Manager: Mary Ellen Black
Production Manager: Larry Main
Senior Project Editor: Christopher Chien
Art and Design Coordinator: David Arsenault
Technology Project Manger: Tom Smith

Contents

3   Digital Communication   83

7 Radio Propagation   219

8 Antennas   257

9 Transmitter and Receiver Circuitry   317

10   Cellular Radio   365

11 Personal Communication Systems   417

Dedication

This book is dedicated to my wife Penny and children Shira and Adam, who have to put up with a lot, especially me monopolizing the family computer.

Communication by radio is one of the oldest uses of electronics. It is also one of the newest, with new developments occurring almost daily. The past several years have seen the introduction of cellular telephony, satellite telephony, and mobile data transmission, to name only a few new systems. The new systems are based on the same principles that govern more traditional forms of communication like radio and television broadcasting and point-to-point microwave links, but the emphasis is different, and so are many of the details. No book can be completely up-to-date in this field, because it changes so quickly, but this one does try. It differs from more traditional texts on radio communication, including my own, in that its emphasis is on modern, mobile communication systems using radio rather than on broadcasting, wireline, or fiber-optic networks or fixed-link services. Mobile and portable wireless communication systems constitute the fastest growing area in the whole communications field, so this approach is justified.

Intended Audience

It is expected that many readers will have little or no prior knowledge of communication systems, so the basics, common to all types of electronic communication, are covered. The reader who has already taken at least one course in communication systems will find these chapters useful for review and a good preparation for the more specific coverage of modern systems beginning in Chapter 10. For those with little or no communication background, the book can be used for a two-course sequence.

The present text is intended to be suitable for students in two-, three- and four-year electronics programs. A knowledge of electronics fundamentals (both analog and digital) is assumed, as is mathematics at the level of basic algebra and trigonometry. A basic knowledge of logarithms and decibels is required. Calculus is not needed.

Textbook Organization

Of course there are similarities between the techniques used in wireless communication and those used in other branches of communication, so it is necessary to look at many traditional topics in radio-frequency communication, such as signal types, modulation schemes, and the effects of noise. Chapter One includes basic material on noise and on signals in the frequency domain. Chapter Two deals with traditional analog modulation schemes. Here the emphasis is on FM, which is of much greater importance than traditional AM and its variants in modern wireless systems. Material on transmission lines, antennas, and radio propagation is found later in the book in Chapters Six and Seven. The emphasis is placed on techniques that are used in modern wireless systems.

A knowledge of basic digital communication theory is essential for the student of wireless communication: many wireless systems use digital modulation schemes and digitized voice, and almost all are capable of transmitting data, whether directly or with the use of a modem. Digital communication is covered in Chapters Three and Four. Since many wireless systems are connected to the wireline telephone system, a knowledge of basic telephony is essential to anyone involved in wireless communication. This subject is covered in Chapter Five. The preparatory work is completed in Chapters Eight, on antennas, and Nine, which deals with transmitter and receiver circuitry.

Starting with cellular telephony in Chapter Ten, specific systems are examined, with the basics kept in mind. The emphasis throughout is on systems rather than individual components and specific circuits, which become obsolete very quickly (the author's three-year-old cell phone is already an antique). Chapter Ten concerns analog and digital cellular telephone systems; Chapter 11 introduces personal communication systems and compares the several ways in which these services are being implemented. Chapter 12 discusses the use of satellites for wireless communication. The following chapter deals with paging systems and wireless data communication. Finally, the book concludes with a look at the future of wireless communication.

Features The emphasis in this book is more on systems than on circuit details. It is assumed that the reader is already familiar with basic analog and digital circuitry, and some circuits specific to wireless applications are introduced as needed. The emphasis in the circuit coverage is on explaining how the circuits work; detailed circuit analysis is left to the many books on circuit theory. Similarly, the emphasis in the coverage of topics such as waveguides and antennas is on practical applications rather than on a detailed theoretical treatment.

Objectives are included at the beginning of each chapter, so that the student can easily refer to them while reading the material. Each chapter also contains a summary and a list of important equations, as well as numerous questions and problems. Answers to odd-numbered problems are provided at the back of the book, and an Instructor's Guide with complete solutions to all the problems is available (ISBN: 0-7668-1267-7). Any book on a subject as dynamic as this requires constant updating: for text updates and additional resources, go to the Delmar Electronics Technology website at

www.electronictech.com

Acknowledgments

The author and Delmar would like to thank the following reviewers who contributed to the development of this book:

Al Brown, ECPI, Hampton, VA

Robert Diffenderfer, DeVry Institute of Technology, Kansas City, MO

John Floyd, Eastern Shore Community College, Melfa, VA

Bruce Frederick, DeVry Institute of Technology, Phoenix, AZ

Carlos Sapijaszko, DeVry Institute of Technology, Alberta, Canada

Norman Zhou, University of Wisconsin-Stout, Menomonie, WI

About the Author

Roy Blake is a Professor of Electronic Technology at Niagara College of Applied Arts and Technology in Welland, Ontario, Canada. He teaches courses in electronic communication at the technician and technologist level. He has written two other textbooks, *Basic Electronic Communication* (West, 1993) and *Comprehensive Electronic Communication* (West, 1997, distributed by Delmar). He received his B.A. and B.A.Sc. (electrical engineering) degrees from the University of Toronto, and an M.Ed. from Brock University. He is a member of the Institute of Electronic Engineers (IEEE) and several of its constituent societies. He lives in Welland with his wife and two children, aged seven and thirteen.

The author can be contacted by e-mail at

rblake@niagarac.on.ca

or through his web page at

http://www.technology.niagarac.on.ca/people/rblake

Fundamental Concepts

1

Objectives

After studying this chapter, you should be able to:

- Explain the nature and importance of wireless communication.
- Outline the history of wireless communication.
- Explain the necessity for modulation in a radio communication system.
- Outline the roles of the transmitter, receiver, and channel in a radio communication system.
- Describe and explain the differences among simplex, half-duplex, and full-duplex communication systems.
- Describe the need for wireless networks and explain the use of repeaters.
- List and briefly describe the major types of modulation.
- State the relationship between bandwidth and information rate for any communication system.
- Calculate thermal noise power in a given bandwidth at a given temperature.
- Explain the concept of signal-to-noise ratio and its importance to communication systems.
- Describe the radio-frequency spectrum and convert between frequency and wavelength.

1.1   Introduction

This is a book on wireless communication. That usually means communication by radio, though ultrasound and infrared light are also used occasionally. The term "wireless" has come to mean nonbroadcast communication, usually between individuals who very often use portable or mobile equipment. The term is rather vague, of course, and there are certainly borderline applications that are called wireless without falling exactly into the above definition.

Wireless communication is the fastest-growing part of the very dynamic field of electronic communication. It is an area with many jobs that go unfilled due to a shortage of knowledgeable people. It is the author's hope that this book will help to remedy that situation.

1.2   Brief History of Wireless Telecommunication

Most of this book is concerned with the present state of wireless communication, with some speculation as to the future. However, in order to understand the present state of the art, a brief glimpse of the past will be useful. Present-day systems have evolved from their predecessors, some of which are still very much with us. Similarly, we can expect that future systems will be developed from current ones.

The Beginning   Wireless telecommunication began only a little later than the wired variety. Morse's telegraph (1837) and Bell's telephone (1876) were soon followed by Hertz's first experiments with radio (1887). Hertz's system was a laboratory curiosity, but Marconi communicated across the English Channel in 1899 and across the Atlantic Ocean in 1901. These successes led to the widespread use of radio for ship-to-ship and ship-to-shore communication using Morse code.

Early wireless systems used crude, though often quite powerful, spark-gap transmitters, and were suitable only for radiotelegraphy. The invention of the triode vacuum tube by De Forest in 1906 allowed for the modulation of a continuous-wave signal and made voice transmission practical. There is some dispute about exactly who did what first, but it appears likely that Reginald Fessenden made the first public broadcast of voice and music in late 1906. Commercial radio broadcasting in both the United States and Canada began in 1920.

Early radio transmitters were too cumbersome to be installed in vehicles. In fact, the first mobile radio systems, for police departments, were one-way, with only a receiver in the police car. The first such system to be considered practical was installed in Detroit in 1928. Two-way police radio, with the equipment occupying most of the car trunk, began in the mid-1930s. Amplitude modulation (AM) was used until the late 1930s, when frequency modulation (FM) began to displace it.

World War II provided a major incentive for the development of mobile and portable radio systems, including two-way systems known as "walkie-talkies" that could be carried in the field and might be considered the distant ancestors of today's cell phones. FM proved its advantages over AM in the war.

Postwar Expansion Soon after the end of World War II, two systems were developed that presaged modern wireless communication. AT&T introduced its **Improved Mobile Telephone Service (IMTS)** in 1946, featuring automatic connection of mobile subscribers to the **public switched telephone network (PSTN)**. This was an expensive service with limited capacity, but it did allow true mobile telephone service. This system is still in use in some remote areas, where, for instance, it allows access to the PSTN from summer cottages.

The next year, in 1947, the American government set up the **Citizens' Band (CB) radio** service. Initially it used frequencies near 460 MHz, but in that respect it was ahead of its time, since equipment for the UHF range was prohibitively expensive. Frequencies in the 27-MHz band were allocated in 1958, and CB radio immediately became very popular. The service was short-range, had no connection to the PSTN, and offered users no privacy, but it was (and still is) cheap and easy to set up. The popularity of CB radio has declined in recent years but it is still useful in applications where its short range and lack of connectivity to the rest of the world are not disadvantages. For example, it serves very well to disseminate information about traffic problems on the highway.

Meanwhile another rather humble-appearing appliance has become ubiquitous: the cordless phone. Usually intended for very short-range communication within a dwelling and its grounds, the system certainly lacks range and drama, but it does have connectivity with the PSTN. Most cordless phones use analog FM in the 46- and 49-MHz bands, but some of the latest models are digital and operate at either 900 MHz or 2.4 GHz. Cordless phones are cheap and simple to use, but their range is limited and, except for the digital models, they offer little privacy.

Pagers were introduced in 1962. The first models merely signaled the user to find a telephone and call a prearranged number. More recent models can deliver an alphanumeric message and even carry a reply. Though relatively limited in function, pagers remain very popular due to their low cost and small size.

The Cellular Revolution   The world's first cellular radio service was installed in Japan in 1979, followed in 1983 by North American services. Cellular systems are quite different from previous radiotelephone services such as IMTS in that, instead of using a single powerful transmitter located on a tall tower for wide coverage, the power of each transmitter is deliberately kept relatively small so that the coverage area, called a cell, will also be small. Many small cells are used so that frequencies can be reused at short distances. Of course, a portable or mobile telephone may move from one cell to another cell during the course of a conversation. In fact, this **handoff** may occur several times during a conversation. Practical cellular systems had to await the development of computers fast enough and cheap enough to keep track of all this activity. Theoretically at least, the number of users in a cellular system can be increased indefinitely, simply by making the cells smaller.

The first cellular systems used analog FM transmission, but digital modulation schemes, which provide greater privacy and can use bandwidth more efficiently, are used in all the new systems. These **personal communication systems (PCS)** usually operate in a higher frequency range (about 1.9 GHz compared with 800 MHz for North American cellular service).

Current cellular systems are optimized for voice but can also transmit data. In the near future, high-speed data transmission using PCS is expected to become a reality. At this point, however, the past merges into the future, and we'll resume the discussion later in this book.

1.3 Elements of a Wireless Communication System

The most basic possible wireless system consists of a transmitter, a receiver, and a channel, usually a radio link, as shown in Figure 1.1. Since radio cannot be used directly with low frequencies such as those in a human voice, it is necessary to superimpose the information content onto a higher frequency **carrier** signal at the transmitter, using a process called **modulation**. The use of modulation also allows more than one information signal to use

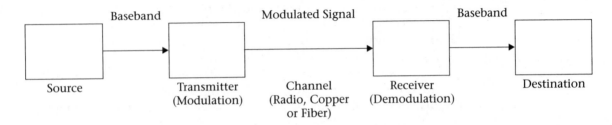

FIGURE 1.1   Elements of a communication system

the radio channel by simply using a different carrier frequency for each. The inverse process, **demodulation**, is performed at the receiver in order to recover the original information.

The information signal is also sometimes called the **intelligence**, the **modulating signal**, or the **baseband** signal. An ideal communication system would reproduce the information signal exactly at the receiver, except for the inevitable time delay as it travels between transmitter and receiver, and except, possibly, for a change in amplitude. Any other changes constitute distortion. Any real system will have some distortion, of course: part of the design process is to decide how much distortion, and of what types, is acceptable.

Simplex and Duplex Communication

Figure 1.1 represents a **simplex** communication system. The communication is one way only, from transmitter to receiver. Broadcasting systems are like this, except that there are many receivers for each transmitter.

Most of the systems we discuss in this book involve two-way communication. Sometimes communication can take place in both directions at once. This is called **full-duplex communication**. An ordinary telephone call is an example of full-duplex communication. It is quite possible (though perhaps not desirable) for both parties to talk at once, with each hearing the other. Figure 1.2 shows full-duplex communication. Note that it simply doubles the previous figure: we need two transmitters, two receivers, and, usually, two channels.

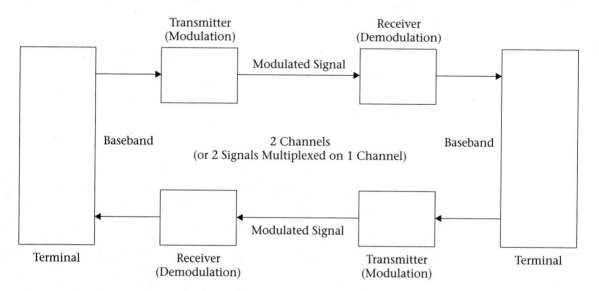

FIGURE 1.2 Full-duplex communication system

Some two-way communication systems do not require simultaneous communication in both directions. An example of this **half-duplex** type of communication is a conversation over citizens' band (CB) radio. The operator pushes a button to talk and releases it to listen. It is not possible to talk and listen at the same time, as the receiver is disabled while the transmitter is activated. Half-duplex systems save bandwidth by allowing the same channel to be used for communication in both directions. They can sometimes save money as well by allowing some circuit components in the transceiver to be used for both transmitting and receiving. They do sacrifice some of the naturalness of full-duplex communication, however. Figure 1.3 shows a half-duplex communication system.

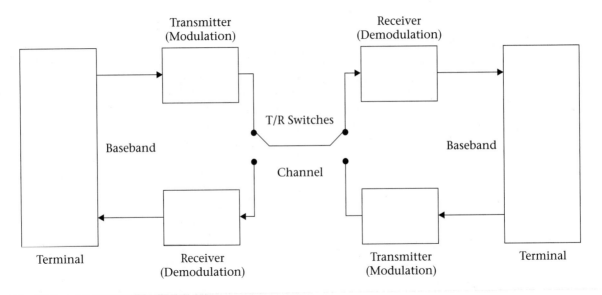

FIGURE 1.3   Half-duplex communication system

Wireless Networks   The full- and half-duplex communication systems shown so far involve communication between only two users. Again, CB radio is a good example of this. When there are more than two simultaneous users, or when the two users are too far from each other for direct communication, some kind of **network** is required. Networks can take many forms, and several will be examined in this book. Probably the most common basic structure in wireless communication is the classic **star network**, shown in Figure 1.4.

The central hub in a radio network is likely to be a **repeater**, which consists of a transmitter and receiver, with their associated antennas, located in

FIGURE 1.4
Star network

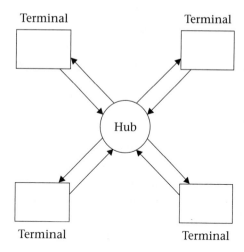

a good position from which to relay transmissions from and to mobile radio equipment. The repeater may also be connected to wired telephone or data networks. The cellular and PCS telephone systems that we look at later in the book have an elaborate network of repeater stations.

1.4   Signals and Noise

The communication systems described in this book differ in many ways, but they all have two things in common. In every case we have a signal, which is used to carry useful information; and in every case there is **noise**, which enters the system from a variety of sources and degrades the signal, reducing the quality of the communication. Keeping the ratio between signal and noise sufficiently high is the basis for a great deal of the work that goes into the design of a communication system. This **signal-to-noise ratio**, abbreviated S/N and almost always expressed in decibels, is an important specification of virtually all communication systems. Let us first consider signal and noise separately, and then take a preliminary look at S/N.

Modulated Signals   Given the necessity for modulating a higher-frequency signal with a lower-frequency baseband signal, it is useful to look at the equation for a sine-wave carrier and consider what aspects of the signal can be varied. A general equation for a sine wave is:

$$e(t) = E_c \sin(\omega_c t + \theta) \tag{1.1}$$

where

$e(t)$ = instantaneous voltage as a function of time

E_c = peak voltage of the carrier wave

ω_c = carrier frequency in radians per second

t = time in seconds

θ = phase angle in radians

It is common to use radians and radians per second, rather than degrees and hertz, in the equations dealing with modulation, because it makes the mathematics simpler. Of course, practical equipment uses hertz for frequency indications. The conversion is easy. Just remember from basic ac theory that

$$\omega = 2\pi f \qquad (1.2)$$

where

ω = frequency in radians per second

f = frequency in hertz

A look at Equation (1.1) shows us that there are only three parameters of a sine wave that can be varied: the amplitude E_c, the frequency ω, and the phase angle θ. It is also possible to change more than one of these parameters simultaneously; for example, in digital communication it is common to vary both the amplitude and the phase of the signal.

Once we decide to vary, or modulate, a sine wave, it becomes a complex waveform. This means that the signal will exist at more than one frequency; that is, it will occupy **bandwidth**. Bandwidth is a concept that will be explored in more detail later in this chapter and will recur often in this book.

Noise   It is not sufficient to transmit a signal from transmitter to receiver if the noise that accompanies it is strong enough to prevent it from being understood. All electronic systems are affected by noise, which has many sources. In most of the systems discussed in this book, the most important noise component is thermal noise, which is created by the random motion of molecules that occurs in all materials at any temperature above absolute zero (0 K or –273° C). We shall have a good deal to say about noise and the ratio between signal and noise power (S/N) in later chapters. For now let us just note that thermal noise power is proportional to the bandwidth over which a system operates. The equation is very simple:

$$P_N = kTB \qquad (1.3)$$

where

$$P_N = \text{noise power in watts}$$
$$k = \text{Boltzmann's constant, } 1.38 \times 10^{-23} \text{ joules/kelvin (J/K)}$$
$$T = \text{temperature in kelvins}$$
$$B = \text{noise power bandwidth in hertz}$$

Note the recurrence of the term *bandwidth*. Here it refers to the range of frequencies over which the noise is observed. If we had a system with infinite bandwidth, theoretically the noise power would be infinite. Of course, real systems never have infinite bandwidth.

A couple of other notes are in order. First, kelvins are equal to degrees Celsius in size; only the zero point on the scale is different. Therefore, converting between degrees Celsius and kelvins is easy:

$$T(K) = T(°C) + 273 \tag{1.4}$$

where

$$T(K) = \text{absolute temperature in kelvins}$$
$$T(°C) = \text{temperature in degrees Celsius}$$

Also, the official terminology is "degrees Celsius" or °C but just "kelvins" or K.

EXAMPLE 1.1 ▼

A resistor at a temperature of 25 °C is connected across the input of an amplifier with a bandwidth of 50 kHz. How much noise does the resistor supply to the input of the amplifier?

SOLUTION

First we have to convert the temperature to kelvins.
From Equation (1.4),

$$
\begin{aligned}
T(K) &= T(°C) + 273 \\
&= 25 + 273 \\
&= 298 \text{ K}
\end{aligned}
$$

Now substitute into Equation (1.3),

$$
\begin{aligned}
P_N &= kTB \\
&= 1.38 \times 10^{-23} \times 298 \times 50 \times 10^3 \\
&= 2.06 \times 10^{-16} \text{ W} \\
&= 0.206 \text{ fW}
\end{aligned}
$$

This is not a lot of power to be sure, but received signal levels in radio communication systems are also very small.

▲

Signal-to-Noise Ratio

Maintaining an adequate ratio of signal power to noise power is essential for any communication system, though the exact definition of "adequate" varies greatly. Obviously there are two basic ways to improve S/N: increase the signal power or reduce the noise power. Increasing signal power beyond a certain point can cause problems, particularly where portable, battery-powered devices are concerned. Reducing noise power requires limiting bandwidth and, if possible, reducing the noise temperature of a system. The system bandwidth must be large enough to accommodate the signal bandwidth, but should be no larger than that. Some modulation schemes are more efficient than others at transmitting information with a given power and bandwidth.

Noise Figure and Noise Temperature

The noise temperature of a complex system is not necessarily equal to the actual operating temperature, but may be higher or lower. The noise temperature for electronic systems is often found by way of the noise figure, so let us look briefly at that specification.

Noise figure describes the way in which a device adds noise to a signal and thereby degrades the signal-to-noise ratio. It is defined as follows:

$$NF = \frac{(S/N)_i}{(S/N)_o} \tag{1.5}$$

where

$(S/N)_i$ = signal-to-noise ratio at the input

$(S/N)_o$ = signal-to-noise ratio at the output

All of the above are expressed as power ratios, not in decibels. When a device has multiple stages, each stage contributes noise, but the first stage is the most important because noise inserted there is amplified by all other stages. The equation that expresses this is:

$$NF_T = NF_1 + \frac{NF_2 - 1}{A_1} + \frac{NF_3 - 1}{A_1 A_2} + \cdots \tag{1.6}$$

where

NF_T = total noise figure for the system

NF_1 = noise figure of the first stage

NF_2 = noise figure of the second stage

A_1 = gain of the first stage

A_2 = gain of the second stage

Again, all these are ratios, not in decibels. The noise figure for the system is usually specified in dB in the usual way:

$$NF(\text{dB}) = 10 \log NF \tag{1.7}$$

Converting noise figure to noise temperature is quite easy:

$$T_{eq} = 290(NF - 1) \tag{1.8}$$

where

T_{eq} = equivalent noise temperature in kelvins

NF = noise figure as a ratio (not in dB)

The noise temperature due to the equipment must be added to the noise temperature contributed by the antenna and its transmission line to find the total system noise temperature. We'll see how that is done after we have looked at receivers, antennas, and transmission lines separately.

EXAMPLE 1.2

A three-stage amplifier has stages with the following specifications. Gain and noise figure are given as ratios.

Stage	Power Gain	Noise Figure
1	10	2
2	25	4
3	30	5

Calculate the power gain in decibels, noise figure in decibels, and equivalent noise temperature for the whole amplifier.

SOLUTION

The power gain is the product of the individual gains:

$$A_T = A_1 A_2 A_3 = 10 \times 25 \times 30 = 7500 = 38.8 \text{ dB}$$

The noise figure is found from

$$NF_T = NF_1 + \frac{NF_2 - 1}{A_1} + \frac{NF_3 - 1}{A_1 A_2} + \cdots$$

$$= 2 + \frac{4-1}{10} + \frac{5-1}{10 \times 25}$$

$$= 2.316$$

$$= 3.65 \text{ dB}$$

The equivalent noise temperature is:

$$T_{eq} = 290(NF - 1)$$

$$= 290(2.316 - 1)$$

$$= 382 \text{ K}$$

1.5 The Frequency Domain

The reader is probably familiar with the **time-domain** representation of signals. An ordinary oscilloscope display, showing amplitude on one scale and time on the other, is a good example.

Signals can also be described in the **frequency domain**. In a frequency-domain representation, amplitude or power is shown on one axis and frequency is displayed on the other. A **spectrum analyzer** gives a frequency-domain representation of signals.

Any signal can be represented either way. For example, a 1-kHz sine wave is shown in both ways in Figure 1.5. The time-domain representation should need no explanation. As for the frequency domain, a sine wave has energy only at its fundamental frequency, so it can be shown as a straight line at that frequency.

Notice that our way of representing the signal in the frequency domain does not show the phase of the signal. The signal in Figure 1.5(b) could be a cosine wave just as easily as a sine wave.

One example of a frequency-domain representation with which the reader will already be familiar is the tuning dial of an ordinary broadcast radio receiver. Each station is assigned a different carrier frequency. Provided that these frequencies are far enough apart, the signals will not interfere with each other. Dividing up the spectrum in this way is known as **frequency-division multiplexing (FDM)**, and can only be understood by referring to the frequency domain.

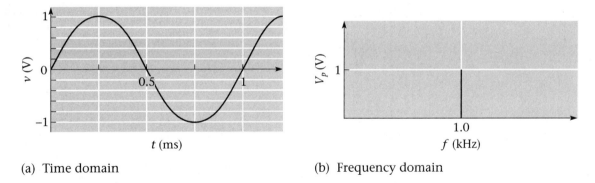

(a) Time domain   (b) Frequency domain

FIGURE 1.5   Sine wave in time and frequency domains

When we move into the study of individual radio signals, frequency-domain representations are equally useful. For instance, the bandwidth of a modulated signal generally has some fairly simple relationship to that of the baseband signal. This bandwidth can easily be found if the baseband signal can be represented in the frequency domain. As we proceed, we will see many other examples in which the ability to work with signals in the frequency domain will be required.

Fourier Series   It should be obvious by now that we need a way to move freely between the two domains. Any well-behaved periodic waveform can be represented as a series of sine and/or cosine waves at multiples of its fundamental frequency plus (sometimes) a dc offset. This is known as a **Fourier series**. This very useful (and perhaps rather surprising) fact was discovered in 1822 by Joseph Fourier, a French mathematician, in the course of research on heat conduction. Not all signals used in communication are strictly periodic, but they are often close enough for practical purposes.

Fourier's discovery, applied to a time-varying signal, can be expressed mathematically as follows:

$$f(t) = \frac{A_o}{2} + A_1 \cos \omega t + B_1 \sin \omega t + A_2 \cos 2\omega t + B_2 \sin 2\omega t \qquad (1.9)$$
$$+ A_3 \cos 3\omega t + B_3 \sin 3\omega t + \cdots$$

where

$f(t)$ = any well-behaved function of time. For our purposes, $f(t)$ will generally be either a voltage $v(t)$ or a current $i(t)$.

A_n and B_n = real-number coefficients; that is, they can be positive, negative, or zero.

ω = radian frequency of the fundamental.

The radian frequency can be found from the time-domain representation of the signal by finding the period (that is, the time T after which the whole signal repeats exactly) and using the equations:

$$f = \frac{1}{T}$$

and

$$\omega = 2\pi f$$

The simplest ac signal is a sinusoid. The frequency-domain representation of a sine wave has already been described and is shown in Figure 1.5 for a voltage sine wave with a period of 1 ms and a peak amplitude of 1 V. For this signal, all the Fourier coefficients are zero except for B_1, which has a value of 1 V. The equation becomes:

$$v(t) = \sin(2000\pi t) \text{ V}$$

which is certainly no surprise. The examples below use the equations for the various waveforms shown in Table 1.1 on page 15.

EXAMPLE 1.3

Find and sketch the Fourier series corresponding to the square wave in Figure 1.6(a).

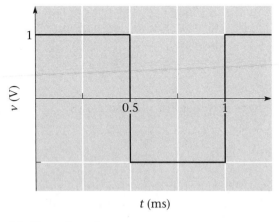

(a) Time domain

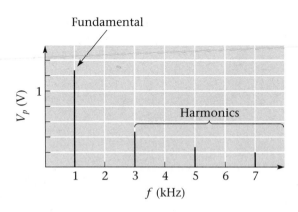

(b) Frequency domain

FIGURE 1.6

TABLE 1.1   Fourier Series for Common Repetitive Waveforms

1. Half-wave rectified sine wave

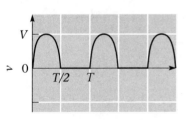

$$v(t) = \frac{V}{\pi} + \frac{V}{2} \sin \omega t - \frac{2V}{\pi} \left(\frac{\cos 2\omega t}{1 \times 3} + \frac{\cos 4\omega t}{3 \times 5} + \frac{\cos 6\omega t}{5 \times 7} + \cdots \right)$$

2. Full-wave rectified sine wave
   (a) With time zero at voltage zero

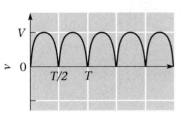

$$v(t) = \frac{2V}{\pi} - \frac{4V}{\pi} \left(\frac{\cos 2\omega t}{1 \times 3} + \frac{\cos 4\omega t}{3 \times 5} + \frac{\cos 6\omega t}{5 \times 7} + \cdots \right)$$

   (b) With time zero at voltage peak

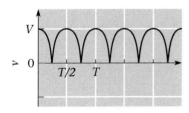

$$v(t) = \frac{2V}{\pi} \left(1 + \frac{2 \cos 2\omega t}{1 \times 3} - \frac{2 \cos 4\omega t}{3 \times 5} + \frac{2 \cos 6\omega t}{5 \times 7} - \cdots \right)$$

3. Square wave
   (a) Odd function

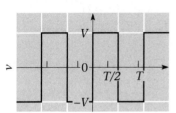

$$v(t) = \frac{4V}{\pi} \left(\sin \omega t + \frac{1}{3} \sin 3\omega t + \frac{1}{5} \sin 5\omega t + \cdots \right)$$

(continues)

TABLE 1.1   (*continued*)

(b) Even function

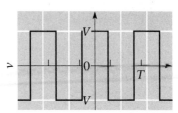

$$v(t) = \frac{4V}{\pi}\left(\cos\ \omega t - \frac{1}{3}\cos\ 3\omega t + \frac{1}{5}\cos\ 5\omega t - \cdots\right)$$

4. Pulse train

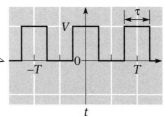

$$v(t) = \frac{V\tau}{T} + \frac{2V\tau}{T}\left(\frac{\sin\ \pi\tau/T}{\pi\tau/T}\cos\ \omega t + \frac{\sin\ 2\pi\tau/T}{2\pi\tau/T}\cos\ 2\omega t + \frac{\sin\ 3\pi\tau/T}{3\pi\tau/T}\cos\ 3\omega t + \cdots\right)$$

5. Triangle wave

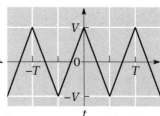

$$v(t) = \frac{8V}{\pi^2}\left(\cos\ \omega t + \frac{1}{3^2}\cos\ 3\omega t + \frac{1}{5^2}\cos\ 5\omega t + \cdots\right)$$

6. Sawtooth wave
    (a) With no dc offset

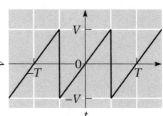

$$v(t) = \frac{2V}{\pi}\left(\sin\ \omega t - \frac{1}{2}\sin\ 2\omega t + \frac{1}{3}\sin\ 3\omega t - \cdots\right)$$

(*continues*)

TABLE 1.1 *(continued)*

(b) Positive-going

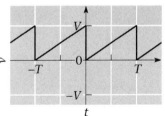

$$v(t) = \frac{V}{2} - \frac{V}{\pi}\left(\sin \omega t + \frac{1}{2}\sin 2\omega t + \frac{1}{3}\sin 3\omega t + \cdots\right)$$

SOLUTION

A square wave is another signal with a simple Fourier representation, although not quite as simple as for a sine wave. For the signal shown in Figure 1.6(a), the frequency is 1 kHz, as before, and the peak voltage is 1 V.

According to Table 1.1, this signal has components at an infinite number of frequencies: all odd multiples of the fundamental frequency of 1 kHz. However, the amplitude decreases with frequency, so that the third harmonic has an amplitude one-third that of the fundamental, the fifth harmonic an amplitude of one-fifth that of the fundamental, and so on. Mathematically, a square wave of voltage with a rising edge at $t = 0$ and no dc offset can be expressed as follows (see Table 1.1):

$$v(t) = \frac{4V}{\pi}\left(\sin \omega t + \frac{1}{3}\sin 3\omega t + \frac{1}{5}\sin 5\omega t + \cdots\right)$$

where

V = peak amplitude of the square wave
ω = radian frequency of the square wave
t = time in seconds

For this example, the above equation becomes:

$$v(t) = \frac{4}{\pi}\left(\sin(2\pi \times 10^3 t) + \frac{1}{3}\sin(6\pi \times 10^3 t) + \frac{1}{5}\sin(10\pi \times 10^3 t) + \cdots\right)V$$

This equation shows that the signal has frequency components at odd multiples of 1 kHz, that is, at 1 kHz, 3 kHz, 5 kHz, and so on. The 1-kHz component has a peak amplitude of

$$V_1 = \frac{4}{\pi} = 1.27 \text{ V}$$

The component at three times the fundamental frequency (3 kHz) has an amplitude one-third that of the fundamental, that at 5 kHz has an amplitude one-fifth that of the fundamental, and so on.

$$V_3 = \frac{4}{3\pi} = 0.424 \text{ V}$$

$$V_5 = \frac{4}{5\pi} = 0.255 \text{ V}$$

$$V_7 = \frac{4}{7\pi} = 0.182 \text{ V}$$

The result for the first four components is sketched in Figure 1.6(b). Theoretically, an infinite number of components would be required to describe the square wave completely, but as the frequency increases, the amplitude of the components decreases rapidly.

The representations in Figures 1.6(a) and 1.6(b) are not two different signals but merely two different ways of looking at the same signal. This can be shown graphically by adding the instantaneous values of several of the sine waves in the frequency-domain representation. If enough of these components are included, the result begins to look like the square wave in the time-domain representation. Figure 1.7 shows the results for two, four, and ten components. It was created by taking the instantaneous values of all the components at the same time and adding them algebraically. This was done for a large number of time values. Doing these calculations by hand would be simple but rather tedious, so a computer was used to perform the calculations and plot the graphs. A perfectly accurate representation of the square wave would require an infinite number of components, but we can see from the figure that using ten terms gives a very good representation because the amplitudes of the higher-frequency components of the signal are very small.

It is possible to go back and forth at will between time and frequency domains, but it should be apparent that information about the relative phases of the frequency components in the Fourier representation of the signal is required to reconstruct the time-domain representation. The Fourier equations do have this information, but the sketch in Figure 1.6(b) does not. If the phase relationships between frequency components are changed in a communication system, the signal will be distorted in the time domain.

Figure 1.8 illustrates this point. The same ten coefficients were used as in Figure 1.7, but this time the waveforms alternated between sine and

FIGURE 1.7
Construction of a square
wave from Fourier
components

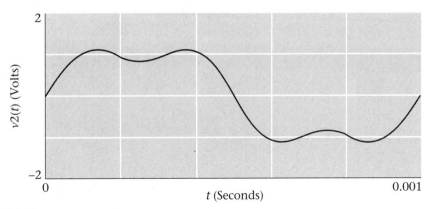

(a) Two components

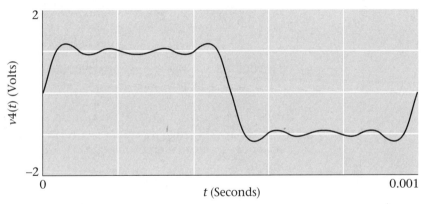

(b) Four components

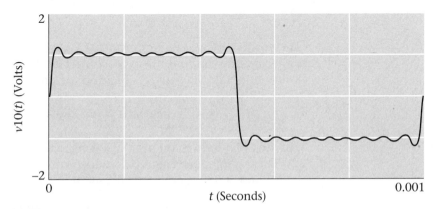

(c) Ten components

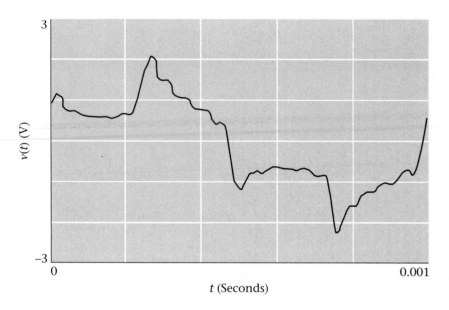

FIGURE 1.8 Addition of square-wave Fourier components with wrong phase angles

cosine: sine for the fundamental, cosine for the third harmonic, sine for the fifth, and so on. The result is a waveform that looks the same on the frequency-domain sketch of Figure 1.6(b) but very different in the time domain.

EXAMPLE 1.4

Find the Fourier series for the signal in Figure 1.9(a).

SOLUTION

The positive-going sawtooth wave of Figure 1.9(a) has a Fourier series with a dc term and components at all multiples of the fundamental frequency. From Table 1.1, the general equation for such a wave is

$$v(t) = \frac{A}{2} - \left(\frac{A}{\pi}\right)\left(\sin \omega t + \frac{1}{2}\sin 2\omega t + \frac{1}{3}\sin 3\omega t + \cdots\right)$$

The first (dc) term is simply the average value of the signal.

For the signal in Figure 1.9, which has a frequency of 1 kHz and a peak amplitude of 5 V, the preceding equation becomes:

$$v(t) = 2.5 - 1.59\ (\sin(2\pi \times 10^3 t) + 0.5\sin\ (4\pi \times 10^3 t) \\ + 0.33\sin\ (6\pi \times 10^3 t) + \cdots)\ \text{V}$$

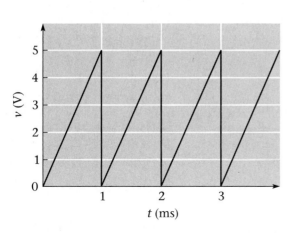

(a) Time domain

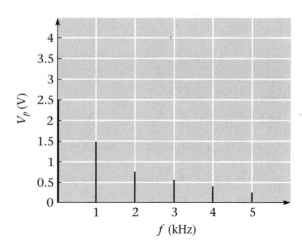

(b) Frequency domain

FIGURE 1.9

The first four voltage components are:

dc component: $V_0 = 2.5$ V

1-kHz component: $V_1 = -1.59$ V (the minus sign represents a phase angle of 180 degrees. A graph of peak values will not usually indicate signs, and a spectrum analyzer will not show phase angles)

2-kHz component: $V_2 = -1.59/2 = -0.795$ V

3-kHz component: $V_3 = -1.59/3 = -0.53$ V

The spectrum is shown in Figure 1.9(b).

Effect of Filtering on Signals   As we have seen, many signals have a bandwidth that is theoretically infinite. Limiting the frequency response of a channel removes some of the frequency components and causes the time-domain representation to be distorted. An uneven frequency response will emphasize some components at the expense of others, again causing distortion. Nonlinear phase shift will also affect the time-domain representation. For instance, shifting the phase angles of some of the frequency components in the square-wave representation of Figure 1.8 changed the signal to something other than a square wave.

However, Figure 1.7 shows that while an infinite bandwidth may theoretically be required, for practical purposes quite a good representation of a square wave can be obtained with a band-limited signal. In general, the wider the bandwidth, the better, but acceptable results can be obtained with a band-limited signal. This is welcome news, because practical systems always have finite bandwidth.

Noise in the Frequency Domain

It was pointed out earlier, in Section 1.4, that noise power is proportional to bandwidth. That implies that there is equal noise power in each hertz of bandwidth. Sometimes this kind of noise is called **white noise**, since it contains all frequencies just as white light contains all colors. In fact, we can talk about a **noise power density** in watts per hertz of bandwidth. The equation for this is very simply derived. We start with Equation (1.3):

$$P_N = kTB$$

This gives the total noise power in bandwidth, B. To find the power per hertz, we just divide by the bandwidth to get an even simpler equation:

$$N_0 = kT \tag{1.10}$$

where

N_0 = noise power density in watts per hertz
k = Boltzmann's constant, 1.38×10^{-23} joules/kelvin (J/K)
T = temperature in kelvins

EXAMPLE 1.5

(a) A resistor has a noise temperature of 300 K. Find its noise power density and sketch the noise spectrum.

(b) A system with a noise temperature of 300 K operates at a frequency of 100 MHz with a bandwidth of 1 MHz. Sketch the noise spectrum.

SOLUTION

(a) From Equation (1.10):

$$
\begin{aligned}
N_0 &= kT \\
&= 1.38 \times 10^{-23} \text{ J/K} \times 300 \text{ K} \\
&= 4.14 \times 10^{-21} \text{ W/Hz}
\end{aligned}
$$

The spectrum is shown in Figure 1.10(a). Note that the spectrum is a straight line, showing that the noise power density is the same at all frequencies. The frequency scale shown runs from 0 to 1 GHz, but theoretically the spectrum remains flat indefinitely. Real systems, of course, never have infinite frequency response.

FIGURE 1.10

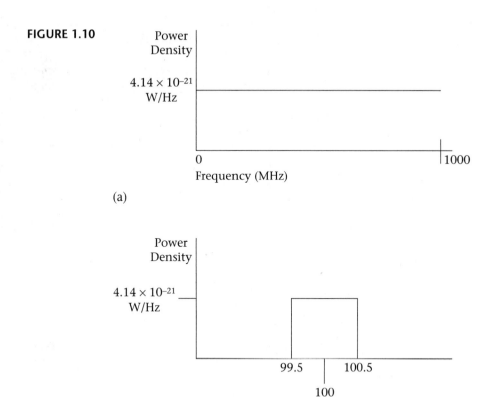

(a)

(b)

(b) Here the noise power density is the same as in part (a) but only over the 1-MHz bandwidth illustrated. Hence the band-limited spectrum of Figure 1.10(b). The exact shape of the pattern will depend on the type of filter used. In the sketch an ideal filter, with complete attenuation outside the passband, is assumed.

1.6 The Radio-Frequency Spectrum

Radio waves are a form of electromagnetic radiation, as are infrared, visible light, ultraviolet light, and gamma rays. The major difference is in the frequency of the waves. The portion of the frequency spectrum that is useful for radio communication at present extends from roughly 100 kHz to about 50 GHz. Table 1.2 shows the conventional designations of the various

TABLE 1.2   The Radio-Frequency Spectrum

Frequency Designation	Frequency Range	Wavelength Range	Wavelength Designation
Extremely High Frequency (EHF)	30–300 GHz	1 mm–1 cm	Millimeter Waves
Super High Frequency (SHF)	3–30 GHz	1–10 cm	Microwaves (microwave region conventionally starts at 1 GHz)
Ultra High Frequency (UHF)	300 Mhz–3 GHz	10 cm–1 m	
Very High Frequency (VHF)	30–300 MHz	1–10 m	
High Frequency (HF)	3–30 MHz	10–100 m	Short Waves
Medium Frequency (MF)	300 kHz–3 MHz	100 m–1 km	Medium Waves

frequency ranges and their associated wavelength ranges. Note that micro-waves and millimeter waves are wavelength designations and fit only approximately into the frequency designations. Wireless communication as described in this book occupies mainly the VHF, UHF, and SHF portions of the spectrum. Lower-frequency systems need inconveniently large antennas and involve methods of signal propagation that are undesirable for the systems we look at. Extremely high frequencies are still difficult to generate and amplify at reasonable cost, though that may well change in the future.

Conversion between frequency and wavelength is quite easy. The general equation that relates frequency to wavelength for any wave is

$$v = f\lambda \tag{1.11}$$

where

v = velocity of propagation of the wave in meters per second
f = frequency of the wave in hertz
λ = wavelength in meters

For radio waves in free space (and air is generally a reasonable approximation to free space) the velocity is the same as that of light: 300×10^6 m/s. The usual symbol for this quantity is c. Equation (1.11) then becomes:

$$c = f\lambda \tag{1.12}$$

EXAMPLE 1.6 ▼

Find the wavelength of a signal at each of the following frequencies:

(a) 850 MHz (cell phone range)

(b) 1.9 GHz (Personal Communication Systems range)

(c) 28 GHz (used for Local Multipoint Distribution Systems (LMDS) for local delivery of television signals by microwave)

SOLUTION

For all of these the method is the same. The problem is repeated to give the reader a feeling for some of the frequencies and wavelengths used in wireless communication. Simply rewrite Equation (1.12) in the form

$$\lambda = \frac{c}{f}$$

(a) $\lambda = \dfrac{300 \times 10^6}{850 \times 10^6} = 0.353 \text{ m} = 353 \text{ mm}$

(b) $\lambda = \dfrac{300 \times 10^6}{1.9 \times 10^9} = 0.158 \text{ m} = 158 \text{ mm}$

(c) $\lambda = \dfrac{300 \times 10^6}{28 \times 10^9} = 0.0107 \text{ m} = 10.7 \text{ mm}$

▲

Bandwidth Requirements

The carrier wave is a sine wave for almost any communication system. A sine wave, of course, exists at only one frequency and therefore occupies zero bandwidth. As soon as the signal is modulated to transmit information, however, the bandwidth increases. A detailed knowledge of the bandwidth of various types of modulated signals is essential to the understanding of the communication systems to be described in this book. Thorough study of signal bandwidths will have to wait until we know more about the modulation schemes referred to above. However, at this time it would be well to look at the concept of bandwidth in more general terms.

First, bandwidth in radio systems is always a scarce resource. Not all frequencies are useful for a given communication system, and there is often competition among users for the same part of the spectrum. In addition, as we have seen, the degrading effect of noise on signals increases with bandwidth. Therefore, in most communication systems it is important to conserve bandwidth to the extent possible.

There is a general rule known as Hartley's Law which relates bandwidth, time, and information content. We will not yet be able to use it for actual calculations, but it would be well to note it for future reference, as Hartley's Law applies to the operation of all communication systems. Here it is:

$$I = ktB \tag{1.13}$$

where

I = amount of information to be transmitted in bits

k = a constant that depends on the modulation scheme and the signal-to-noise ratio

t = time in seconds

B = bandwidth in hertz

Our problem thus far is that we do not have precise ways of quantifying either the amount of information I or the constant k. However, the general form of the equation is instructive. It tells us that the rate at which information is transmitted is proportional to the bandwidth occupied by a communication system. To transmit more information in a given time requires more bandwidth (or a more efficient modulation scheme).

EXAMPLE 1.7

Telephone voice transmission requires transmission of frequencies up to about 3.4 kHz. Broadcast video using the ordinary North American standard, on the other hand, requires transmission of frequencies up to 4.2 MHz. If a certain modulation scheme needs 10 kHz for the audio transmission, how much bandwidth will be needed to transmit video using the same method?

SOLUTION

Hartley's Law states that bandwidth is proportional to information rate, which in this case is given by the baseband bandwidth. Assuming that audio needs a bandwidth from dc to 3.4 kHz, while video needs dc to 4.2 MHz, the bandwidth for video will be

$$B_{TV} = B_{TA} \times \frac{B_V}{B_A}$$

where

$$B_{TV} = \text{transmission bandwidth for video}$$
$$B_{TA} = \text{transmission bandwidth for audio}$$
$$B_{V} = \text{baseband bandwidth for video}$$
$$B_{A} = \text{baseband bandwidth for audio}$$

Substituting the numbers from the problem we get

$$B_{TV} = B_{TA} \times \frac{B_V}{B_A}$$
$$= 10 \text{ kHz} \times \frac{4.2 \text{ MHz}}{3.4 \text{ kHz}}$$
$$= 12.3 \text{ MHz}$$

Obviously the type of baseband signal to be transmitted makes a great deal of difference to any spectrum management plan.

Hartley's Law also shows that it is possible to reduce the bandwidth required to transmit a given amount of information by using more time for the process. This is an important possibility where data must be sent, but of course it is not practical when real-time communication is required—in a telephone call, for instance. The reader may have experienced this trade-off of time for bandwidth in downloading an audio or video file from the internet. If the bandwidth of the connection is low, such a file may take much longer to download than it does to play.

Frequency Reuse   Spectrum space in wireless systems is nearly always in short supply. Even with the communication bandwidth restricted as much as possible, there is often more traffic than can be accommodated. Of course the spectrum used for a given purpose in one area can be reused for a different purpose in another area that is physically far enough away that signals do not travel from one area to the other with sufficient strength to cause unacceptable interference levels. How far that is depends on many factors such as transmitter power, antenna gain and height, and the type of modulation used. Many recent systems, such as cellular telephony, automatically reduce transmitter power to the minimum level consistent with reliable communication, thereby allowing frequencies to be reused at quite small distances. Such schemes can use spectrum very efficiently.

1.7   "Convergence" and Wireless Communication

There has been much talk recently about "convergence," the merger of all kinds of previously separate electronic systems, for example, telephony (both wireline and wireless), broadcast and cable television, and data communication (most notably the internet). Convergence does seem to be beginning to happen, though more slowly than many had anticipated. The process is slowed both by technical problems involving the very different types of signals and media used in fields that have evolved separately for many years, and by more mundane but equally serious problems of regulatory jurisdiction and commercial interests. It is by no means clear exactly how wireless communication will fit into the final picture, even if a field as dynamic as this can be imagined to have a final state. Some people (many of whom seem to work for wireless phone companies) have suggested that eventually wireless phones will replace wireline equipment, and everyone will have one phone (with one number) which they will carry with them everywhere. Wired communication will then do what it does best: carry high-bandwidth signals like television to fixed locations. On the other hand, there is very serious development work underway involving high-bandwidth wireless communication for world-wide web access from portable devices, for instance. If it is unclear even to the experts what the future holds, we must be careful in our predictions. This much is certain though: wireless communication will be a large part of the total communication picture, and a knowledge of the technologies involved will certainly help a technologist to understand future developments as they occur.

Summary

The main points to remember from this chapter are:

- Any wireless communication system requires a transmitter and a receiver connected by a channel.

- Simplex communication systems allow communication in one direction only. Half-duplex systems are bidirectional, but work in only one direction at a time. Full-duplex systems can communicate in both directions simultaneously.

- Most wireless networks are variations of the star network configuration, often with radio repeaters at the hub.

- Radio systems transmit information by modulating a sine-wave carrier signal. Only three basic parameters can be modulated: amplitude, frequency, and phase. Many variations are possible, however.

- The ratio of signal power to noise power is one of the most important specifications for any communication system. Thermal noise is the most important type of noise in most wireless systems.
- The frequency domain is useful for the observation and understanding of both signals and noise in communication systems.
- Many signals can be analyzed in the frequency domain with the aid of Fourier series.
- The bandwidth required by a system depends on the modulation scheme employed and the information transmission rate required. Bandwidth should be kept to the minimum necessary to reduce noise problems and to conserve radio-frequency spectrum.
- Convergence is a term describing the possible merger of many different kinds of communication and related technologies.

● Equation List

$$e(t) = E_c \sin(\omega_c t + \theta) \tag{1.1}$$

$$P_N = kTB \tag{1.3}$$

$$T(K) = T(°C) + 273 \tag{1.4}$$

$$NF = \frac{(S/N)_i}{(S/N)_o} \tag{1.5}$$

$$NF_T = NF_1 + \frac{NF_2 - 1}{A_1} + \frac{NF_3 - 1}{A_1 A_2} + \cdots \tag{1.6}$$

$$T_{eq} = 290(NF - 1) \tag{1.8}$$

$$f(t) = \frac{A_o}{2} + A_1 \cos \omega t + B_1 \sin \omega t + A_2 \cos 2\omega t + B_2 \sin 2\omega t \\ + A_3 \cos 3\omega t + B_3 \sin 3\omega t + \cdots \tag{1.9}$$

$$N_0 = kT \tag{1.10}$$

$$v = f\lambda \tag{1.11}$$

$$c = f\lambda \tag{1.12}$$

$$I = ktB \tag{1.13}$$

● Key Terms

bandwidth portion of frequency spectrum occupied by a signal

baseband information signal

carrier high-frequency signal which is modulated by the baseband signal in a communication system

Citizens' Band (CB) radio short-distance unlicensed radio communication system

demodulation recovery of a baseband signal from a modulated signal

Fourier series expression showing the structure of a signal in the frequency domain

frequency domain method of analyzing signals by observing them on a power-frequency plane

frequency-division multiplexing combining of several signals into one communication channel by assigning each a different carrier frequency

full-duplex communication two-way communication in which both terminals can transmit simultaneously

half-duplex communication two-way communication system in which only one station can transmit at a time

handoff transfer of a call in progress from one cell site to another

Improved Mobile Telephone Service (IMTS) a mobile telephone service, now obsolescent, using trunked channels but not cellular in nature

intelligence information to be communicated

modulating signal the information signal that is used to modulate a carrier for transmission

network an organized system for communicating among terminals

noise an unwanted random signal that extends over a considerable frequency spectrum

noise power density the power in a one-hertz bandwidth due to a noise source

personal communication system (PCS) a cellular telephone system designed mainly for use with portable (hand-carried) telephones

public switched telephone network (PSTN) the ordinary public wireline phone system

repeater a transmitter-receiver combination used to receive and retransmit a signal

signal-to-noise ratio   ratio between the signal power and noise power at some point in a communication system

simplex   a unidirectional communication system; for example, broadcasting

spectrum analyzer   test instrument that typically displays signal power as a function of frequency

star network   a computer network topology in which each terminal is connected to a central mainframe or server

time domain   representation of a signal as a function of time and some other parameter, such as voltage

white noise   noise containing all frequencies with equal power in every hertz of bandwidth

● Questions

1. Why were the first radio communication systems used for telegraphy only?

2. When were the first two-way mobile radio communication systems installed, and for what purpose?

3. What characteristics of CB radio led to its great popularity?

4. Why are cellular radio systems more efficient in their use of spectrum than earlier systems?

5. What types of modulation are used with cellular phones?

6. Explain the differences among simplex, half-duplex, and full-duplex communication.

7. Identify each of the following communication systems as simplex, half-duplex, or full-duplex.
   (a) cordless telephone
   (b) television broadcast
   (c) intercom with push-to-talk bar

8. Why is it necessary to use a high-frequency carrier with a radio communication system?

9. Name the three basic modulation methods.

10. Suppose that a voice frequency of 400 Hz is transmitted using a transmitter operating at 800 MHz. Which of these is:
    (a) the information frequency?
    (b) the carrier frequency?

(c) the baseband frequency?

(d) the modulating frequency?

11. What effect does doubling the bandwidth of a system have on its noise level?

12. What is the meaning of the signal-to-noise ratio for a system, and why is it important?

13. What is the difference between the kelvin and Celsius temperature scales?

14. State whether the time or frequency domain would be more appropriate for each of the following:

    (a) a display of all UHF television channels

    (b) measuring the peak voltage of a waveform

    (c) measuring the bandwidth of a waveform

    (d) determining the rise time of a signal

15. What is meant by the term *frequency-division multiplexing*?

16. Why is thermal noise sometimes called *white noise*?

17. Give the frequency designation for each of the following systems:

    (a) marine radio at 160 MHz

    (b) cell phones at 800 MHz

    (c) direct-to-home satellite television at 12 GHz

    (d) CB radio at 27 Mhz

● Problems

1. Express the frequency of a 10-kHz signal in radians per second.

2. Find the noise power produced by a resistor at a temperature of 60 °C in a bandwidth of 6 MHz in

   (a) watts

   (b) dBm

   (c) dBf

3. If the signal power at a certain point in a system is 2 W and the noise power is 50 mW, what is the signal-to-noise ratio, in dB?

4. Sketch the spectrum for the half-wave rectified signal in Figure 1.11, showing harmonics up to the fifth. Show the voltage and frequency scales and indicate whether your voltage scale shows peak or RMS voltage.

FIGURE 1.11

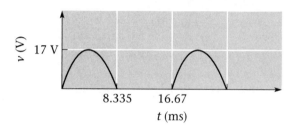

5. Sketch the frequency spectrum for the triangle wave shown in Figure 1.12 for harmonics up to the fifth. Show the voltage and frequency scales.

FIGURE 1.12

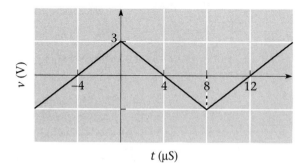

6. A 1-kHz square wave passes through each of three communication channels whose bandwidths are given below. Sketch the output in the time domain for each case.
   (a) 0 to 10 kHz
   (b) 2 kHz to 4 kHz
   (c) 0 to 4 kHz

7. Sketch the spectrum for the pulse train shown in Figure 1.13.

FIGURE 1.13

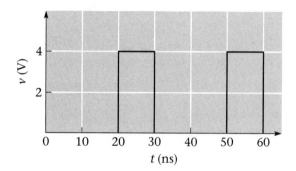

8. Sketch the spectrum for the sawtooth waveform in Figure 1.14. Explain why this waveform has no dc component, unlike the sawtooth waveform in Example 1.3.

FIGURE 1.14

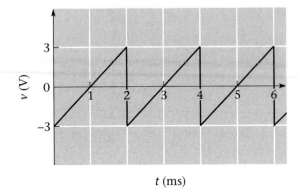

9. Visible light consists of electromagnetic radiation with free-space wavelengths between 400 and 700 nanometers (nm). Express this range in terms of frequency.

10. Equation (1.11) applies to any kind of wave. The velocity of sound waves in air is about 344 m/s. Calculate the wavelength of a sound wave with a frequency of 1 kHz.

Analog Modulation Schemes

Objectives

After studying this chapter, you should be able to:

- Explain the concept of modulation.
- Describe the differences among analog modulation schemes.
- Analyze amplitude-modulated signals in the time and frequency domains.
- Analyze frequency-modulated signals in the frequency domain.
- Describe phase modulation.
- Explain the need for pre-emphasis and de-emphasis with FM signals.

2.1 Introduction

In Chapter 1 we saw that modulation is necessary in order to transmit intelligence over a radio channel. A radio-frequency signal can be modulated by either analog or digital information. In either case, the information signal must change one or more of three parameters: amplitude, frequency, and phase.

With the exception of Morse code transmission, which is digital though not binary, the earliest wireless communication systems used analog modulation, and these schemes are still very popular in such diverse areas as broadcasting and cellular telephony. Analog modulation schemes tend to be more intuitive and hence easier to understand than their digital variants, so they will be considered first. Of the analog schemes, amplitude modulation (AM) is simplest and was first historically, therefore, it seems logical to begin with it. Frequency modulation (FM) is more common in modern systems, so it will be discussed next. Finally, phase modulation (PM) is seen less often than the others in analog systems, but it is very common in digital communication, so we will introduce it here but leave the details for later.

2.2 Amplitude Modulation

An amplitude-modulated signal can be produced by using the instantaneous amplitude of the information signal (the baseband or modulating signal) to vary the peak amplitude of a higher-frequency signal. Figure 2.1(a) shows a baseband signal consisting of a 1-kHz sine wave, which can be combined with the 10-kHz carrier signal shown in Figure 2.1(b) to produce the amplitude-modulated signal of Figure 2.1(c). If the peaks of the individual waveforms of the modulated signal are joined, an **envelope** results that resembles the original modulating signal. It repeats at the modulating frequency, and the shape of each "half" (positive or negative) is the same as that of the modulating signal.

Figure 2.1(c) shows a case where there are only ten cycles of the carrier for each cycle of the modulating signal. In practice, the ratio between carrier frequency and modulating frequency is usually much greater. For instance, an AM citizens' band (CB) station would have a carrier frequency of about 27 MHz and a modulating frequency on the order of 1 kHz. A waveform like this is shown in Figure 2.2. Since there are thousands of cycles of the carrier for each cycle of the envelope, the individual RF cycles are not visible, and only the envelope can be seen.

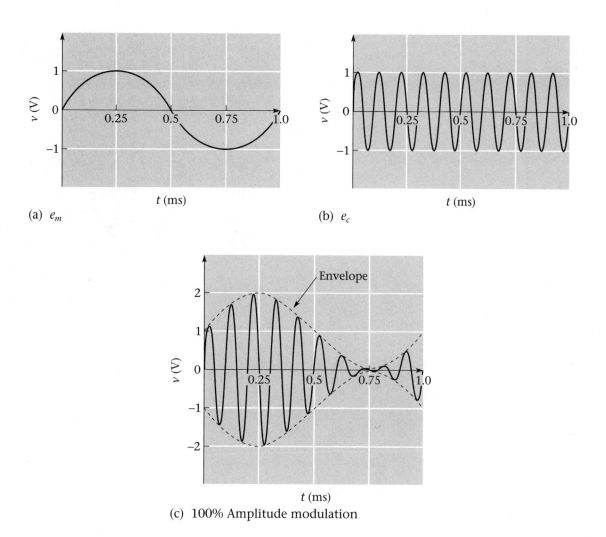

(a) e_m

(b) e_c

(c) 100% Amplitude modulation

FIGURE 2.1 Amplitude modulation

The AM envelope allows for very simple demodulation. All that is necessary is to rectify the signal to remove one-half of the envelope, then low-pass filter the remainder to recover the modulation. A simple but quite practical AM demodulator is shown in Figure 2.3.

Because AM relies on amplitude variations, it follows that any amplifier used with an AM signal must be linear, that is, it must reproduce amplitude variations exactly. This principle can be extended to any signal that has an envelope. This point is important, because nonlinear amplifiers are typically less expensive and more efficient than linear amplifiers.

FIGURE 2.2

AM envelope

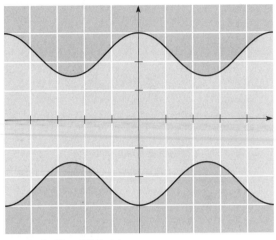

Vertical: 25 mV/division
Horizontal: 200 μs/division

FIGURE 2.3

AM demodulator

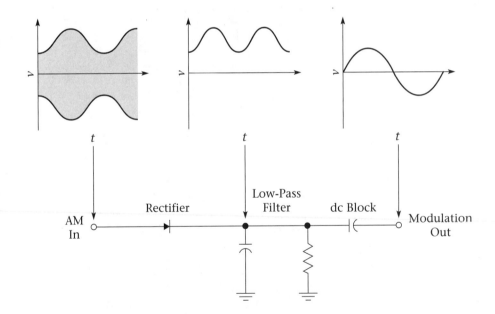

Time-Domain Analysis   Now that we understand the general idea of AM, it is time to examine the system in greater detail. We shall look at the modulated signal in both the time and frequency domains, as each method emphasizes some of the important characteristics of AM. The time domain is probably more familiar, so we begin there.

AM is created by using the instantaneous modulating signal voltage to vary the amplitude of the modulated signal. The carrier is almost always a sine wave. The modulating signal can be a sine wave, but is more often an arbitrary waveform, such as an audio signal. However, an analysis of sine-wave modulation is very useful, since Fourier analysis often allows complex signals to be expressed as a series of sinusoids.

We can express the above relationship by means of an equation:

$$v(t) = (E_c + e_m) \sin \omega_c t \tag{2.1}$$

where

$v(t)$ = instantaneous amplitude of the modulated signal in volts

E_c = peak amplitude of the carrier in volts

e_m = instantaneous amplitude of the modulating signal in volts

ω_c = the frequency of the carrier in radians per second

t = time in seconds

If the modulating (baseband) signal is a sine wave, Equation (2.1) has the following form:

$$v(t) = (E_c + E_m \sin \omega_m t) \sin \omega_c t \tag{2.2}$$

where

E_m = peak amplitude of the modulating signal in volts

ω_m = frequency of the modulating signal in radians per second

and the other variables are as defined for Equation (2.1).

EXAMPLE 2.1

A carrier with an RMS voltage of 2 V and a frequency of 1.5 MHz is modulated by a sine wave with a frequency of 500 Hz and amplitude of 1 V RMS. Write the equation for the resulting signal.

SOLUTION

First, note that Equation (2.2) requires peak voltages and radian frequencies. We can easily get these as follows:

$$E_c = \sqrt{2} \times 2 \text{ V}$$
$$= 2.83 \text{ V}$$

$$E_m = \sqrt{2} \times 1 \text{ V}$$
$$= 1.41 \text{ V}$$

$$\omega_c = 2\pi \times 1.5 \times 10^6$$
$$= 9.42 \times 10^6 \text{ rad/s}$$

$$\omega_m = 2\pi \times 500$$
$$= 3.14 \times 10^3 \text{ rad/s}$$

So the equation is

$$v(t) = (E_c + E_m \sin \omega_m t) \sin \omega_c t$$
$$= [2.83 + 1.41 \sin (3.14 \times 10^3 t)] \sin (9.42 \times 10^6 t) \text{ V}$$

Modulation Index

The ratio between the amplitudes of the modulating signal and the carrier is defined as the **modulation index**, m. Mathematically,

$$m = E_m / E_c \tag{2.3}$$

Modulation can also be expressed as a percentage, by multiplying m by 100. For example, $m = 0.5$ corresponds to 50% modulation.

Substituting m into Equation (2.2) gives:

$$v(t) = E_c(1 + m \sin \omega_m t) \sin \omega_c t \tag{2.4}$$

EXAMPLE 2.2

Calculate m for the signal of Example 2.1 and write the equation for this signal in the form of Equation (2.4).

SOLUTION

To avoid an accumulation of round-off errors we should go back to the original voltage values to find m.

$$m = E_m / E_c$$
$$= 1/2$$
$$= 0.5$$

It is all right to use the RMS values for calculating this ratio, as the factors of $\sqrt{2}$, if used to find the peak voltages, will cancel.

Now we can rewrite the equation:

$$v(t) = E_c(1 + m \sin \omega_m t) \sin \omega_c t$$
$$= 2.83 [1 + 0.5 \sin (3.14 \times 10^3 t)] \sin (9.42 \times 10^6 t) \text{ V}$$

It is worthwhile to examine what happens to Equation (2.4) and to the modulated waveform, as m varies. To start with, when $m = 0$, $E_m = 0$ and we have the original, unmodulated carrier. As m varies between 0 and 1, the changes due to modulation become more pronounced. Resultant waveforms for several values of m are shown in Figure 2.4. Note especially the result for $m = 1$ or 100%. Under these conditions the peak signal voltage will vary between zero and twice the unmodulated carrier amplitude.

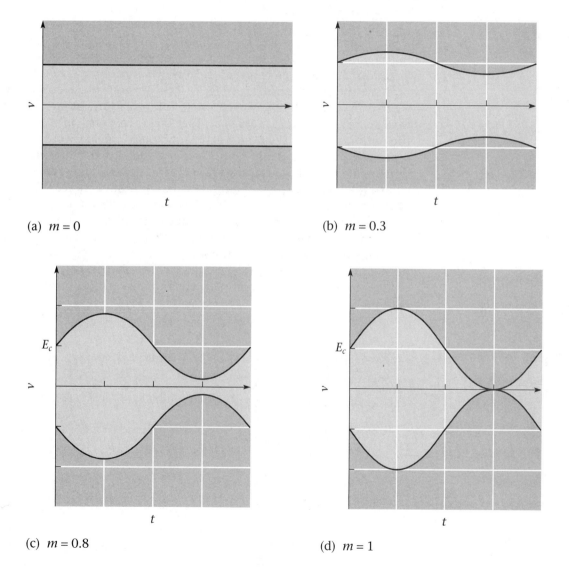

(a) $m = 0$

(b) $m = 0.3$

(c) $m = 0.8$

(d) $m = 1$

FIGURE 2.4 Envelopes for various values of m

Overmodulation   When the modulation index is greater than one, the signal is said to be overmodulated. There is nothing in Equation (2.4) that would seem to prevent E_m from being greater than E_c, that is, m greater than one. There are practical difficulties, however. Figure 2.5(a) shows the result of simply substituting $m = 2$ into Equation (2.4). As you can see, the envelope no longer resembles the modulating signal. Thus the type of demodulator described earlier no longer gives undistorted results, and the signal is no longer a full-carrier AM signal.

Whenever we work with mathematical models, we must remember to keep checking against physical reality. This situation is a good example. It is possible to build a circuit that does produce an output that agrees with Equation (2.4) for m greater than 1. However, most practical AM modulators produce the signal shown in Figure 2.5(b) under these conditions. This waveform is completely useless for communication. In fact, if this signal were subjected to Fourier analysis, the sharp "corners" on the waveform as the output goes to zero on negative modulation peaks would be found to represent high-frequency components added to the original baseband signal. This type of **overmodulation** creates spurious frequencies known as **splatter**, which cause the modulated signal to have increased bandwidth. This can cause interference with a signal on an adjacent channel.

From the foregoing, we can conclude that for full-carrier AM, m must be in the range from 0 to 1. Overmodulation creates distortion in the demodulated signal and may result in the signal occupying a larger bandwidth than normal. Since spectrum space is tightly controlled by law, overmodulation of an AM transmitter is actually illegal, and means must be provided to prevent it.

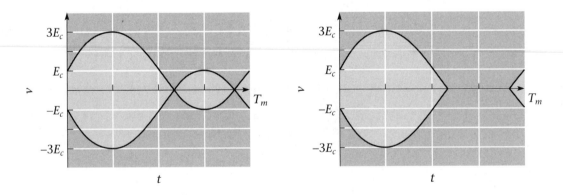

(a) $m = 2$ in Equation (2.4)   (b) $m = 2$ with practical modulator

FIGURE 2.5   Overmodulation

Modulation Index for Multiple Modulating Frequencies

Practical AM systems are seldom used to transmit sine waves, of course. The information signal is more likely to be a voice signal, which contains many frequencies. When there are two or more sine waves of different, uncorrelated frequencies (that is, frequencies that are not multiples of each other) modulating a single carrier, m is calculated by using the equation

$$m_T = \sqrt{m_1^2 + m_2^2 + \cdots}$$
(2.5)

where

m_T = total resultant modulation index

m_1, m_2, etc. = modulation indices due to the individual modulating components.

EXAMPLE 2.3

Find the modulation index if a 10-volt carrier is amplitude modulated by three different frequencies, with amplitudes of 1, 2, and 3 volts respectively.

SOLUTION

The three separate modulation indices are:

m_1 = 1/10 = 0.1
m_2 = 2/10 = 0.2
m_3 = 3/10 = 0.3

$$m_T = \sqrt{m_1^2 + m_2^2 + m_3^2}$$
$$= \sqrt{0.1^2 + 0.2^2 + 0.3^2}$$
$$= 0.374$$

Measurement of Modulation Index

If we let E_m and E_c be the peak modulation and carrier voltages respectively, then we can see, from Equation (2.4), that the maximum envelope voltage is simply

$$E_{max} = E_c(1 + m)$$
(2.6)

and the minimum envelope voltage is

$$E_{min} = E_c(1 - m)$$
(2.7)

Note, by the way, that these results agree with the conclusions expressed earlier: for $m = 0$, the peak voltage is E_c, and for $m = 1$, the envelope voltage ranges from $2E_c$ to zero.

Applying a little algebra to the above expressions, it is easy to show that

$$m = \frac{E_{max} - E_{min}}{E_{max} + E_{min}} \qquad (2.8)$$

Of course, doubling both E_{max} and E_{min} will have no effect on this equation, so it is quite easy to find m by displaying the envelope on an oscilloscope and measuring the maximum and minimum peak-to-peak values for the envelope voltage.

EXAMPLE 2.4

Calculate the modulation index for the waveform shown in Figure 2.2.

SOLUTION

It is easiest to use peak-to-peak values with an oscilloscope. From the figure we see that:

$$E_{max} = 150 \text{ mV}_{p\text{-}p} \qquad E_{min} = 70 \text{ mV}_{p\text{-}p}$$

$$m = \frac{E_{max} - E_{min}}{E_{max} + E_{min}}$$
$$= \frac{150 - 70}{150 + 70}$$
$$= 0.364$$

Frequency-Domain Analysis

So far we have looked at the AM signal exclusively in the time domain, that is, as it can be seen on an oscilloscope. In order to find out more about this signal, however, it is necessary to consider its spectral makeup. We could use Fourier methods to do this, but for a simple AM waveform it is easier, and just as valid, to use trigonometry.

To start, we should observe that although both the carrier and the modulating signal may be sine waves, the modulated AM waveform is *not* a sine wave. This can be seen from a simple examination of the waveform of Figure 2.1(c). It is important to remember that the modulated waveform is not a sine wave when, for instance, trying to find RMS from peak voltages. The usual formulas, so laboriously learned in fundamentals courses, do not apply here!

If an AM signal is not a sine wave, then what is it? We already have a mathematical expression, given by Equation (2.4):

$$v(t) = E_c(1 + m \sin \omega_m t) \sin \omega_c t$$

Expanding it and using a trigonometric identity will prove useful. Expanding gives

$$v(t) = E_c \sin \omega_c t + mE_c \sin \omega_c t \sin \omega_m t$$

The first term is just the carrier. The second can be expanded using the trigonometric identity

$$\sin A \sin B = \frac{1}{2}[\cos(A - B) - \cos(A + B)]$$

to give

$$v(t) = E_c \sin \omega_c t + \frac{mE_c}{2}[\cos(\omega_c - \omega_m)t - \cos(\omega_c + \omega_m)t]$$

which can be separated into three distinct terms:

$$v(t) = E_c \sin \omega_c t + \frac{mE_c}{2}\cos(\omega_c - \omega_m)t - \frac{mE_c}{2}\cos(\omega_c + \omega_m)t \qquad (2.9)$$

We now have, in addition to the original carrier, two other sinusoidal waves, one above the carrier frequency and one below. When the complete signal is sketched in the frequency domain as in Figure 2.6, we see the carrier and two additional frequencies, one to each side. These are called, logically enough, **side frequencies**. The separation of each side frequency from the carrier is equal to the modulating frequency; and the relative amplitude of the side frequency, compared with the carrier, is proportional to m, becoming half the carrier voltage for $m = 1$. In a real situation there is generally more than one set of side frequencies, because there is more than one modulating frequency. Each modulating frequency produces two side frequencies. Those above the carrier can be grouped into a band of frequencies called the upper **sideband**. There is also a lower sideband, which looks like a mirror image of the upper, reflected in the carrier.

FIGURE 2.6

AM in the frequency domain

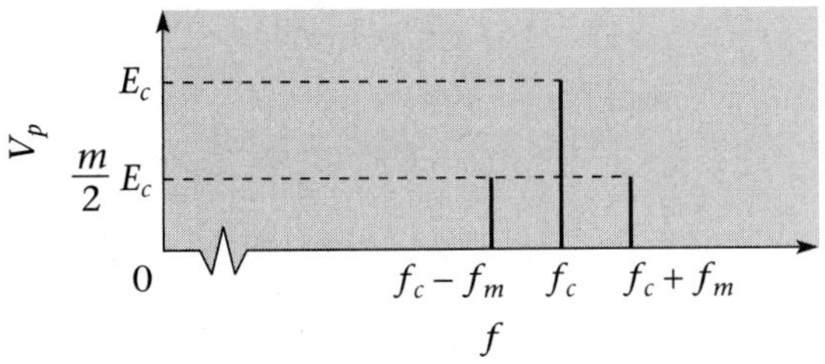

From now on we will generally use the term *sideband,* rather than *side frequency,* even for the case of single-tone modulation, because it is more general and more commonly used in practice.

Mathematically, we have:

$$f_{usb} = f_c + f_m \tag{2.10}$$

$$f_{lsb} = f_c - f_m \tag{2.11}$$

$$E_{lsb} = E_{usb} = \frac{mE_c}{2} \tag{2.12}$$

where

$$
\begin{aligned}
f_{usb} &= \text{upper sideband frequency} \\
f_{lsb} &= \text{lower sideband frequency} \\
E_{usb} &= \text{peak voltage of the upper-sideband component} \\
E_{lsb} &= \text{peak voltage of the lower-sideband component} \\
E_c &= \text{peak carrier voltage}
\end{aligned}
$$

EXAMPLE 2.5

(a) A 1-MHz carrier with an amplitude of 1 volt peak is modulated by a 1-kHz signal with $m = 0.5$. Sketch the voltage spectrum.

(b) An additional 2-kHz signal modulates the carrier with $m = 0.2$. Sketch the voltage spectrum.

SOLUTION

(a) The frequency scale is easy. There are three frequency components. The carrier is at:

$$f_c = 1 \text{ MHz}$$

The upper sideband is at:

$$
\begin{aligned}
f_{usb} &= f_c + f_m \\
&= 1 \text{ MHz} + 1 \text{ kHz} \\
&= 1.001 \text{ MHz}
\end{aligned}
$$

The lower sideband is at:

$$
\begin{aligned}
f_{lsb} &= f_c - f_m \\
&= 1 \text{ MHz} - 1 \text{ kHz} \\
&= 0.999 \text{ MHz}
\end{aligned}
$$

Next we have to determine the amplitudes of the three components. The carrier is unchanged with modulation, so it remains at 1 V peak. The two sidebands have the same peak voltage:

$$E_{lsb} = E_{usb} = \frac{mE_c}{2}$$

$$= \frac{0.5 \times 1}{2}$$

$$= 0.25 \ V$$

Figure 2.7(a) shows the solution.

(b) The addition of another modulating signal at a different frequency simply adds another set of side frequencies. It does not change anything that was done in part (a). The new frequency components are at 1.002 and 0.998 MHz, and their amplitude is 0.1 volt. The result is shown in Figure 2.7(b).

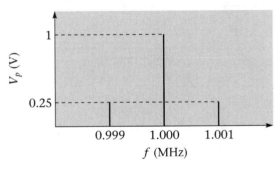

(a) $f_c = 1$ MHz $f_m = 1$ kHz $m = 0.5$ $E_c = 1$ V

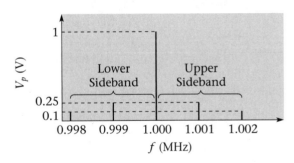

(b) $f_c = 1$ MHz $E_c = 1$ V $f_{m1} = 1$ kHz $m_1 = 0.5$
$f_{m2} = 2$ kHz $m_2 = 0.2$

FIGURE 2.7

Bandwidth Signal bandwidth is one of the most important characteristics of any modulation scheme. In general, a narrow bandwidth is desirable. In any situation where spectrum space is limited, a narrow bandwidth allows more signals to be transmitted simultaneously than does a wider bandwidth. It also allows a narrower bandwidth to be used in the receiver. The receiver must have a wide enough bandwidth to pass the complete signal, including all the sidebands, or distortion will result. Since thermal noise is evenly distributed over the frequency domain, a narrower receiver bandwidth includes less noise and this increases the signal-to-noise ratio, unless there are other factors.

The bandwidth calculation is very easy for AM. The signal extends from the lower side frequency, which is the difference between the carrier frequency and the modulation frequency, to the upper side frequency, at the sum of the carrier frequency and the modulation frequency. The difference between these is simply twice the modulation frequency. If there is more than one modulating frequency, the bandwidth is twice the *highest* modulating frequency. Mathematically, the relationship is:

$$B = 2F_m \qquad (2.13)$$

where

$$B = \text{bandwidth in hertz}$$
$$F_m = \text{the highest modulating frequency in hertz}$$

EXAMPLE 2.6 ▼

Citizens' band radio channels are 10 kHz wide. What is the maximum modulation frequency that can be used if a signal is to remain entirely within its assigned channel?

SOLUTION

From Equation (2.13) we have

$$B = 2 F_m$$

so

$$F_m = \frac{B}{2}$$

$$= \frac{10 \text{ kHz}}{2}$$

$$= 5 \text{ kHz}$$

▲

Power Relationships Power is important in any communication scheme, because the crucial signal-to-noise ratio at the receiver depends as much on the signal power being large as on the noise power being small. The power that is most important, however, is not the total signal power but only that portion that is used to transmit information. Since the carrier in an AM signal remains unchanged with modulation, it contains no information. Its only function is to aid in demodulating the signal at the receiver. This makes AM inherently wasteful of power, compared with some other modulation schemes to be described later.

The easiest way to look at the power in an AM signal is to use the frequency domain. We can find the power in each frequency component, then add to get total power. We shall assume that the signal appears across a resistance R, so that reactive volt-amperes can be ignored. We will also assume that the power required is average power.

Suppose that the modulating signal is a sine wave. Then the AM signal consists of three sinusoids, the carrier and two sidebands, as shown in Figure 2.6.

The power in the carrier is easy to calculate, since the carrier by itself is a sine wave. The carrier is given by the equation

$$e_c = E_c \sin \omega_c t$$

where

$\quad e_c$ = instantaneous carrier voltage

$\quad E_c$ = peak carrier voltage

$\quad \omega_c$ = carrier frequency in radians per second

Since E_c is the peak carrier voltage, the power developed when this signal appears across a resistance R is simply

$$P_c = \frac{\left(\dfrac{E_c}{\sqrt{2}}\right)^2}{R}$$

$$= \frac{E_c^2}{2R}$$

The next step is to find the power in each sideband. The two frequency components have the same amplitude, so they have equal power. Assuming sine-wave modulation, each sideband is a cosine wave whose peak voltage is given by Equation (2.12):

$$E_{lsb} = E_{usb} = mE_c/2$$

Since the carrier and both sidebands are part of the same signal, the sidebands appear across the same resistance, R, as the carrier. Looking at the lower sideband,

$$P_{lsb} = \frac{E_{lsb}^2}{2R}$$

$$= \frac{\left(\dfrac{mE_c}{2}\right)^2}{2R}$$

$$= \frac{m^2 E_c^2}{4 \times 2R}$$

$$= \frac{m^2}{4} \times \frac{E_c^2}{2R}$$

$$P_{lsb} = P_{usb} = \frac{m^2}{4} P_c \tag{2.14}$$

Since the two sidebands have equal power, the total sideband power is given by

$$P_{sb} = \frac{m^2}{2} P_c \tag{2.15}$$

The total power in the whole signal is just the sum of the power in the carrier and the sidebands, so it is

$$P_t = P_c + \left(\frac{m^2}{2}\right) P_c$$

or

$$P_t = P_c\left(1 + \frac{m^2}{2}\right) \tag{2.16}$$

These latest equations tell us several useful things:

- The total power in an AM signal increases with modulation, reaching a value 50% greater than that of the unmodulated carrier for 100% modulation.
- The extra power with modulation goes into the sidebands: the carrier power does not change with modulation.
- The useful power, that is, the power that carries information, is rather small, being a maximum of one-third of the total signal power for 100% modulation and much less at lower modulation indices. For this reason, AM transmission is more efficient when the modulation index is as close to 1 as practicable.

ANALOG MODULATION SCHEMES • 51

EXAMPLE 2.7

An AM transmitter has a carrier power output of 50 W. What would be the total power produced with 80% modulation?

SOLUTION

$$P_t = P_c \left(1 + \frac{m^2}{2} \right)$$

$$= 50 \text{ W} \left(1 + \frac{0.8^2}{2} \right)$$

$$= 66 \text{ W}$$

Measuring Modulation Index in the Frequency Domain

Since the ratio between sideband and carrier power is a simple function of m, it is quite possible to measure modulation index by observing the spectrum of an AM signal. The only complication is that spectrum analyzers generally display power ratios in decibels. The power ratio between sideband and carrier power can easily be found from the relation:

$$\frac{P_{lsb}}{P_c} = \text{antilog} \left(\frac{dB}{10} \right) \tag{2.17}$$

where

P_c = carrier power

P_{lsb} = power in one sideband

dB = difference between sideband and carrier signals, measured in dB (this number will be negative)

Once the ratio between carrier and sideband power has been found, it is easy to find the modulation index from Equation (2.14):

$$P_{lsb} = \frac{m^2}{4} P_c$$

$$m^2 = \frac{4 P_{lsb}}{P_c}$$

$$m = 2 \sqrt{\frac{P_{lsb}}{P_c}} \tag{2.18}$$

Although the time-domain measurement described earlier is simpler and uses less-expensive equipment, frequency-domain measurement enables much smaller values of m to be found. A modulation level of 5%, for instance, would be almost invisible on an oscilloscope, but it is quite obvious, and easy to measure, on a spectrum analyzer. The spectrum analyzer also allows the contribution from different modulating frequencies to be observed and calculated separately.

EXAMPLE 2.8

Calculate the modulation frequency and modulation index for the spectrum analyzer display shown in Figure 2.8.

FIGURE 2.8

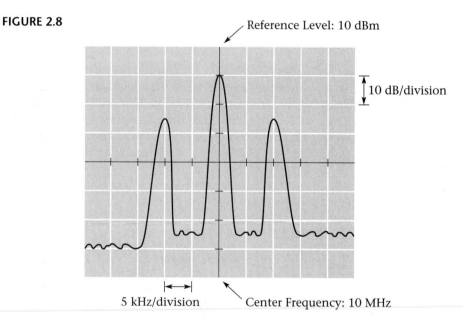

Reference Level: 10 dBm

10 dB/division

5 kHz/division

Center Frequency: 10 MHz

SOLUTION

First let us find f_m. The difference between the carrier and either sideband is 2 divisions at 5 kHz/division, or 10 kHz. So $f_m = 10$ kHz.

Next, we need to find the modulation index. The two sidebands have the same power, so we can use either. The spectrum analyzer is set for 10 dB/division, and each sideband is 1.5 divisions, or 15 dB, below the carrier. This corresponds to a power ratio of

$$\frac{P_{lsb}}{P_c} = \text{antilog}\left(\frac{-15}{10}\right)$$
$$= 0.0316$$

From Equation (2.18),

$$m = 2\sqrt{\frac{P_{lsb}}{P_c}}$$
$$= 2\sqrt{0.0316}$$
$$= 0.356$$

2.3 Suppressed-Carrier AM Systems

It is possible to improve the efficiency and reduce the bandwidth of an AM signal by removing the carrier and/or one of its sidebands. Recall from the previous section that the carrier has at least two-thirds of the power in an AM signal, but none of the information. This can be understood by noting that the presence of modulation has no effect on the carrier. Removing the carrier to create a *double-sideband suppressed-carrier (DSBSC)* AM signal should therefore result in a power gain for the information-carrying part of the signal of at least three (or about 4.8 dB), assuming that the power removed from the carrier could be put into the sidebands. Note also that the upper and lower sidebands are mirror images of each other, containing exactly the same information. Removing one of these sidebands would reduce the signal bandwidth by half. Assuming that the receiver bandwidth is also reduced by half, this should result in a reduction of the noise power by a factor of two (3 dB). Therefore, removing the carrier and one sideband should cause the resulting *single-sideband suppressed-carrier AM (SSBSC or just SSB)* signal to have a signal-to-noise improvement of 7.8 dB or more, compared with full-carrier double-sideband AM.

It is quite practical to remove the carrier from an AM signal, provided it is re-inserted at the receiver. Removing one sideband is also effective, and there is no need to replace it. Single-sideband AM is quite popular for voice communication systems operating in the high-frequency range (3–30 MHz) and has also been used for terrestrial point-to-point microwave links carrying telephone and television signals.

Figure 2.9 shows the idea. Figure 2.9(a) shows the baseband spectrum of a typical voice signal. In Figure 2.9(b) we have double-sideband suppressed-carrier AM (DSBSC). The carrier frequency of 1 MHz is indicated but there is no carrier, just the upper and lower sidebands. In Figure 2.9(c), the lower sideband has been removed and only the upper sideband is transmitted.

FIGURE 2.9

DSB and SSB transmission

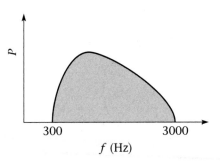

(a) Baseband spectrum

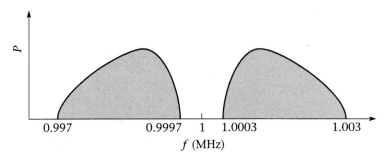

(b) DSBSC spectrum

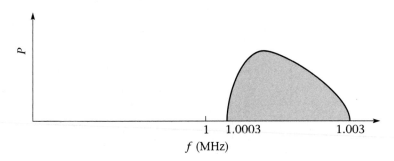

(c) SSB (USB) spectrum

Since single-sideband is a variant of AM, an SSB signal does have an envelope and must be used with linear amplifiers. The envelope is different from that for a full-carrier AM signal, however. Figure 2.10 shows a signal with two modulation frequencies, called a *two-tone test signal*. Note that the envelope is caused by the algebraic addition of the two sideband components. Its frequency is that of the difference between the two modulating signal frequencies, in this case 2 kHz.

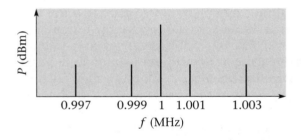

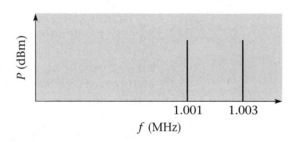

(a) AM with 1 kHz and 3 kHz modulation
(frequency domain)

(b) USB with two-tone modulation
(frequency domain)

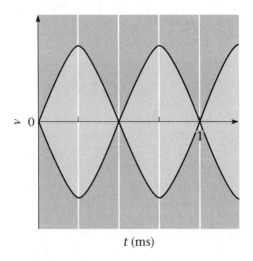

(c) USB with two-tone modulation
(time domain)

FIGURE 2.10 Two-tone modulation

 ## 2.4 Frequency and Phase Modulation

Frequency modulation (FM) is probably the most commonly used analog modulation technique, seeing application in everything from broadcasting to cordless phones. **Phase modulation** (PM) is rarely used in analog systems but is very common in digital communication. Obviously, frequency and phase are very closely related, so it makes sense to discuss the two schemes together. In fact, they are often grouped under the heading of **angle modulation**.

In our discussion of amplitude modulation, we found that the amplitude of the modulated signal varied in accordance with the instantaneous amplitude of the modulating signal. In FM it is the frequency, and in PM the phase of the modulated signal that varies with the amplitude of the modulating signal. This is important to remember: in all types of modulation it is the amplitude, not the frequency, of the baseband signal that does the modulating.

The amplitude and power of an angle-modulation signal do not change with modulation. Thus, an FM signal has no envelope. This is actually an advantage; an FM receiver does not have to respond to amplitude variations, and this lets it ignore noise to some extent. Similarly, FM equipment can use nonlinear amplifiers throughout, since amplitude linearity is not important.

Frequency Modulation

Figure 2.11 demonstrates the concept of frequency modulation. Although a sine wave is mathematically simpler, a square-wave modulating signal is used in the figure to make the process easier to follow by eye. Figure 2.11(a) shows the unmodulated carrier and the modulating signal. Figure 2.11(b) shows the modulated signal in the time domain, as it would appear on an oscilloscope. The amount of frequency change has been exaggerated for clarity. The amplitude remains as before, and the frequency changes can be seen in the changing times between zero crossings for the waveforms. Figure 2.11(c) of the figure shows how the signal frequency varies with time in accordance with the amplitude of the modulating signal. Finally, in Figure 2.11(d) we see how the phase angle varies with time. When the frequency is greater than f_c, the phase angle of the signal gradually increases until it leads that of the carrier, and when the frequency is lower than f_c, the signal phase gradually lags that of the carrier.

The maximum amount by which the transmitted frequency shifts in one direction from the carrier frequency is defined as the **deviation**. The total frequency *swing* is thus twice the deviation. A **frequency modulation index**, m_f, is also defined:

$$m_f = \frac{\delta}{f_m} \tag{2.19}$$

where

$$
\begin{aligned}
m_f &= \text{frequency modulation index} \\
\delta &= \text{peak deviation in hertz} \\
f_m &= \text{modulating frequency in hertz}
\end{aligned}
$$

The FM modulation index varies with the modulating frequency, unlike the case for AM. This choice of a definition for m_f causes the modulation index to be equal to the peak phase deviation in radians, which is inversely proportional to the modulating frequency. The modulation index for phase modulation is also defined as the peak phase deviation.

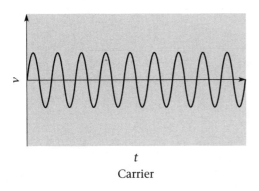

Carrier

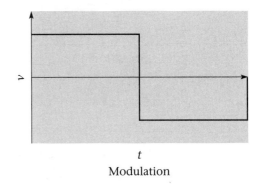

Modulation

(a) Carrier and modulating signals before modulation

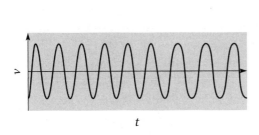

(b) Modulated signal (time domain)

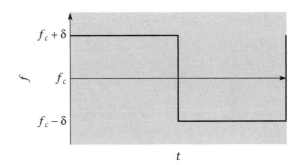

(c) Frequency as a function of time

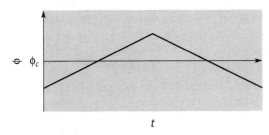

(d) Phase as a function of time

FIGURE 2.11 Frequency modulation

EXAMPLE 2.9

A cell phone transmitter has a maximum frequency deviation of 12 kHz. Calculate the modulation index if it operates at maximum deviation with a voice frequency of

(a) 300 Hz

(b) 2500 Hz

SOLUTION

(a)
$$m_f = \frac{\delta}{f_m}$$
$$= \frac{12 \text{ kHz}}{300 \text{ Hz}}$$
$$= 40$$

(b)
$$m_f = \frac{\delta}{f_m}$$
$$= \frac{12 \text{ kHz}}{2500 \text{ Hz}}$$
$$= 4.8$$

Note that there is no requirement for the FM (or PM) modulation index to be less than 1. When FM modulation is expressed as a percentage, it is the deviation as a percentage of the maximum allowed deviation that is being stated.

The Angle Modulation Spectrum

Frequency modulation produces an infinite number of sidebands, even for single-tone modulation. These sidebands are separated from the carrier by multiples of f_m, but their amplitude tends to decrease as their distance from the carrier frequency increases. Sidebands with amplitude less than about 1% of the total signal voltage can usually be ignored; for practical purposes an angle-modulated signal can be considered to be band-limited. In most cases, though, its bandwidth is much larger than that of an AM signal.

Bessel Functions

The equation for modulation of a carrier with amplitude A and radian frequency ω_c by a single-frequency sinusoid is of the form

$$v(t) = A \sin (\omega_c t + m_f \sin \omega_m t) \tag{2.20}$$

This equation cannot be simplified by ordinary trigonometry, as is the case for amplitude modulation. About the only useful information that can be gained by inspection is the fact that the signal amplitude remains constant regardless of the modulation index. This observation is significant, since it demonstrates one of the major differences between AM and FM or PM, but it provides no information about the sidebands.

This signal can be expressed as a series of sinusoids by using Bessel functions of the first kind. Proving this is beyond the scope of this text, but it can be done. The Bessel functions themselves are rather tedious to evaluate numerically, but that, too, has been done. Some results are presented in Figure 2.12 and Table 2.1 shown on page 60. Bessel functions are equally valid for FM and PM systems, since the modulation index is equal to the peak phase deviation, in radians, for both techniques.

FIGURE 2.12

Bessel Functions

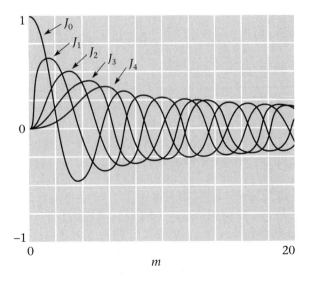

The table and graph of Bessel functions represent normalized voltages for the various frequency components of an FM signal. That is, the numbers in the tables will represent actual voltages if the unmodulated carrier has an amplitude of one volt. J_0 represents the component at the carrier frequency. J_1 represents each of the first pair of sidebands, at frequencies of $f_c + f_m$ and $f_c - f_m$. J_2 represents the amplitude of each of the second pair of sidebands, which are separated from the carrier frequency by twice the modulating frequency, and so on. Figure 2.13 shows this on a frequency-domain plot. All of the Bessel terms should be multiplied by the voltage of the unmodulated carrier to find the actual sideband amplitudes. Of course, the

TABLE 2.1   Bessel Functions

m	J_0	J_1	J_2	J_3	J_4	J_5	J_6	J_7	J_8	J_9	J_{10}	J_{11}	J_{12}	J_{13}	J_{14}	J_{15}	J_{16}	J_{17}	J_{18}	J_{19}	J_{20}
0	1.00																				
0.25	0.98	0.12																			
0.5	0.94	0.24	0.03																		
0.75	0.86	0.35	0.07	0.01																	
1	0.77	0.44	0.11	0.02																	
1.25	0.65	0.51	0.17	0.04	0.01																
1.5	0.51	0.56	0.23	0.06	0.01																
1.75	0.37	0.58	0.29	0.09	0.02																
2	0.22	0.58	0.35	0.13	0.03																
2.25	0.08	0.55	0.40	0.17	0.05	0.01															
2.4	0.00	0.52	0.43	0.20	0.06	0.02															
2.5	−0.05	0.50	0.45	0.22	0.07	0.02															
2.75	−0.16	0.43	0.47	0.26	0.10	0.03	0.01														
3	−0.26	0.34	0.49	0.31	0.13	0.04	0.01														
3.5	−0.38	0.14	0.46	0.39	0.20	0.08	0.03	0.01													
4	−0.40	−0.07	0.36	0.43	0.28	0.13	0.05	0.01													
4.5	−0.32	−0.23	0.22	0.42	0.35	0.20	0.08	0.03	0.01												
5	−0.18	−0.33	0.05	0.36	0.39	0.26	0.13	0.05	0.02	0.01											
5.5	0.00	−0.34	−0.12	0.26	0.40	0.32	0.19	0.09	0.03	0.01											
6	0.15	−0.28	−0.24	0.11	0.36	0.36	0.25	0.13	0.06	0.02	0.01										
6.5	0.26	−0.15	−0.31	−0.03	0.28	0.37	0.30	0.18	0.09	0.04	0.01										
7	0.30	−0.01	−0.30	−0.17	0.16	0.35	0.34	0.23	0.13	0.06	0.02	0.01									
7.5	0.27	0.14	−0.23	−0.26	0.02	0.28	0.35	0.28	0.17	0.09	0.04	0.01	0.01								
8	0.17	0.24	−0.11	−0.29	−0.11	0.19	0.34	0.32	0.22	0.13	0.06	0.03	0.01								
8.5	0.04	0.27	0.02	−0.26	−0.21	0.07	0.29	0.34	0.27	0.17	0.09	0.04	0.02	0.01							
8.65	0.00	0.27	0.06	−0.24	−0.23	0.03	0.27	0.34	0.28	0.18	0.10	0.05	0.02	0.01							
9	−0.09	0.25	0.14	−0.18	−0.27	−0.06	0.20	0.33	0.30	0.21	0.13	0.06	0.03	0.01							
10	−0.25	0.04	0.26	0.06	−0.22	−0.23	−0.01	0.22	0.32	0.29	0.21	0.12	0.06	0.03	0.01						
11	−0.17	−0.18	0.14	0.23	−0.01	−0.24	−0.20	0.02	0.23	0.31	0.28	0.20	0.12	0.06	0.03	0.01					
12	0.05	−0.22	−0.08	0.20	0.18	−0.07	−0.24	−0.17	0.04	0.23	0.30	0.27	0.20	0.12	0.07	0.03	0.01				
13	0.21	−0.07	−0.22	0.00	0.22	0.13	−0.12	−0.24	−0.14	0.07	0.23	0.29	0.26	0.19	0.12	0.07	0.03	0.01			
14	0.17	0.13	−0.15	−0.18	0.08	−0.15	−0.23	−0.11	0.08	0.24	0.29	0.25	0.19	0.12	0.07	0.03	0.02	0.01			
15	−0.01	0.20	0.04	−0.19	−0.12	0.13	0.21	0.03	−0.17	−0.22	−0.09	0.10	0.24	0.28	0.25	0.18	0.12	0.07	0.03	0.02	0.01
16	−0.17	0.09	0.19	−0.04	−0.20	−0.06	0.17	0.18	−0.01	−0.19	−0.21	−0.07	0.11	0.24	0.27	0.24	0.18	0.11	0.07	0.03	0.02
17	−0.17	−0.10	0.16	0.14	−0.11	−0.19	0.00	0.19	0.15	−0.04	−0.20	−0.19	−0.05	0.12	0.24	0.27	0.23	0.17	0.11	0.07	0.04

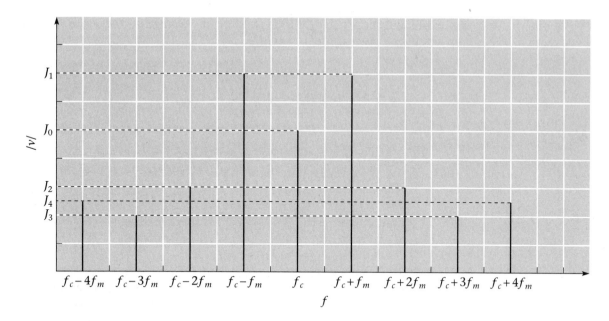

FIGURE 2.13 FM in the frequency domain

Bessel coefficients are equally valid for peak or RMS voltages, but the user should be careful to keep track of which type of measurement is being used.

When Bessel functions are used, the signal of Equation (2.20) becomes

$$
\begin{aligned}
v(t) &= A \sin(\omega_c t + m_f \sin \omega_m t)\\
&= A\, \{J_0(m_f) \sin \omega_c t - J_1(m_f)[\sin(\omega_c - \omega_m)t - \sin(\omega_c + \omega_m)t]\\
&\quad + J_2(m_f)[\sin(\omega_c - 2\omega_m)t + \sin(\omega_c + 2\omega_m)t]\\
&\quad - J_3(m_f)[\sin(\omega_c - 3\omega_m)t + \sin(\omega_c + 3\omega_m)t]\\
&\quad + \cdots\}
\end{aligned}
\tag{2.21}
$$

With angle modulation, the total signal voltage and power do not change with modulation. Therefore, the appearance of power in the sidebands indicates that the power at the carrier frequency must be reduced below its unmodulated value in the presence of modulation. In fact, the carrier-frequency component disappears for certain values of m_f (for example, 2.4 and 5.5).

This constant-power aspect of angle modulation can be demonstrated using the table of Bessel functions. For simplicity, normalized values can be used. Let the unmodulated signal have a voltage of one volt RMS across a resistance of one ohm. Its power is, of course, one watt. When modulation is applied, the carrier voltage will be reduced and sidebands will appear. J_0 from the table will represent the RMS voltage at the carrier frequency and the power at the carrier frequency will be

$$P_c = \frac{V_c^2}{R}$$

$$= \frac{J_0^2}{1}$$

$$= J_0^2$$

Similarly, the power in each of the first pair of sidebands will be

$$P_{SB_1} = J_1^2$$

The combined power in the first set of sidebands will be twice as much as this, of course. The power in the whole signal will then be

$$P_T = J_0^2 + 2J_1^2 + 2J_2^2 + \cdots$$

If the series is carried on far enough, the result will be equal to one watt, regardless of the modulation index.

The bandwidth of an FM or PM signal is to some extent a matter of definition. The Bessel series is infinite, but as can be seen from the table or the graph, the amplitude of the components will gradually diminish until at some point they can be ignored. The process is slower for large values of m, so the number of sets of sidebands that has to be considered is greater for larger modulation indices. A practical rule of thumb is to ignore sidebands with a Bessel coefficient of less than 0.01. The bandwidth, for practical purposes, is equal to twice the number of the highest significant Bessel coefficient, multiplied by the modulating frequency.

EXAMPLE 2.10 ▼

An FM signal has a deviation of 3 kHz and a modulating frequency of 1 kHz. Its total power is 5 W, developed across a 50 Ω resistive load. The carrier frequency is 160 MHz.

(a) Calculate the RMS signal voltage.

(b) Calculate the RMS voltage at the carrier frequency and each of the first three sets of sidebands.

(c) Calculate the frequency of each sideband for the first three sideband pairs.

(d) Calculate the power at the carrier frequency, and in each sideband, for the first three pairs.

(e) Determine what percentage of the total signal power is unaccounted for by the components described above.

(f) Sketch the signal in the frequency domain, as it would appear on a spectrum analyzer. The vertical scale should be power in dBm, and the horizontal scale should be frequency.

SOLUTION

(a) The signal power does not change with modulation, and neither does the voltage, which can easily be found from the power equation.

$$P_T = \frac{V_T^2}{R_L}$$

$$V_T = \sqrt{P_T R_L}$$

$$= \sqrt{5\,\text{W} \times 50\,\Omega}$$

$$= 15.8\ \text{V(RMS)}$$

(b) The modulation index must be found in order to use Bessel functions to find the carrier and sideband voltages.

$$m_f = \frac{\delta}{f_m}$$

$$= \frac{3\,\text{kHz}}{1\,\text{kHz}}$$

$$= 3$$

From the Bessel function table, the coefficients for the carrier and the first three sideband pairs are:

$$J_0 = -0.26 \qquad J_1 = 0.34 \qquad J_2 = 0.49 \qquad J_3 = 0.31$$

These are normalized voltages, so they will have to be multiplied by the total RMS signal voltage to get the RMS sideband and carrier-frequency voltages.

For the carrier,

$$V_c = J_0 V_T$$

J_0 has a negative sign. This simply indicates a phase relationship between the components of the signal. It would be required if we wanted to add together all the components to get the resultant signal. For our present purpose, however, it can be ignored, and we can use

$$V_c = |J_0| V_T$$

$$= 0.26 \times 15.8\ \text{V}$$

$$= 4.11\ \text{V}$$

Similarly we can find the voltage for each of the three sideband pairs. Note that these are voltages for individual components. There will be a lower and an upper sideband with each of these calculated voltages.

$$
\begin{aligned}
V_1 &= J_1 V_T \\
&= 0.34 \times 15.8 \text{ V} \\
&= 5.37 \text{ V}
\end{aligned}
$$

$$
\begin{aligned}
V_2 &= J_2 V_T \\
&= 0.49 \times 15.8 \text{ V} \\
&= 7.74 \text{ V}
\end{aligned}
$$

$$
\begin{aligned}
V_3 &= J_3 V_T \\
&= 0.31 \times 15.8 \text{ V} \\
&= 4.9 \text{ V}
\end{aligned}
$$

(c) The sidebands are separated from the carrier frequency by multiples of the modulating frequency. Here, $f_c = 160$ MHz and $f_m = 1$ kHz, so there are sidebands at each of the following frequencies.

$$
\begin{aligned}
f_{USB1} &= 160 \text{ MHz} + 1 \text{ kHz} = 160.001 \text{ MHz} \\
f_{USB2} &= 160 \text{ MHz} + 2 \text{ kHz} = 160.002 \text{ MHz} \\
f_{USB3} &= 160 \text{ MHz} + 3 \text{ kHz} = 160.003 \text{ MHz} \\
f_{LSB1} &= 160 \text{ MHz} - 1 \text{ kHz} = 159.999 \text{ MHz} \\
f_{LSB2} &= 160 \text{ MHz} - 2 \text{ kHz} = 159.998 \text{ MHz} \\
f_{LSB3} &= 160 \text{ MHz} - 3 \text{ kHz} = 159.997 \text{ MHz}
\end{aligned}
$$

(d) Since each of the components of the signal is a sinusoid, the usual equation can be used to calculate power. All the components appear across the same 50 Ω load.

$$
\begin{aligned}
P_c &= \frac{V_c^2}{R_L} \\
&= \frac{4.11^2}{50} \\
&= 0.338 \text{ W}
\end{aligned}
$$

Similarly, it can be shown that

$$ P_1 = 0.576 \text{ W} \qquad P_2 = 1.2 \text{ W} \qquad P_3 = 0.48 \text{ W} $$

(e) To find the total power in the carrier and the first three sets of sidebands, it is only necessary to add the powers calculated above, counting each of the sideband powers twice, because each of the calculated powers represents one of a pair of sidebands. We only count the carrier once, of course.

$$P_T = P_c + 2(P_1 + P_2 + P_3)$$
$$= 0.338 + 2(0.576 + 1.2 + 0.48) \text{ W}$$
$$= 4.85 \text{ W}$$

This is not quite the total signal power, which was given as 5 W. The remainder is in the additional sidebands. To find how much is unaccounted for by the carrier and the first three sets of sidebands, we can subtract. Call the difference P_x.

$$P_x = 5 - 4.85 = 0.15 \text{ W}$$

As a percentage of the total power this is

$$P_x(\%) = \frac{0.15}{5} \times 100$$
$$= 3\%$$

All the information we need for the sketch is on hand, except that the power values have to be converted to dBm using the equation

$$P(\text{dBm}) = 10 \log \frac{P}{1 \text{ mW}}$$

This gives

$$P_c(\text{dBm}) = 10 \log 338 = 25.3 \text{ dBm}$$
$$P_1(\text{dBm}) = 10 \log 576 = 27.6 \text{ dBm}$$
$$P_2(\text{dBm}) = 10 \log 1200 = 30.8 \text{ dBm}$$
$$P_3(\text{dBm}) = 10 \log 480 = 26.8 \text{ dBm}$$

The sketch is shown in Figure 2.14.

FIGURE 2.14

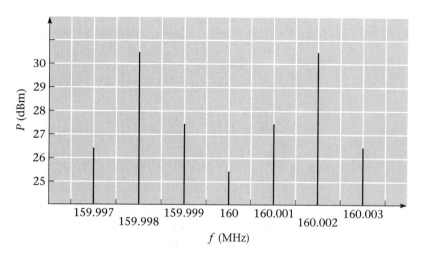

Bandwidth   For PM, the bandwidth varies directly with the modulating frequency, since doubling the frequency doubles the distance between sidebands. It is also roughly proportional to the maximum phase deviation, since increasing m_p increases the number of sidebands. For FM, however, the situation is complicated by the fact that

$$m_f = \frac{\delta}{f_m}$$

For a given amount of frequency deviation, the modulation index is inversely proportional to the modulating frequency. Recall that the frequency deviation is proportional to the amplitude of the modulating signal. Then, if the amplitude of the modulating signal remains constant, increasing its frequency reduces the modulation index. Reducing m_f reduces the number of sidebands with significant amplitude. On the other hand, the increase in f_m means that the sidebands will be further apart in frequency, since they are separated from each other by f_m. These two effects are in opposite directions. The result is that the bandwidth does increase somewhat with increasing modulating-signal frequency, but the bandwidth is not directly proportional to the frequency. Sometimes FM is called a "constant-bandwidth" communication mode for this reason, though the bandwidth is not really constant. Figure 2.15 provides a few examples that show the relationship between modulating frequency and bandwidth. For this example the deviation remains constant at 10 kHz as the modulating frequency varies from 2 kHz to 10 kHz.

One other point must be made about the sidebands. With AM, restricting the bandwidth of the receiver has a very simple effect on the signal. Since the side frequencies farthest from the carrier contain the high-frequency baseband information, restricting the receiver bandwidth reduces its response to high-frequency baseband signals, leaving all else unchanged. When reception conditions are poor, bandwidth can be restricted to the minimum necessary for intelligibility. For FM the situation is more complicated, since even low-frequency modulating signals can generate sidebands that are far removed from the carrier frequency. FM receivers must be designed to include all the significant sidebands that are transmitted; otherwise severe distortion, not just limited frequency response, will result.

Carson's Rule   The calculation of the bandwidth of an FM signal from Bessel functions is easy enough, since the functions are available in a table, but it can be a bit tedious. There is an approximation, known as Carson's rule, that can be used to find the bandwidth of an FM signal. It is not as accurate as using Bessel functions, but can be applied almost instantly, without using tables or even a calculator.

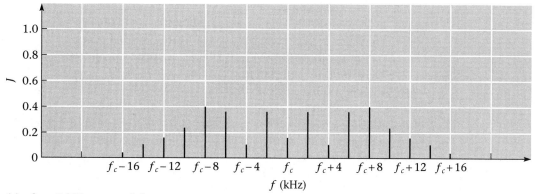

(a) $f_m = 2$ kHz $m_f = 5.0$

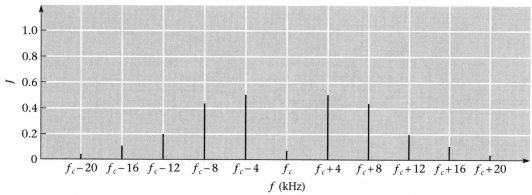

(b) $f_m = 4$ kHz $m_f = 2.5$

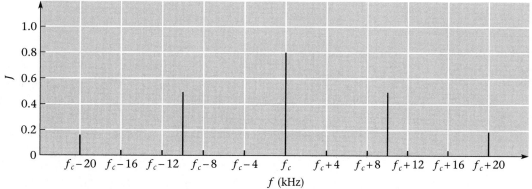

(c) $f_m = 10$ kHz $m_f = 1.0$

FIGURE 2.15 Variation of FM bandwidth with modulating frequency

Here is Carson's rule:

$$B \cong 2[\delta_{(max)} + f_{m(max)}] \tag{2.22}$$

Equation (2.22) assumes that the bandwidth is proportional to the sum of the deviation and the modulating frequency. This is not strictly true. Carson's rule also makes the assumption that maximum deviation occurs with the maximum modulating frequency. Sometimes this leads to errors in practical situations, where often the highest baseband frequencies have much less amplitude than lower frequencies, and therefore do not produce as much deviation.

EXAMPLE 2.11 ▼

Use Carson's rule to calculate the bandwidth of the signal used in Example 2.10.

SOLUTION

Here there is only one modulating frequency, so

$$
\begin{aligned}
B &\cong 2(\delta + f_m) \\
&= 2(3 \text{ kHz} + 1 \text{ kHz}) \\
&= 8 \text{ kHz}
\end{aligned}
$$

In the previous example we found that 97% of the power was contained in a bandwidth of 6 kHz. An 8-kHz bandwidth would contain more of the signal power. Carson's rule gives quite reasonable results in this case, with very little work.

Narrowband and Wideband FM

We mentioned earlier that there are no theoretical limits to the modulation index or the frequency deviation of an FM signal. The limits are practical and result from a compromise between signal-to-noise ratio and bandwidth. In general, larger values of deviation result in an increased signal-to-noise ratio, while also resulting in greater bandwidth. The former is desirable, but the latter is not, especially in regions of the spectrum where frequency space is in short supply. It is also necessary to have some agreement about deviation, since receivers must be designed for a particular signal bandwidth.

For these reasons, the bandwidth of FM transmissions is generally limited by government regulations that specify the maximum frequency deviation and the maximum modulating frequency, since both of these affect bandwidth. In general, relatively narrow bandwidth (on the order of 10 to

30 kHz) is used for voice communication, with wider bandwidths for such services as FM broadcasting (about 200 kHz) and satellite television (36 MHz for one system).

FM and Noise   The original reason for developing FM was to give improved performance in the presence of noise, and that is still one of its main advantages over AM. This improved noise performance can actually result in a better signal-to-noise ratio at the output of a receiver than is found at its input.

One way to approach the problem of FM and noise is to think of the noise voltage as a phasor having random amplitude and phase angle. The noise adds to the signal, causing random variations in both the amplitude and phase angle of the signal as seen by the receiver. Figure 2.16 shows this vector addition.

FIGURE 2.16
Effect of noise on an
FM signal

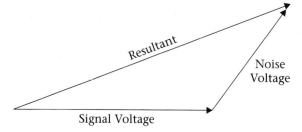

The amplitude component of noise is easily dealt with in a well-designed FM system. Since FM signals do not depend on an envelope for detection, the receiver can employ limiting to remove any amplitude variations from the signal. That is, it can use amplifiers whose output amplitude is the same for a wide variety of input signal levels. The effect of this on the amplitude of a noisy signal is shown in Figure 2.17. As long as the signal amplitude is

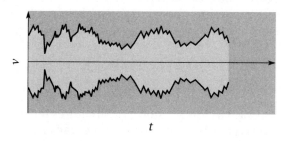

(a) Signal before limiting

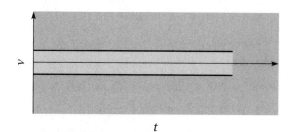

(b) Signal after limiting

FIGURE 2.17   Limiting

considerably larger than the noise to begin with, the amplitude component of the noise will not be a problem.

It is not possible for the receiver to ignore phase shifts, however. A PM receiver obviously must respond to phase changes, but so will an FM receiver because, as we have seen, phase shifts and frequency shifts always occur together. Therefore, phase shifts due to noise are associated with frequency shifts that will be interpreted by the receiver as part of the modulation.

Figure 2.18 shows the situation at the input to the receiver. The circle represents the fact that the noise phasor has a constantly changing angle with respect to the signal. Its greatest effect, and thus the peak phase shift to the signal, will occur when the noise phasor is perpendicular to the resultant. At that time, the phase shift due to noise is

$$\phi_N = \sin^{-1}\left(\frac{E_N}{E_S}\right) \tag{2.23}$$

FIGURE 2.18
Phase shift due to noise

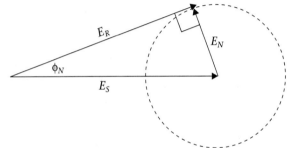

(E_N/E_S) is the reciprocal of the voltage signal-to-noise ratio at the input. A little care is needed here, as S/N is usually given as a power ratio, in decibels, and will have to be converted to a voltage ratio before being used in Equation (2.23).

Equation (2.23) can be simplified as long as the signal is much larger than the noise. This will cause the phase deviation to be small, and for small angles the sine of the angle is approximately equal to the angle itself, in radians. Thus in a practical situation we can use

$$\phi_N \approx \frac{E_N}{E_S} \tag{2.24}$$

The phase shift due to noise can be reduced by making the signal voltage, relative to the noise voltage, as large as possible. This requires increased transmission power, a better receiver noise figure, or both. Perhaps less obvious is the fact that the relative importance of phase shifts due to noise can be reduced by having the phase shifts in the signal as large as possible. This is accomplished by keeping the value of m_f high, since m_f represents the peak phase shift in radians. It would seem that the ratio of signal voltage to noise voltage at the output would be proportional to m_f, and this is approximately true under strong-signal conditions.

EXAMPLE 2.12

An FM signal has a frequency deviation of 5 kHz and a modulating frequency of 1 kHz. The signal-to-noise ratio at the input to the receiver detector is 20 dB. Calculate the approximate signal-to-noise ratio at the detector output.

SOLUTION

First, notice the word "approximate." Our analysis is obviously a little simplistic, since noise exists at more than one frequency. We are also going to assume that the detector is completely unresponsive to amplitude variations and that it adds no noise of its own. Our results will not be precise but they will show the process that is involved.

First, let us convert 20 dB to a voltage ratio.

$$\frac{E_S}{E_N} = \log^{-1} \frac{(S/N)(\text{dB})}{20}$$

$$= \log^{-1} \frac{20}{20}$$

$$= 10$$

$$\frac{E_N}{E_S} = \frac{1}{10}$$

$$= 0.1$$

Since $E_S \gg E_N$, we can use Equation (2.24).

$$\phi_N \approx \frac{E_N}{E_S}$$

$$= 0.1 \text{ rad}$$

Remembering that the receiver will interpret the noise as an FM signal with a modulation index equal to ϕ_N, we find

$$m_{fN} = 0.1$$

This can be converted into an equivalent frequency deviation δ_N due to the noise.

$$\delta_N = m_f f_m$$

$$= 0.1 \times 1 \text{ kHz}$$

$$= 100 \text{ Hz}$$

The frequency deviation due to the signal is given as 5 kHz, and the receiver output voltage is proportional to the deviation. Therefore, the output S/N as a voltage ratio will be equal to the ratio between the deviation due to the signal and that due to the noise.

$$\left(\frac{E_s}{E_N}\right) = \frac{\delta_s}{\delta_N}$$

$$= \frac{5 \text{ kHz}}{100 \text{ Hz}}$$

$$= 50$$

Since S/N is nearly always expressed in decibels, change this to dB.

$$(S/N)_o \text{ (dB)} = 20 \log 50$$
$$= 34 \text{ dB}$$

This is an improvement of 14 dB over the S/N at the input.

Threshold Effect and Capture Effect

An FM signal can produce a better signal-to-noise ratio at the output of a receiver than an AM signal with a similar input S/N, but this is not always the case. The superior noise performance of FM depends on there being a sufficient input S/N ratio. There exists a threshold S/N below which the performance is no better than AM. In fact, it is worse, because the greater bandwidth of the FM signal requires a wider receiver noise bandwidth. When the signal strength is above the threshold, the improvement in noise performance for FM can be more than 20 dB compared with AM.

The noise-rejection characteristic of FM applies equally well to interference. As long as the desired signal is considerably stronger than the interference, the ratio of desired to interfering signal strength will be greater at the output of the detector than at the input. We could say that the stronger signal "captures" the receiver, and in fact this property of FM is usually called the **capture effect**. It is very easy to demonstrate with any FM system. For example, it is the reason that there is less interference between cordless telephones, which share a few channels in the 46- and 49-MHz bands, than one might expect.

Pre-emphasis and De-emphasis

An FM receiver interprets the phase shifts due to noise as frequency modulation. Phase and frequency deviation are related by Equation (2.19):

$$m_f = \frac{\delta}{f_m}$$

which can be restated

$$\delta = m_f f_m$$

The modulation index m_f is simply the peak phase deviation in radians. The frequency deviation is proportional to the modulating frequency. This tells us, if the phase deviation due to thermal noise is randomly distributed over the baseband spectrum, the amplitude of the demodulated noise will be proportional to frequency. This relationship between noise voltage and frequency is shown in Figure 2.19. Since power is proportional to the square of voltage, the noise power will have the parabolic spectrum shown in Figure 2.19. An improvement in *S/N* can be made by boosting (pre-emphasizing) the higher frequencies in the baseband signal before modulation, with a corresponding cut in the receiver after demodulation. Obviously it is necessary to use similar filter characteristics for pre- and de-emphasis. Usually simple first-order filters are used.

FIGURE 2.19
Spectrum of demodulated noise

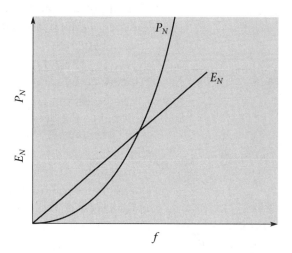

Note that pre-emphasis and de-emphasis are unnecessary with phase modulation. Since the phase deviation due to noise is converted directly into baseband noise output without the intermediate step of conversion into an equivalent frequency deviation, the output noise will have a flat spectrum, assuming thermal noise at the input to the demodulator.

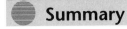 **Summary**   The main points to remember from this chapter are:

- In the time domain, the process of amplitude modulation creates a signal with an envelope that closely resembles the original information signal.

- In the frequency domain, an amplitude-modulated signal consists of the carrier, which is unchanged from its unmodulated state, and two sidebands. The total bandwidth of the signal is twice the maximum modulating frequency.

- An amplitude-modulated signal can be demodulated by an envelope detector, which consists of a diode followed by a lowpass filter.

- The peak voltage of an amplitude-modulated signal varies with the modulation index, becoming twice that of the unmodulated carrier for the maximum modulation index of 1.

- The power in an amplitude-modulated signal increases with modulation. The extra power goes into the sidebands. At maximum modulation, the total power is 50% greater than the power in the unmodulated carrier.

- Angle modulation includes frequency and phase modulation, which are closely related.

- Frequency modulation is widely used for analog communication, while phase modulation sees greatest application in data communication.

- The power of an angle-modulation signal does not change with modulation, but the bandwidth increases due to the generation of multiple sets of sidebands.

- The voltage and power of each sideband can be calculated using Bessel functions. An approximate bandwidth is given by Carson's rule.

- Frequency modulation has a significant advantage compared with AM in the presence of noise or interference, provided the deviation is relatively large and the signal is reasonably strong.

- The signal-to-noise ratio for FM can be improved considerably by using pre-emphasis and de-emphasis. This involves greater gain for the higher baseband frequencies before modulation, with a corresponding reduction after demodulation.

● Equation List

$$v(t) = (E_c + E_m \sin \omega_m t) \sin \omega_c t \tag{2.2}$$

$$m = E_m/E_c \tag{2.3}$$

$$v(t) = E_c(1 + m \sin \omega_m t) \sin \omega_c t \tag{2.4}$$

$$m_T = \sqrt{m_1^2 + m_2^2 + \cdots} \qquad (2.5)$$

$$E_{max} = E_c(1 + m) \qquad (2.6)$$

$$E_{min} = E_c(1 - m) \qquad (2.7)$$

$$m = \frac{E_{max} - E_{min}}{E_{max} + E_{min}} \qquad (2.8)$$

$$v(t) = E_c \sin \omega_c t + \frac{mE_c}{2} \cos(\omega_c - \omega_m)t - \frac{mE_c}{2} \cos(\omega_c + \omega_m)t \qquad (2.9)$$

$$f_{usb} = f_c + f_m \qquad (2.10)$$

$$f_{lsb} = f_c - f_m \qquad (2.11)$$

$$E_{lsb} = E_{usb} = \frac{mE_c}{2} \qquad (2.12)$$

$$B = 2f_m \qquad (2.13)$$

$$P_{lsb} = P_{usb} = \frac{m^2}{4} P_c \qquad (2.14)$$

$$P_{SB} = \frac{m^2}{2} P_c \qquad (2.15)$$

$$P_t = P_c\left(1 + \frac{m^2}{2}\right) \qquad (2.16)$$

$$m = 2\sqrt{\frac{P_{lsb}}{P_c}} \qquad (2.18)$$

$$m_f = \frac{\delta}{f_m} \qquad (2.19)$$

$$
\begin{aligned}
v(t) = {} & A \sin (\omega_c t + m_f \sin \omega_m t) \\
= {} & A \{J_0(m_f) \sin \omega_c t - J_1(m_f)[\sin (\omega_c - \omega_m)t - \sin (\omega_c + \omega_m)t] \\
& + J_2(m_f) [\sin (\omega_c - 2\omega_m)t + \sin (\omega_c + 2\omega_m)t] \\
& - J_3(m_f)[\sin (\omega_c - 3\omega_m)t + \sin (\omega_c + 3 \omega_m)t] \\
& + \cdots\}
\end{aligned}
\qquad (2.21)
$$

$$B \cong 2[\delta_{(max)} + f_{m(max)}] \tag{2.22}$$

$$\phi_N \approx \frac{E_N}{E_S} \tag{2.24}$$

● Key Terms

angle modulation   term that applies to both frequency modulation (FM) and phase modulation (PM) of a transmitted signal

capture effect   tendency of an FM receiver to receive the strongest signal and reject others

deviation   in FM, the peak amount by which the instantaneous signal frequency differs from the carrier frequency in each deviation

envelope   imaginary pattern formed by connecting the peaks of individual RF waveforms in an amplitude-modulated signal

frequency modulation   modulation scheme in which the transmitted frequency varies in accordance with the instantaneous amplitude of the information signal

frequency modulation index   peak phase shift in a frequency-modulated signal, in radians

modulation index   number indicating the degree to which a signal is modulated

overmodulation   modulation to an extent greater than that allowed for either technical or regulatory reasons

phase modulation   communication system in which the phase of a high-frequency carrier is varied according to the amplitude of the baseband (information) signal

side frequencies   frequency components produced above and below the carrier frequency by the process of modulation

sideband   a group of side frequencies above or below the carrier frequency

splatter   frequency components produced by a transmitter that fall outside its assigned channel

● Questions

1. What is meant by the "envelope" of an AM waveform, and what is its significance?

2. Although amplitude modulation certainly involves changing the amplitude of the signal; it is not true to say that the amplitude of the carrier is modulated. Explain this statement.

3. Why is it desirable to have the modulation index of an AM signal as large as possible, without overmodulating?

4. Describe what happens when a typical AM modulator is overmodulated, and explain why overmodulation is undesirable.

5. How does the bandwidth of an AM signal relate to the information signal?

6. Describe two ways in which the modulation index of an AM signal can be measured.

7. By how much does the power in an AM signal increase with modulation, compared to the power of the unmodulated carrier?

8. What two types of modulation are included in the term "angle modulation"?

9. Compare, in general terms, the bandwidth and signal-to-noise ratio of FM and AM.

10. Describe and compare two ways to determine the practical bandwidth of an FM signal.

11. What is pre-emphasis and how is it used to improve the signal-to-noise ratio of FM transmissions?

12. For FM, what characteristic of the modulating signal determines the instantaneous frequency deviation?

13. What is the capture effect?

14. Where is phase modulation used?

15. Explain why the signal-to-noise ratio of FM can increase with the bandwidth. Is this always true for FM? Compare with the situation for AM.

16. Compare the effects of modulation on the carrier power and the total signal power in FM and AM.

17. What is the threshold effect?

18. Explain how limiting reduces the effect of noise on FM signals.

19. Explain how noise affects FM signals even after limiting.

20. Explain the fact that there is no simple relationship between modulating frequency and bandwidth for an FM signal.

21. Why does limiting the receiver bandwidth to less than the signal bandwidth cause more problems with FM than with AM?

Problems

1. An AM signal has the equation:

$$v(t) = (15 + 4 \sin 44 \times 10^3 t) \sin 46.5 \times 10^6 t \text{ volts}$$

   (a) Find the carrier frequency.
   (b) Find the frequency of the modulating signal.
   (c) Find the value of m.
   (d) Find the peak voltage of the unmodulated carrier.
   (e) Sketch the signal in the time domain showing voltage and time scales.

2. An AM signal has a carrier frequency of 3 MHz and an amplitude of 5 V peak. It is modulated by a sine wave with a frequency of 500 Hz and a peak voltage of 2 V. Write the equation for this signal and calculate the modulation index.

3. An AM signal consists of a 10-MHz carrier modulated by a 5-kHz sine wave. It has a maximum positive envelope voltage of 12 V and a minimum of 4 V.

   (a) Find the peak voltage of the unmodulated carrier.
   (b) Find the modulation index and percent.
   (c) Sketch the envelope.
   (d) Write the equation for the signal voltage as a function of time.

4. An AM transmitter is modulated by two sine waves, at 1 kHz and 2.5 kHz, with a modulation due to each of 25% and 50% respectively. What is the effective modulation index?

5. For the AM signal sketched in Figure 2.20, calculate:
   (a) the modulation index
   (b) the peak carrier voltage
   (c) the peak modulating-signal voltage

6. For the signal of Figure 2.21, calculate:
   (a) the index of modulation
   (b) the RMS voltage of the carrier without modulation

7. An audio system requires a frequency response from 50 Hz to15 kHz for high fidelity. If this signal were transmitted using AM, what bandwidth would it require?

8. A transmitter operates with a carrier frequency of 7.2 MHz. It is amplitude modulated by two tones with frequencies of 1500 Hz and 3000 Hz. What frequencies are produced at its output?

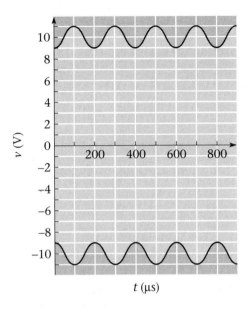

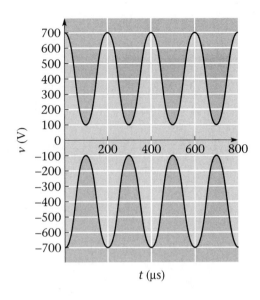

FIGURE 2.20   **FIGURE 2.21**

9. Sketch the signal whose equation is given in Problem 1 in the frequency domain, showing frequency and voltage scales.

10. An AM signal has the following characteristics:

$$f_c = 150 \text{ MHz} \qquad f_m = 3 \text{ kHz} \qquad E_c = 50 \text{ V} \qquad E_m = 40 \text{ V}$$

For this signal, find:

(a) the modulation index

(b) the bandwidth

(c) the peak voltage of the upper sideband

11. An AM signal observed on a spectrum analyzer shows a carrier at +12 dBm, with each of the sidebands 8 dB below the carrier. Calculate:

(a) the carrier power in milliwatts

(b) the modulation index

12. An AM transmitter with a carrier power of 10 W at a frequency of 25 MHz operates into a 50-Ω load. It is modulated at 60% by a 2-kHz sine wave.

(a) Sketch the signal in the frequency domain. Show power and frequency scales. The power scale should be in dBm.

(b) What is the total signal power?

(c) What is the RMS voltage of the signal?

13. A 5-MHz carrier is modulated by a 5-kHz sine wave. Sketch the result in both frequency and time domains for each of the following types of modulation. Time and frequency scales are required, but amplitude scales are not.

    (a) DSB full-carrier AM

    (b) DSBSC AM

    (c) SSBSC AM (USB)

14. If a transmitter power of 100 W is sufficient for reliable communication over a certain path using SSB, approximately what power level would be required using:

    (a) DSBSC

    (b) full-carrier AM

15. An AM transmitter has a carrier power of 50 W at a carrier frequency of 12 MHz. It is modulated at 80% by a 1-kHz sine wave.

    (a) How much power is contained in the sidebands?

    (b) Suppose the transmitter in part (a) can also be used to transmit a USB signal with an average power level of 50 watts. By how much (in dB) will the signal-to-noise ratio be improved when the transmitter is used in this way, compared with the situation in part (a)?

16. An FM signal has a deviation of 10 kHz and a modulating frequency of 2 kHz. Calculate the modulation index.

17. Calculate the frequency deviation for an FM signal with a modulating frequency of 5 kHz and a modulation index of 2.

18. A sine-wave carrier at 100 MHz is modulated by a 1-kHz sine wave. The deviation is 100 kHz. Draw a graph showing the variation of instantaneous modulated signal frequency with time.

19. A phase-modulated signal has a modulation index of 2 with a modulating signal having an amplitude of 100 mV and a frequency of 4 kHz. What would be the effect on the modulation index of:

    (a) changing the frequency to 5 kHz

    (b) changing the voltage to 200 mV

20. An FM signal has a deviation of 10 kHz and is modulated by a sine wave a frequency of 5 kHz. The carrier frequency is 150 MHz, and the signal has a total power of 12.5 W, operating into an impedance of 50 Ω.

    (a) What is the modulation index?

    (b) How much power is present at the carrier frequency?

(c) What is the voltage level of the second sideband below the carrier frequency?

(d) What is the bandwidth of the signal, ignoring all components which have less than 1% of the total signal voltage?

21. An FM transmitter operates with a total power of 10 watts, a deviation of 5 kHz, and a modulation index of 2.

(a) What is the modulating frequency?

(b) How much power is transmitted at the carrier frequency?

(c) If a receiver has a bandwidth sufficient to include the carrier and the first two sets of sidebands, what percentage of the total signal power will it receive?

22. An FM transmitter has a carrier frequency of 220 MHz. Its modulation index is 3 with a modulating frequency of 5 kHz. The total power output is 100 watts into a 50 Ω load.

(a) What is the deviation?

(b) Sketch the spectrum of this signal, including all sidebands with more than 1% of the signal voltage.

(c) What is the bandwidth of this signal according to the criterion used in part (b)?

(d) Use Carson's rule to calculate the bandwidth of this signal, and compare with the result found in part (c).

23. An FM transmitter has a carrier frequency of 160 MHz. The deviation is 10 kHz and the modulation frequency is 2 kHz. A spectrum analyzer shows that the carrier-frequency component of the signal has a power of 5 W. What is the total signal power?

24. Use Carson's rule to compare the bandwidth that would be required to transmit a baseband signal with a frequency range from 300 Hz to 3 kHz using:

(a) narrowband FM with maximum deviation of 5 kHz

(b) wideband FM with maximum deviation of 75 kHz

25. An FM receiver operates with a signal-to-noise ratio of 30 dB at its detector input and is operating with $m_f = 10$.

(a) If the received signal has a voltage of 10 mV, what is the amplitude of the noise voltage?

(b) Find the maximum phase shift that could be given to the signal by the noise voltage.

(c) Calculate the signal-to-noise ratio at the detector output, assuming the detector is completely insensitive to amplitude variations.

26. A certain full-carrier DSB AM signal has a bandwidth of 20 kHz. What would be the approximate bandwidth required if the same information signal were to be transmitted using:

    (a) DSBSC AM

    (b) SSB

    (c) FM with 10 kHz deviation

    (d) FM with 50 kHz deviation

27. Using the table of Bessel functions, demonstrate that the total power in an FM signal is equal to the power in the unmodulated carrier for $m = 2$. Compare with the situation for full-carrier AM and for SSBSC AM.

28. Suppose you were called upon to recommend a modulation technique for a new communication system for voice frequencies. State which of the techniques studied so far you would recommend, and why, in each of the following situations:

    (a) simple, cheap receiver design is of greatest importance

    (b) narrow signal bandwidth is of greatest importance

    (c) immunity to noise and interference is of greatest importance

Digital Communication

3

Objectives

After studying this chapter, you should be able to:

- Compare analog and digital communication techniques and discuss the advantages of each.

- Calculate the minimum sampling rate for a signal and explain the necessity for sampling at that rate or above.

- Find the spurious frequencies produced by aliasing when the sample rate is too low.

- Describe the common types of analog pulse modulation.

- Describe pulse-code modulation and calculate the number of quantizing levels, the bit rate, and the dynamic range for PCM systems.

- Explain companding, show how it is accomplished, and explain its effects.

- Describe the coding and decoding of a PCM signal.

- Describe differential PCM and explain its operation and advantages.

- Describe delta modulation and explain the advantages of adaptive delta modulation.

- Distinguish between lossless and lossy compression and provide examples of each.

- Describe the operation of common types of vocoders.

3.1 Introduction

In the previous chapter we looked at analog modulation techniques. These were historically the earliest ways to transmit information by radio, and they are still very popular. However, recent developments have made digital methods more important, to the point where it is expected that most new wireless communication systems will be digital.

In this chapter we will look at the digital encoding of analog signals. Voice is a good example of such a signal; it begins and ends as analog information. Once coded digitally, the analog signal is indistinguishable from any other data stream, such as a word-processing document. In Chapter 4 we will look at the transmission of this data by radio.

While the advantage of digital transmission for data is self-evident, it is not immediately obvious why it is desirable to go to the trouble and expense of converting analog signals to digital form and back again. There are in fact several very good reasons for doing so. Here are some of the most important for wireless communication systems.

- Digital signals can be manipulated more easily than analog signals. They are easier to multiplex, for instance.

- Digital signals can easily be encrypted to ensure privacy.

- When an analog signal goes through a chain of signal processors, such as transmitters, receivers, and amplifiers, noise and distortion accumulate. This process can be made much less severe in digital systems by regenerating the signal and by using various types of error control. These concepts will be explained in more detail later in this chapter.

- Data compression can be used with a digital signal to reduce its bandwidth to less than that required to transmit the original analog signal. Until recently the use of digital techniques was restricted because digital communication required more bandwidth than analog. That situation has changed quite radically in recent years.

You might have noticed that "higher fidelity" and "improved frequency response" do not appear in the above list. It is certainly true that some digital schemes, such as compact disc audio, have better fidelity than some analog schemes, like FM broadcasts; but this is not always the case. The distortion and frequency response of a digital audio system are fixed by the sampling process, which will be described in the next section. The quality of the reproduction can be predicted quite accurately, and the use of regeneration and error correction can prevent it from deteriorating very much, but not all digital transmission is of high quality. For instance, whether you know it or not you have been communicating digitally for years: practically all modern telephone switches are digital. Digital telephony sounds pretty good, for a telephone call, but does not compare with compact disc audio. We will find out why in the next sections.

 ## 3.2 Sampling

An analog signal varies continuously with time. If we want to transmit such a signal digitally, that is, as a series of numbers, we must first sample the signal. This involves finding its amplitude at discrete time intervals. Only in this way can we arrive at a finite series of numbers to transmit.

Sampling Rate In 1928, Harry Nyquist showed mathematically that it is possible to reconstruct a band-limited analog signal from periodic samples, as long as the sampling rate is at least twice the frequency of the highest-frequency component of the signal. This assumes that an ideal low-pass filter prevents higher frequencies from entering the sampler. Since real filters are not ideal, in practice the sampling rate must be considerably more than twice the maximum frequency to be transmitted.

If the sampling rate is too low, a form of distortion called **aliasing** or **foldover distortion** is produced. In this form of distortion, frequencies in the sampled signal are translated downward. Figure 3.1 shows what happens.

In Figure 3.1(a) the sampling rate is adequate and the signal can be reconstructed. In Figure 3.1(b), however, the rate is too low and the attempt to reconstruct the original signal results in a lower-frequency output signal. Once aliasing is present, it cannot be removed.

The frequency of the interference generated by aliasing is easier to see by looking at the frequency domain. For simplicity, assume that the signal to be sampled, which we will call the baseband signal, is a sine wave:

$$e_b = E_b \sin \omega_b t \tag{3.1}$$

where

e_b = instantaneous baseband signal voltage
E_b = peak baseband signal voltage
ω_b = the radian frequency of the baseband signal

The sampling process is equivalent to multiplying the baseband signal by a pulse train at the sampling frequency. See Figure 3.2. For simplicity, assume that the pulses have an amplitude of one volt. Then, when a pulse is

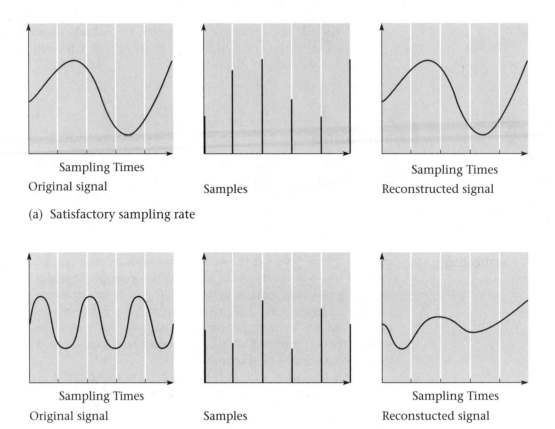

Sampling Times
Original signal

Samples

Sampling Times
Reconstructed signal

(a) Satisfactory sampling rate

Sampling Times
Original signal

Samples

Sampling Times
Reconstucted signal

(b) Sampling rate too low

FIGURE 3.1 Aliasing

present, the output of the sampler will be the same as the baseband signal amplitude, and when there is no pulse, the sampler output will be zero.

The spectrum for the pulse train in Figure 3.2 is given in Chapter 1 as:

$$e_s = \frac{1}{T} + 2\frac{1}{T}\left(\frac{\sin \pi\tau/T}{\pi\tau/T}\cos \omega_s t + \frac{\sin 2\pi\tau/T}{2\pi\tau/T}\cos 2\omega_s t + \frac{\sin 3\pi\tau/T}{3\pi\tau/T}\cos 3\omega_s t + \cdots\right) \quad (3.2)$$

$$= \frac{1}{T} + \frac{2\sin \pi\tau/T}{\pi\tau}\cos \omega_s t + \frac{\sin 2\pi\tau/T}{\pi\tau}\cos 2\omega_s t + \frac{2\sin 3\pi\tau/T}{3\pi\tau}\cos 3\omega_s t + \cdots$$

where

e_s = instantaneous voltage of the sampling pulse
τ = pulse duration
T = pulse period
ω_s = radian frequency of the pulse train

FIGURE 3.2
Sampling pulses

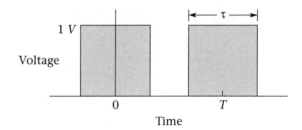

Multiplying the two signals given in Equations (3.1) and (3.2) together gives the following output:

$$v(t) = \frac{E_b}{T} \sin \omega_b t + \frac{2E_b \sin \pi\tau/T}{\pi\tau} \sin \omega_b t \cos \omega_s t \qquad (3.3)$$

$$+ \frac{E_b \sin 2\pi\tau/T}{\pi\tau} \sin \omega_b t \cos 2\omega_s t$$

$$+ \frac{2E_b \sin 3\pi\tau/T}{3\pi\tau} \sin \omega_b t \cos 3\omega_s t$$

As is often the case with equations of this type, we do not have to "solve" anything to understand what is happening. The first term is simply the original baseband signal multiplied by a constant. If only this term were present, the original signal could be recovered from the sampled signal. However, we need to look at the other terms to see whether they will interfere with the signal recovery.

The second term contains the product of $\sin \omega_b t$ and $\cos \omega_s t$. Recall the trigonometric identities:

$$\sin A \cos B = \frac{1}{2} [\sin(A - B) + \sin(A + B)] \qquad (3.4)$$

and

$$\sin(-A) = -\sin A \qquad (3.5)$$

Equation (3.5) can be used to express Equation (3.4) in a more convenient form for the problem at hand. Substituting Equation (3.5) into Equation (3.4) gives:

$$\sin A \cos B = \frac{1}{2} [\sin(B + A) - \sin(B - A)] \qquad (3.6)$$

This new identity can be used to expand the second term of Equation (3.3) as follows:

$$\frac{2E_b \sin \pi\tau/T}{\pi\tau} \sin \omega_b t \cos \omega_s t \tag{3.7}$$

$$= \frac{E_b \sin \pi\tau/T}{\pi\tau} [\sin(\omega_s + \omega_b)t - \sin(\omega_s - \omega_b)t]$$

$$= \frac{E_b \sin \pi\tau/T}{\pi\tau} \sin(\omega_s + \omega_b)t - \frac{E_b \sin \pi\tau/T}{\pi\tau} \sin(\omega_s - \omega_b)t$$

Now we can see that this term consists of components at the sum and difference of the baseband and sampling frequencies. The sum term can easily be eliminated by a low-pass filter, since its frequency is obviously much higher than the baseband. The difference term is more interesting. If $w_s > 2\omega_b$, then $\omega_s - \omega_b > \omega_b$ and the difference part of this term can also be removed by a low-pass filter, at least in theory. However, if $\omega_s < 2\omega_b$, the difference will be less than ω_b. An aliased component will appear as

$$f_a = f_s - f_b \tag{3.8}$$

and low-pass filtering will not be effective in removing it.

The other terms in Equation (3.3) are not interesting here because they all represent frequencies greater than $\omega_s - \omega_b$. Therefore, if we make sure that $(\omega_s - \omega_b) > \omega_b$, these other terms will not be a problem. Let us, then, rewrite Equation (3.3), including only the first term and the expanded second term:

$$v(t) = \frac{E_b}{T} \sin \omega_b t + \frac{E_b \sin \pi\tau/T}{\pi\tau} \sin(\omega_s + \omega_b)t \tag{3.9}$$

$$- \frac{E_b \sin \pi\tau/T}{\pi\tau} \sin(\omega_s - \omega_b)t$$

An example will further clarify the problem.

EXAMPLE 3.1 ▼

A digital communication system uses sampling at 10 kilosamples per second (kSa/s). The receiver filters out all frequencies above 5 kHz. What frequencies appear at the receiver for each of the following signal frequencies at the input to the transmitter?

(a) 1 kHz

(b) 5 kHz

(c) 6 kHz

SOLUTION

(a) The first term in Equation (3.9) is simply the input frequency, which is, of course, the only one we want to see in the output. The second term is the sum of the input and the sampling frequencies, and the third is the difference. In this case, the frequencies generated are:

$$f_b = 1 \text{ kHz} \qquad f_s + f_b = 11 \text{ kHz} \qquad f_s - f_b = 9 \text{ kHz}$$

However, only the 1-kHz component passes through the filter and the system operates correctly.

(b) $\qquad f_b = 5 \text{ kHz} \qquad f_s + f_b = 15 \text{ kHz} \qquad f_s - f_b = 5 \text{ kHz}$

Again, the system works properly and the 15-kHz component is removed by the filter, and only the input frequency of 5 kHz appears at the output.

(c) $\qquad f_b = 6 \text{ kHz} \qquad f_s + f_b = 16 \text{ kHz} \qquad f_s - f_b = 4 \text{ kHz}$

Here we have a serious problem. The 16-kHz component is removed by the filter but the 4-kHz component is not. With 6 kHz at the input we produce 6 kHz and 4 kHz at the output.

Natural and Flat-Topped Sampling

The equations in the previous section assumed that a sample consisted of the baseband signal multiplied by a rectangular pulse. To simplify the mathematics, we assumed that the pulse had an amplitude of 1 V. These assumptions yield a sample pulse whose shape follows that of the original signal, as shown in Figure 3.3(a). This technique is called **natural sampling**.

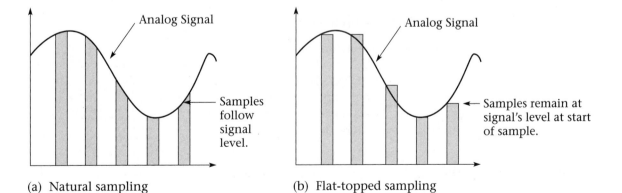

(a) Natural sampling

(b) Flat-topped sampling

FIGURE 3.3 Natural and flat-topped sampling

Practical systems generally sample by using a sample-and-hold circuit, which maintains the signal level at the start of the sample pulse. The results of such a method are shown in Figure 3.3(b). This technique is known as **flat-topped sampling**. If the samples were to be transmitted directly as analog pulses of different amplitudes, there would be some small difference in frequency response between the two techniques. However, for digital transmission there is no practical difference between the two sampling methods.

Analog Pulse Modulation

As previously mentioned, it would be possible to transmit the samples directly as analog pulses. This technique, called **pulse-amplitude modulation (PAM)**, does not offer any great advantage over conventional analog transmission. In current systems, PAM is used as an intermediate step; before being transmitted, the PAM signal is digitized. Similarly, at the receiver, the digital signal is converted back to PAM as part of the demodulation process. The original signal can then be recovered using a low-pass filter.

Some improvement in noise performance can be made by transmitting pulses of equal amplitude but variable length (with the duration of the pulses corresponding to the amplitude of the samples). This technique is called **pulse-duration modulation (PDM)** or **pulse-width modulation (PWM)**. PDM has uses in communication, in some telemetry systems for instance; but it is not likely to be seen in modern wireless systems. Similarly, it is possible to transmit the information signal by using pulses of equal amplitude and duration but changing their timing in accordance with the sample amplitude. This system, called **pulse-position modulation (PPM)**, is mentioned only for completeness, as it is rarely seen. Figure 3.4 shows the basic nature of all these systems.

3.3 Pulse-Code Modulation

Pulse-code modulation (PCM) is the most commonly used digital modulation scheme. In PCM the available range of signal voltages is divided into levels, and each is assigned a binary number. Each sample is then represented by the binary number representing the level closest to its amplitude, and this number is transmitted in serial form. In *linear* PCM, levels are separated by equal voltage gradations.

Quantization and Quantizing Noise

The number of levels available depends on the number of bits used to express the sample value. The number of levels is given by

$$N = 2^m \tag{3.10}$$

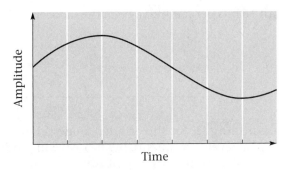

(a) Original signal

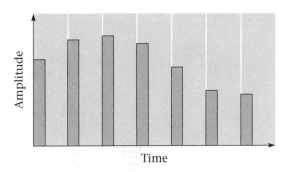

(b) Pulse-amplitude modulation (PAM)

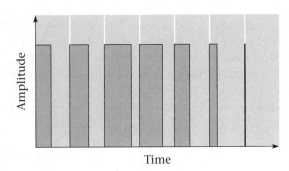

(c) Pulse-duration modulation (PDM)

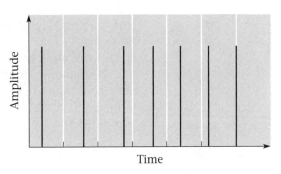

(d) Pulse-position modulation (PPM)

FIGURE 3.4   Analog pulse modulation

where

N = number of levels

m = number of bits per sample

EXAMPLE 3.2

Calculate the number of levels if the number of bits per sample is:

(a) 8 (as used in telephony)

(b) 16 (as used in the compact disc audio system)

SOLUTION

(a) The number of levels with 8 bits per sample is, from Equation (3.10),

$$N = 2^m$$
$$= 2^8$$
$$= 256$$

(b) The number of levels with 16 bits per sample is, from the same equation,

$$N = 2^m$$
$$= 2^{16}$$
$$= 65536$$

This process is called **quantizing**. Since the original analog signal can have an infinite number of signal levels, the quantizing process will produce errors called **quantizing errors** or often **quantizing noise**.

Figure 3.5 shows how quantizing errors arise. The largest possible error is one-half the difference between levels. Thus the error is proportionately greater for small signals. This means that the signal-to-noise ratio varies with the signal level and is greatest for large signals. The level of quantizing noise can be decreased by increasing the number of levels, which also increases the number of bits that must be used per sample.

The dynamic range of a system is the ratio between the strongest possible signal that can be transmitted and the weakest discernible signal. For a linear PCM system, the maximum dynamic range in decibels is given approximately by

$$DR = 1.76 + 6.02m \text{ dB} \tag{3.11}$$

where

$$DR = \text{dynamic range in decibels}$$
$$m = \text{number of bits per sample}$$

This equation ignores any noise contributed by the analog portion of the system.

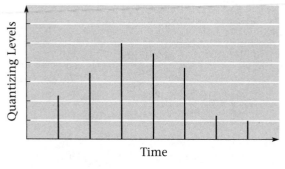

(a) Samples before quantizing

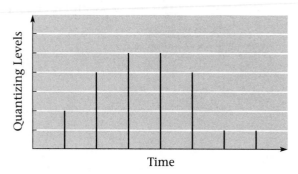

(b) Samples after quantizing

FIGURE 3.5   Quantizing error

EXAMPLE 3.3

Find the maximum dynamic range for a linear PCM system using 16-bit quantizing.

SOLUTION

From Equation (3.11)

$$DR = 1.76 + 6.02m \text{ Db}$$
$$= 1.76 + 6.02 \times 16$$
$$= 98.08 \text{ dB}$$

Bit Rate Increasing the number of bits per sample increases the bit rate, which is given very approximately by

$$D = f_s m \tag{3.12}$$

where

D = data rate in bits per second
f_s = sample rate in samples per second
m = number of bits per sample

Extra bits are often included to detect and correct errors. A few bits, called **framing bits**, are also needed to ensure that the transmitter and receiver agree on which bits constitute one sample. The actual bit rate will therefore be somewhat higher than calculated above.

EXAMPLE 3.4

Calculate the minimum data rate needed to transmit audio with a sampling rate of 40 kHz and 14 bits per sample.

SOLUTION

From Equation (3.12)

$$D = f_s m$$
$$= 40 \times 10^3 \times 14$$
$$= 560 \times 10^3 \text{ b/s}$$
$$= 560 \text{ kb/s}$$

Companding The transmission bandwidth varies directly with the bit rate. In order to keep the bit rate and thus the required bandwidth low, **companding** is often used. This system involves using a compressor amplifier at the transmitter, with greater gain for low-level than for high-level signals. The compressor will reduce the quantizing error for small signals. The effect of compression on the signal can be reversed by using an expander at the receiver, with a gain characteristic that is the inverse of that at the transmitter.

It is necessary to follow the same standards at both ends of the circuit so that the dynamics of the output signal are the same as at the input. The system used in the North American telephone system uses a characteristic known as the μ (*mu*) *law*, which has the following equation for the compressor:

$$v_o = \frac{V_o \ln(1 + \mu v_i / V_i)}{\ln(1 + \mu)} \tag{3.13}$$

where

v_o = actual output voltage from the compressor
V_o = maximum output voltage
V_i = the maximum input voltage
v_i = the actual input voltage
μ = a parameter that defines the amount of compression
(contemporary systems use $\mu = 255$)

European telephone systems use a similar but not identical scheme called *A-law compression*.

Figure 3.6 on page 95 shows the μ-255 curve. The curve is a transfer function for the compressor, relating input and output levels. It has been normalized, that is, v_i/V_i and v_o/V_o are plotted, rather than v_i and v_o.

EXAMPLE 3.5

A signal at the input to a mu-law compressor is positive, with its voltage one-half the maximum input voltage. What proportion of the maximum output voltage is produced?

SOLUTION

From Equation (3.13)

$$v_o = \frac{V_o \ln(1 + \mu v_i / V_i)}{\ln(1 + \mu)}$$

$$= \frac{V_o \ln(1 + 255 \times 0.5)}{\ln(1 + 255)}$$

$$= 0.876 \; V_o$$

This problem can also be solved graphically, using Figure 3.6.

FIGURE 3.6
Mu-Law
compression

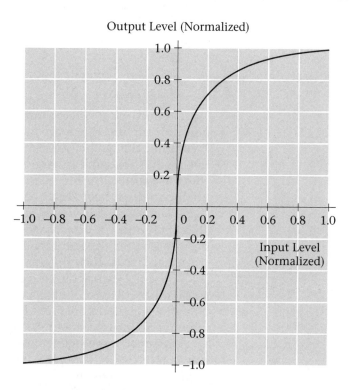

Digital companding is also possible. The method is to quantize a signal using a greater number of bits than will be transmitted, and then perform arithmetic on the samples to reduce the number of bits. This is the way companding is done in most modern telephone equipment. This type of companding is part of the coding and decoding process, which is the topic of the next section.

Coding and Decoding

The process of converting an analog signal into a PCM signal is called **coding**, and the inverse operation, converting back from digital to analog, is known as **decoding**. Both procedures are often accomplished in a single integrated-circuit device called a **codec**.

Figure 3.7 is a block diagram showing the steps for converting an analog signal into a PCM code. The first block is a low-pass filter, required to prevent aliasing. As shown in section 3.2, the filter must block all frequency components above one-half the sampling rate. This requires a high-order filter.

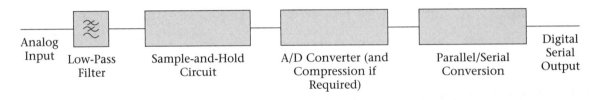

FIGURE 3.7 PCM coding

The next step is to sample the incoming waveform using a sample-and-hold circuit. There are many such circuits; a simple one is shown in Figure 3.8. The field-effect transistor Q turns on during the sample time, allowing the capacitor to charge to the amplitude of the incoming signal. The transistor then turns off, and the capacitor stores the signal value until the analog-to-digital converter has had time to convert the sample to digital form. The two operational amplifiers, connected as voltage followers, isolate the circuit from the other stages. The low output impedance of the first stage ensures that the capacitor quickly charges or discharges to the value of the incoming signal when the transistor conducts.

The samples must now be coded as binary numbers. If we are using linear PCM, all that is required is a standard analog to digital (A/D) converter. Compression, if required, can be applied to the analog signal, but it is more common to incorporate the compression into the coding process.

The codecs used in telephony generally accomplish compression by using a piecewise-linear approximation to the mu-law curve shown earlier in Figure 3.6. The positive- and negative-going parts of the curve are each divided into seven segments, with an additional segment centered around zero, resulting in a total of fifteen segments. Figure 3.9 shows the segmented curve. Segments 0 and 1 have the same slope and do not compress the segment. For each higher-numbered segment, the step size is double that of the previous segment. Each segment has sixteen steps. The result is a close approximation to the actual curve.

The binary number produced by the codec in a telephone system has eight bits. The first is a sign bit, which is a one for a positive voltage and a zero for negative. Bits 2, 3, and 4 represent the segment number, from zero to seven. The last four bits determine the step within the segment. If we normalize the signal, that is, set the maximum input level equal to one volt, the

FIGURE 3.8
Sample-and-hold circuit

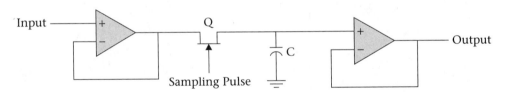

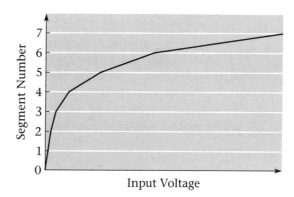

FIGURE 3.9
Segmented mu-law curve

step sizes can easily be calculated as follows: let the step size for the 0 and 1 segments be x mV. Then segment 2 has a step size of $2x$, segment 3 a step size of $3x$, and so on. Since each segment has 16 steps, the value of x can be found as follows.

$$16(x + x + 2x + 4x + 8x + 16x + 32x + 64x) = 1000 \text{ mV}$$
$$x = 0.488 \text{ mV}$$

The relationship between input voltage and segment is shown in Table 3.1.

TABLE 3.1 Mu-Law Compressed PCM Coding

Segment	Voltage Range (mV)	Step Size (mV)
0	0–7.8	0.488
1	7.8–15.6	0.488
2	15.6–31.25	0.9772
3	31.25–62.5	1.953
4	62.5–125	3.906
5	125–250	7.813
6	250–500	15.625
7	500–1000	31.25

EXAMPLE 3.6

Code a positive-going signal with amplitude 30% of the maximum allowed as a PCM sample.

SOLUTION

The signal is positive, so the first bit is a one. On the normalized voltage scale, the amplitude is 300 mV. A glance at Table 3.1 shows that the signal is in segment 6. That means the next three bits are 110 (6 in binary). This segment starts at 250 mV and increases 15.625 mV per step. The signal voltage is 50 mV above the lower limit, which translates into 50/15.625 = 3.2 steps. This is less than halfway from step 3 to step 4, so it will be quantized as step 3, making the last four bits 0011 (3 in binary). Therefore the code representing this sample is 11100011.

In operation, many modern codecs achieve compression by first encoding the signal using a 12-bit linear PCM code, then converting the 12-bit linear code into an 8-bit compressed code by discarding some of the bits. This is a simple example of **digital signal processing (DSP)**. Once an analog signal has been digitized, it can be manipulated in a great many ways simply by performing arithmetic with the bits that make up each sample. In the case of the 12-to-8 bit conversion described here, some precision will be lost for large-amplitude samples, but the data rate needed to transmit the information will be much less than for 12-bit PCM. Since most of the samples in an audio signal have amplitudes much less than the maximum, there is a gain in accuracy compared with 8-bit linear PCM.

Briefly, the conversion works as follows. The 12-bit PCM sample begins with a sign bit, which is retained. The other eleven bits describe the amplitude of the sample, with the most significant bit first. For low-level samples, the last few bits and the sign bit may be the only non-zero bits. The segment number for the 8-bit code can be determined by subtracting the number of leading zeros (not counting the sign bit) in the 12-bit code from 7. The next four bits after the first 1 are counted as the level number within the segment. Any remaining bits are discarded.

EXAMPLE 3.7

Convert the 12-bit sample 100110100100 into an 8-bit compressed code.

SOLUTION

Copy the sign bit to the 8-bit code. Next count the leading zeros (2) and subtract from 7 to get 5 (101 in binary). The first four bits of the 8-bit code are thus 1101. Now copy the next four bits after the first 1 (not counting the sign bit) to the 8-bit code. Thus the next four bits are 1010. Discard the rest. The corresponding 8-bit code is 11011010.

The decoding process is the reverse of coding. It is illustrated in the block diagram in Figure 3.10. The expansion process follows an algorithm analogous to that used in the compressor. The low-pass filter at the output removes the high-frequency components in the PAM signal that exits from the digital-to-analog converter.

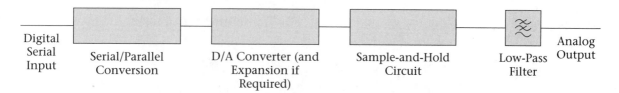

FIGURE 3.10 PCM decoding

Differential PCM   Instead of coding the entire sample amplitude for each sample, it is possible to code and transmit only the difference between the amplitude of the current sample and that of the previous sample. Since successive samples often have similar amplitudes, it should be possible to use fewer bits to encode the changes. The most common (and most extreme) example of this process is **delta modulation**, which is discussed in the next section.

3.4   Delta Modulation

In delta modulation, only one bit is transmitted per sample. That bit is a one if the current sample is more positive than the previous sample, zero if the current sample is more negative. Since only a small amount of information about each sample is transmitted, delta modulation requires a much higher sampling rate than PCM for equal quality of reproduction. Nyquist did not say that transmitting samples at twice the maximum signal frequency would always give undistorted results, only that it could, provided the samples were transmitted accurately.

Figure 3.11 shows how delta modulation generates errors. In region (i), the signal is not varying at all; the transmitter can only send ones and zeros, however, so the output waveform has a triangular shape, producing a noise signal called granular noise. On the other hand, the signal in region (iii) changes more rapidly than the system can follow, creating an error in the output called **slope overload**.

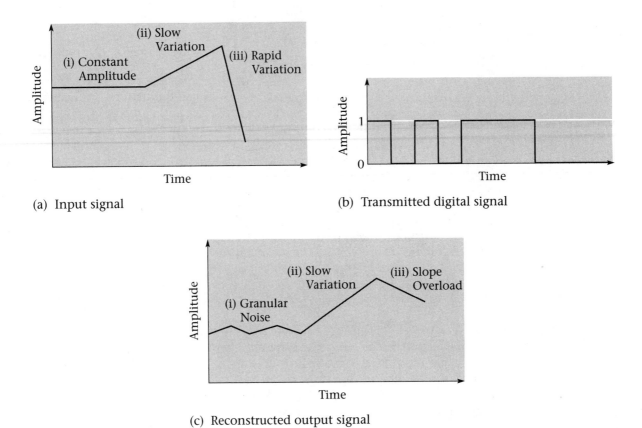

(a) Input signal

(b) Transmitted digital signal

(c) Reconstructed output signal

FIGURE 3.11 Delta modulation

Adaptive Delta Modulation Adaptive delta modulation, in which the step size varies according to previous values, is more efficient. Figure 3.12 shows how it works. After a number of steps in the same direction, the step size increases. A well-designed adaptive delta modulation scheme can transmit voice at about half the bit rate of a PCM system, with equivalent quality.

3.5 Data Compression

In the previous section we looked at companding, which we noticed consisted of compression at the transmitter and expansion at the receiver. Essentially this is an analog technique for improving dynamic range by increasing the signal-to-noise ratio for low-level signals although, as we

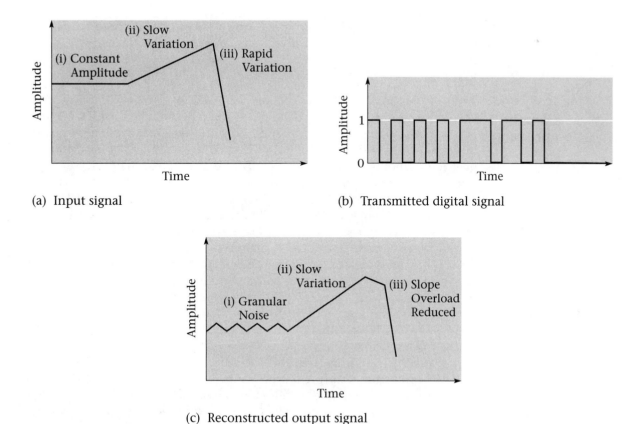

(a) Input signal

(b) Transmitted digital signal

(c) Reconstructed output signal

FIGURE 3.12 Adaptive delta modulation

saw, it can be implemented using digital signal processing. We now turn our attention to the bits that result from the analog-to-digital conversion just discussed and consider whether there is any way to reduce the number of bits that have to be transmitted per second. This reduction is also called compression, but it is really a completely different process from the one just described. We shall call it data compression to emphasize the difference.

Generally, without data compression more bandwidth is required to transmit an analog signal in digital form. For instance, analog telephony requires less than 4 kHz per channel with single-sideband AM transmission. Digital telephony conventionally operates at 64 kb/s. The exact bandwidth requirement for this depends on the modulation scheme but is likely to be much more than 4 kHz unless the channel has a very high signal-to-noise radio and an elaborate modulation scheme is used. In order to use digital techniques in wireless communication, it is very desirable to reduce the bandwidth to no more, and preferably less, than that needed for analog transmission.

Lossy and Lossless Compression

There are two main categories of data compression. Lossless compression involves transmitting all of the data in the original signal but using fewer bits. Lossy compression, on the other hand, allows for some reduction in the quality of the transmitted signal. Obviously there has to be some limit on the loss in quality, depending on the application. For instance, up until now the expectation of voice quality has been less for a mobile telephone than for a wireline telephone. This expectation is now changing as wireless telephones become more common. People are no longer impressed with the fact that wireless telephony works at all; they want it to work as well as a fixed telephone.

Lossless compression schemes generally look for redundancies in the data. For instance, a string of zeros can be replaced with a code that tells the receiver the length of the string. This technique is called **run-length encoding**. It is very useful in some applications: facsimile (fax) transmission, for instance, where it is unnecessary to transmit as much data for white space on the paper as for the message.

In voice transmission it is possible to greatly reduce the bit rate, or even stop transmitting altogether, during time periods in which there is no speech. For example, during a typical conversation each person generally talks for less than half the time. Taking advantage of this to increase the bandwidth for transmission in real time requires there to be more than one signal multiplexed. When the technique is applied to a radio system, it also allows battery-powered transmitters to conserve power by shutting off or reducing power during pauses in speech.

Lossy compression can involve reducing the number of bits per sample or reducing the sampling rate. As we have seen, the first reduces the signal-to-noise ratio and the second limits the high-frequency response of the signal, so there are limits to both methods. Other lossy compression methods rely on knowledge of the type of signal, and often, on knowledge of human perception. This means that voice, music, and video signals would have to be treated differently. These more advanced methods often involve the need for quite extensive digital signal processing. Because of this, they have only recently become practical for real-time use with portable equipment. A couple of brief examples will show the sort of thing that is possible.

Vocoders

A **vocoder** (*voice coder*) is an example of lossy compression applied to human speech. A typical vocoder tries to reduce the amount of data that needs to be transmitted by constructing a model for the human vocal system. Human sounds are produced by emitting air from the lungs at an adjustable rate. For voiced sounds this air causes the vocal cords to vibrate at an adjustable frequency; for unvoiced sounds the air passes the vocal cords without vibrating them. In either case, the sound passes through the larynx and mouth, which act as filters, changing the frequency response of the system at frequent intervals. Typically there are from three to six resonant peaks in the frequency response of the vocal tract.

Vocoders can imitate the human voice with an electronic system. Modern vocoders start with the vocal-tract model above. There is an excitation function, followed by a multi-pole bandpass filter. Parameters for the excitation and the filter response must be transmitted at intervals of about 20 ms, depending on the system. Vocoders of this type are known as *linear predictive coders* because of the mathematical process used to generate the filter parameters from an analysis of the voice signal.

The first step in transmitting a signal using a vocoder is to digitize it in the usual way, using PCM, generally at 64 kb/s. Then the signal is analyzed and the necessary excitation and filter parameters extracted. Only these parameters need to be sent to the receiver where the signal is reconstructed. The transmitted data rate is typically in the range of about 2.4 to 9.6 kb/s, allowing a much smaller transmission bandwidth than would be required for the original 64 kb/s rate.

There are two main ways of generating the excitation signal in a linear predictive vocoder. In *pulse excited linear predictive* (*PELP* or sometimes *RPELP,* for *regular pulse excited linear predictive*) vocoders, a white noise generator is used for unvoiced sounds, and a variable-frequency pulse generator produces the voiced sounds. The pulse generator creates a tone rich in harmonics, as is the sound produced by human vocal cords. Both sources have variable amplitudes. Figure 3.13 illustrates the process at the receiver.

Residual excited linear predictive (*RELP*) vocoders, on the other hand, apply the inverse of the filter that will be used at the receiver to the voice signal. The output of this filter is a signal that, when applied to the receiver filter, will reproduce the original signal exactly. Figure 3.14 shows how this

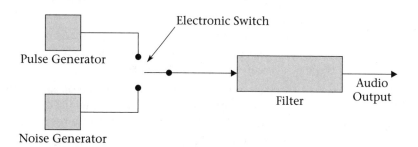

FIGURE 3.13   PELP vocoder

FIGURE 3.14   Generation of excitation signal using codebook

process works at the transmitter. The residual signal is too complex to transmit exactly with the available bit rate, so it must be represented in a more economical way. One method is to compare it with values in a table, called a *codebook*, and transmit the number of the closest codebook entry. The receiver looks up the codebook entry, generates the corresponding signal, and uses it instead of the pulse and noise generators shown in Figure 3.13. Many other vocoder variations are possible as well.

Reasonable quality can be achieved with vocoders using data rates much lower than those required for PCM. So far, the quality is not quite as good as for straightforward PCM, however.

It should be obvious that vocoders are intended for use with voice only; whereas, the PCM system described above can be used to send any 64 kb/s data stream, including music, fax, or computer files. None of these will work properly with a vocoder. Vocoders even tend to give a somewhat unnatural quality to human speech. Still, the gain in bit rate and hence bandwidth, compared to PCM, is so great that vocoders are very common in digital wireless voice communication.

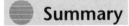 **Summary**   The main points to remember from this chapter are:

- Modern communication systems are often a mixture of analog and digital sources and transmission techniques. The trend is toward digital systems.

- Modern digital systems have better performance and use less bandwidth than equivalent analog systems.

- An analog signal that is to be transmitted digitally must be sampled at least twice per cycle of its highest-frequency component. Failure to do so creates undesirable aliasing.

- PCM requires that the amplitude of each sample of a signal be converted to a binary number. The more bits used for the number, the greater the accuracy, but the greater the bit rate required.

- Delta modulation transmits only one bit per sample, indicating whether the signal level is increasing or decreasing, but it needs a higher sampling rate than PCM for equivalent results.

- The signal-to-noise ratio for either PCM or delta modulation signals can often be improved by using companding.
- Lossless compression eliminates redundant data bits, thereby reducing the bit rate with no effect on signal quality.
- Lossy compression compromises signal quality in order to reduce the bit rate. For voice transmissions, vocoders are often used to achieve great reductions in bit rate.

● Equation List

$$f_a = f_s - f_b \tag{3.8}$$

$$v(t) = \frac{E_b}{T} \sin \omega_b t + \frac{E_b \sin \pi\tau/T}{\pi\tau} \sin(\omega_s + \omega_b)t$$
$$- \frac{E_b \sin \pi\tau/T}{\pi\tau} \sin(\omega_s - \omega_b)t \tag{3.9}$$

$$N = 2^m \tag{3.10}$$

$$DR = 1.76 + 6.02m \text{ dB} \tag{3.11}$$

$$D = f_s m \tag{3.12}$$

$$v_o = \frac{V_o \ln(1 + \mu v_i/V_i)}{\ln(1 + \mu)} \tag{3.13}$$

● Key Terms

aliasing   distortion created by using too low a sampling rate when coding an analog signal for digital transmission

codec   device that converts sampled analog signal to and from its PCM or delta modulation equivalent

coding   conversion of a sampled analog signal into a PCM or delta modulation bitstream

companding   combination of compression at the transmitter and expansion at the receiver of a communication system

decoding   conversion of a PCM or delta modulation bitstream to analog samples

delta modulation   a coding scheme that records the change in signal level since the previous sample

digital signal processing (DSP)   filtering of signals by converting them to digital form, performing arithmetic operations on the data bits, then converting back to analog form

flat-topped sampling   sampling of an analog signal using a sample-and-hold circuit, such that the sample has the same amplitude for its whole duration

foldover distortion   see **aliasing**

framing bits   bits added to a digital signal to help the receiver to detect the beginning and end of data frames

natural sampling   sampling of an analog signal so that the sample amplitude follows that of the original signal for the duration of the sample

pulse-amplitude modulation (PAM)   a series of pulses in which the amplitude of each pulse represents the amplitude of the information signal at a given time

pulse-code modulation (PCM)   a series of pulses in which the amplitude of the information signal at a given time is coded as a binary number

pulse-duration modulation (PDM)   a series of pulses in which the duration of each pulse represents the amplitude of the information signal at a given time

pulse-position modulation (PPM)   a series of pulses in which the timing of each pulse represents the amplitude of the information signal at a given time

pulse-width modulation (PWM)   see **pulse-duration modulation (PDM)**

quantizing   representation of a continuously varying quantity as one of a number of discrete values

quantizing errors   inaccuracies caused by the representation of a continuously varying quantity as one of a number of discrete values

quantizing noise   see **quantizing errors**

run-length encoding   method of data compression by encoding the length of a string of ones or zeros instead of transmitting all the one or zero bits individually

slope overload in delta modulation, an error condition that occurs when the analog signal to be digitized varies too quickly for the system to follow

vocoder circuit for digitizing voice at a low data rate by using knowledge of the way in which voice sounds are produced

● Questions

1. Give four advantages and one disadvantage of using digital (rather than analog) techniques for the transmission of voice signals.

2. Explain the necessity for sampling an analog signal before transmitting it digitally.

3. What is the Nyquist rate? What happens when a signal is sampled at less than the Nyquist rate?

4. Explain the difference between natural and flat-topped sampling.

5. (a) List three types of analog pulse modulation.

    (b) Which pulse modulation scheme is used as an intermediate step in the creation of PCM?

    (c) Which pulse modulation scheme also finds use in audio amplifiers and motor speed-control systems?

6. What is meant by the term *quantizing noise*?

7. For a PCM signal, describe the effects of:

    (a) increasing the sampling rate

    (b) increasing the number of bits per sample

8. (a) Briefly explain what is meant by companding.

    (b) What advantage does companded PCM have over linear PCM for voice communication?

9. How does differential PCM differ from standard PCM?

10. Explain why the sampling rate must be greater for delta modulation than for PCM.

11. What is meant by slope overload in a delta modulation system? How can this problem be reduced?

12. What are the two functions of a codec? Where in a telephone system is it usually located?

13. Explain briefly how μ-law compression is implemented in a typical codec.

14. Explain the difference between lossless and lossy data compression. Give an example of each.

15. How do vocoders model the human vocal cords? How do they model the mouth and larynx?

16. What gives vocoders their somewhat artificial voice quality?

17. Does digital audio always have higher quality than analog audio? Explain.

● Problems

1. It is necessary to transmit the human voice using a frequency range from 300 Hz to 3.5 kHz using a digital system.
   (a) What is the minimum required sampling rate, according to theory?
   (b) Why would a practical system need a higher rate than the one you calculated in part (a)?

2. The human voice actually has a spectrum that extends to much higher frequencies than are necessary for communication. Suppose a frequency of 5 kHz was present in a sampler that sampled at 8 kHz.
   (a) What would happen?
   (b) How can the problem described in part (a) be prevented?

3. A 1-kHz sine wave with a peak value of 1 volt and no dc offset is sampled every 250 microseconds. Assume the first sample is taken as the voltage crosses zero in the upward direction. Sketch the results over 1 ms using:
   (a) PAM with all pulses in the positive direction
   (b) PDM
   (c) PPM

4. The compact disc system of digital audio uses two channels with TDM. Each channel is sampled at 44.1 kHz and coded using linear PCM with sixteen bits per sample. Find:
   (a) the maximum audio frequency that can be recorded (assuming ideal filters)
   (b) the maximum dynamic range in decibels
   (c) the bit rate, ignoring error correction and framing bits
   (d) the number of quantizing levels

5. Suppose an input signal to a μ-law compressor has a positive voltage and an amplitude 25% of the maximum possible. Calculate the output voltage as a percentage of the maximum output.

6. Suppose a composite video signal with a baseband frequency range from dc to 4 MHz is transmitted by linear PCM, using eight bits per sample and a sampling rate of 10 MHz.

   (a) How many quantization levels are there?

   (b) Calculate the bit rate, ignoring overhead.

   (c) What would be the maximum signal-to-noise ratio, in decibels?

   (d) What type of noise determines the answer to part (c)?

7. How would a signal with 50% of the maximum input voltage be coded in 8-bit PCM, using digital compression?

8. Convert a sample coded (using mu-law compression) as 11001100 to a voltage with the maximum sample voltage normalized as 1 V.

9. Convert the 12-bit PCM sample 110011001100 to an 8-bit compressed sample.

10. A typical PCS system using a vocoder operates at 9600 b/s. By what factor has the amount of data required been reduced, compared with standard digital telephony?

Digital Modulation

Objectives

After studying this chapter, you should be able to:

- Describe the basic types of digital modulation.
- Calculate the maximum data rate for a channel with a given modulation scheme and signal-to-noise ratio.
- Explain the use of eye diagrams and constellation diagrams.
- Explain the difference between bit rate and baud rate and calculate both for typical digital modulation systems.
- Describe and compare FSK, PSK, and QAM and perform simple calculations with each.
- Explain the concepts of multiplexing and multiple access using frequency and time division.
- Describe the principles of spread-spectrum communication and distinguish between frequency-hopping and direct-sequence systems.
- Calculate spreading gain and signal-to-noise ratio for spread-spectrum systems.
- Describe code-division multiple access and compare with FDMA and TDMA.

4.1 Introduction

Digital signals have become very important in wireless communication. Some of the information to be transmitted is already in digital form. In Chapter 3, we examined the ways in which analog signals such as voice can be converted to digital form for transmission.

In order to send data by radio, it is necessary to use a higher frequency carrier wave, just as for analog communication. Since the high-frequency carrier is a sine wave for digital as well as analog signals, the same three basic parameters are available for modulation: amplitude, frequency, and phase. All three, singly and in combination, are used in digital systems.

Often the modulator and demodulator are collectively described as a **modem**. You are probably familiar with telephone modems. These modulate digital data onto an audio-frequency carrier. Radio-frequency modems are similar in principle though quite a bit different in construction due to the much higher carrier frequencies used.

In a holdover from the days of Morse code (a digital, though not a binary, communication scheme), the word *keying* is still often used to denote digital modulation schemes. Thus we have **frequency-shift keying (FSK)** and **phase-shift keying (PSK)**. Straightforward **amplitude-shift keying (ASK)** is rare in digital communication unless we count Morse code, but **quadrature AM (QAM)** is very common. In this chapter we shall look at all of these schemes as they are used in wireless transmission.

In Chapter 1 we noted that information capacity is proportional to bandwidth. This fundamental relationship is given by Hartley's law, which is repeated here for convenience:

$$I = ktB \qquad (4.1)$$

where

I = amount of information to be transmitted in bits

k – a constant that depends on the modulation scheme and the signal-to-noise ratio

t = time in seconds

B = bandwidth in hertz

We noted in Chapter 1 that the constant k is important. In Chapter 2, when we studied analog signals, we saw that the bandwidth required for the same information rate varied greatly with the modulation scheme. In that chapter we used baseband bandwidth as a rough measure of information content. We are now in a position to discuss the value of k for different digital modulation schemes. The information rate for digital communication will be expressed in bits per second.

Digital signals result in discrete, rather than continuous, changes in the modulated signal. The receiver examines the signal at specified times, and the state of the signal at each such time is called a **symbol**. The eventual output will be binary, but it is certainly possible to use more than two states for the transmitted symbol. Complex schemes using several levels can send more data in a given bandwidth.

There is a theoretical limit to the maximum data rate that can be transmitted in a channel with a given bandwidth. The Shannon-Hartley theorem states:

$$C = 2B \log_2 M \qquad\qquad (4.2)$$

where

C = information capacity in bits per second
B = channel bandwidth in hertz
M = number of possible states per symbol

For instance, if an FSK modulator can transmit either of two frequencies, $M = 2$. A more elaborate modulator, using four different phase angles, each of which can have two different amplitudes, has $M = 4 \times 2 = 8$.

The limiting effect of bandwidth on data rate is understood most easily by looking at a low-pass, rather than a bandpass, channel. Suppose that the channel can pass all frequencies from zero to some maximum frequency B. Then, of course, the highest frequency that can be transmitted is B. Suppose that a simple binary signal consisting of alternate ones and zeros is transmitted through the channel. This time let a logic 1 be 1 V and a logic 0 be –1 V. The input signal will look like Figure 4.1(a): it will be a square wave with a frequency one-half the bit rate (since there are two bits, a one and a zero, for each cycle). This signal, which is a square wave, has harmonics at all odd

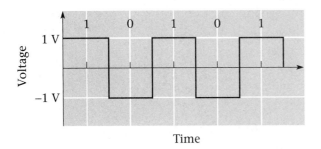

(a) Input square-wave signal

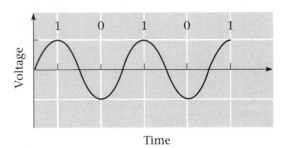

(b) Output signal at maximum rate

FIGURE 4.1 Digital transmission through a low-pass channel

multiples of its fundamental frequency, with declining amplitude as the frequency increases. This was shown in Chapter 1, using Fourier analysis. At very low bit rates the output signal after passage through the channel will be similar to the input, but as the bit rate increases, the frequency of the square wave also increases and more of its harmonics are filtered out. Therefore, the output will become more and more distorted. Finally, for a bit rate of $2B$ the frequency of the input signal will be B, and only the fundamental of the square wave will pass through the channel, as shown in Figure 4.1(b). The receiver will still be able to distinguish a one from a zero, and the information will be transmitted. Thus, with binary information, the channel capacity will be $2B$, as predicted by Equation (4.2).

Often the ratio of bit rate to channel bandwidth, expressed in bits per second per hertz, is used as a figure of merit in digital radio systems. In the preceding paragraph, we can see that for a two-level code the theoretical maximum bandwidth efficiency is 2 b/s/Hz.

If a random pattern of ones and zeros were transmitted using the code of Figure 4.1 and the signal were applied to an oscilloscope sweeping at the bit rate, we would see the patterns shown in Figure 4.2. These are called "eye diagrams." The reason should be obvious from Figure 4.2(b). Because some sweeps of the scope take place during a high- and some during a low-signal level, both levels appear on the scope. A diagram like that in Figure 4.2(a) indicates a data rate much lower than can be carried by the channel. Figure 4.2(b) indicates an optimal system. If an attempt were made to transmit data too quickly, the eye would begin to close due to intersymbol interference, as shown in Figure 4.2(c).

Now suppose that four voltage levels, each corresponding to a two-bit sequence, are used, as in Figure 4.3. The bandwidth required for the fundamental of this signal is the same as before. We have, it seems, managed to transmit twice as much information in the same bandwidth. Again, this agrees with Equation (4.2).

Unfortunately, the information capacity of a channel cannot be increased without limit by increasing the number of states because noise makes it difficult to distinguish between signal states. The ultimate limit is called the Shannon limit:

$$C = B \log_2(1 + S/N) \tag{4.3}$$

where

C = information capacity in bits per second

B = bandwidth in hertz

S/N = signal-to-noise ratio (as a power ratio, not in decibels)

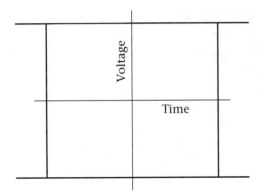

(a) Eye diagram for ideal digital signal

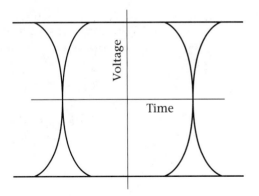

(b) Eye diagram for low-pass channel

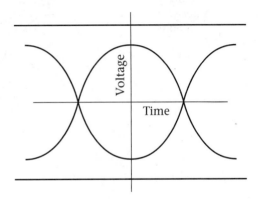

(c) Eye diagram showing inter-symbol interference

FIGURE 4.2 Eye diagrams

FIGURE 4.3
Four-level code

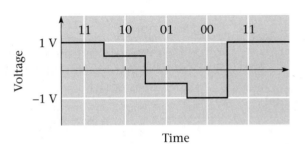

By the way, it is quite easy to find logs to the base 2, even if your calculator lacks this function. Simply find the log to the base 10 (the common log) of the given number and divide by the log of 2, that is:

$$\log_2 N = \frac{\log_{10} N}{\log_{10} 2} \tag{4.4}$$

The effect of noise on a signal can be seen in the eye diagram of Figure 4.4. The noise causes successive oscilloscope traces to be at different amplitudes. If the noise is severe enough, the eye closes and data recovery is unreliable.

FIGURE 4.4
Eye diagram showing inter-symbol interference and noise

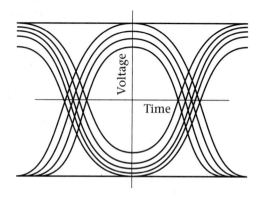

EXAMPLE 4.1

A radio channel has a bandwith of 10 kHz and a signal-to-noise ratio of 15 dB. What is the maximum data rate than can be transmitted:

(a) Using any system?

(b) Using a code with four possible states?

SOLUTION

(a) We can find the theoretical maximum data rate for this channel from Equation (4.3). First, though, we need the signal-to-noise ratio as a power ratio. We can convert the given decibel value as follows:

$$\frac{S}{N} = \log^{-1}\left(\frac{15}{10}\right)$$

$$= 31.6$$

Now we can use Equation (4.3):

$$\begin{aligned}
C &= B \log_2(1 + S/N) \\
&= 10 \times 10^3 \log_2(1 + 31.6) \\
&= 10 \times 10^3 \times 5.03 \\
&= 50.3 \text{ kb/s}
\end{aligned}$$

(b) We can use Equation (4.2) to find the maximum possible bit rate given the specified code and bandwidth. We will then have to compare this answer with that of part (a). From Equation (4.2),

$$\begin{aligned}
C &= 2B \log_2 M \\
&= 2 \times 10 \times 10^3 \times \log_2 4 \\
&= 2 \times 10 \times 10^3 \times 2 \\
&= 40 \text{ kb/s}
\end{aligned}$$

Since this is less than the maximum possible for this channel, it should be possible to transmit over this channel, with a four-level scheme, at 40 kb/s. A more elaborate modulation scheme would be required to attain the maximum data rate of 50.3 kb/s for the channel.

At this point we should distinguish between **bit rate** and **baud rate**. The bit rate is simply the number of bits transmitted per second (C in the preceding two equations), while the baud rate is the number of symbols per second. Therefore, if we let the baud rate be S (for symbols, since B is already being used for bandwidth), then:

$$C = S \log_2 M \tag{4.5}$$

where

C = capacity in bits per second

S = baud rate in symbols per second

M = number of possible states per symbol

EXAMPLE 4.2

A modulator transmits symbols, each of which has 64 different possible states, 10,000 times per second. Calculate the baud rate and bit rate.

SOLUTION

The baud rate is simply the symbol rate, or 10 kbaud. The bit rate is given by Equation (4.5):

$$
\begin{aligned}
C &= S \log_2 M \\
&= 10 \times 10^3 \times \log_2 64 \\
&= 60 \text{ kb/s}
\end{aligned}
$$

4.2 Frequency-Shift Keying

Probably the simplest digital modulation scheme in current use is *frequency-shift keying (FSK)*. In its simplest form two frequencies are transmitted, one corresponding to binary one, the other to zero. In digital communication systems, a one is often denoted by the term **mark**, and a zero is called a **space**. This is another holdover from telegraphy in which Morse code messages were often recorded as marks on a paper tape. FSK is a robust scheme; that is, like analog FM, it tends to be reliable in the presence of noise. Its disadvantage is: since each symbol has only two possible states, it is not very efficient in terms of bandwidth. It tends to be used for low-data-rate applications, such as pagers, and for transmitting bursts of data over systems that are mainly analog.

FSK is also used extensively in high-frequency radio systems for radio-teletype transmission. High-frequency radio channels tend to be very noisy, and phase shifts induced into the signal by travel through the ionosphere make the use of any scheme requiring accurate phase information impractical. Data rates for HF communication are very low (on the order of 100 b/s) and frequency shifts between mark and space vary from 170 Hz to 850 Hz.

Usually HF FSK is actually transmitted and received using SSBSC AM equipment. Two different audio tones, one for mark and one for space, with the required frequency difference are chosen. These are applied one at a time to the microphone input of the transmitter. This results in the transmission of RF frequencies that differ by the same amount as the audio frequencies. At the receiver two audio tones differing by the required frequency shift will be produced. Figure 4.5 shows the spectrum of a typical HF FSK transmission, as well as a typical pair of audio tones used to modulate the transmitter.

A variant of FSK is **AFSK (audio frequency-shift keying)**, in which two different audio frequencies, corresponding to mark and space, are used to modulate a carrier by any of the analog methods discussed in Chapter 2. AFSK is often combined with FM modulation for the audio tones. Figure 4.6 shows how this can work using a VHF amateur radio link as an example. There is an extensive amateur packet-radio network, using AFSK transmission with a frequency shift of 1 kHz between mark and space, operating at a data rate of 1200 bits per second.

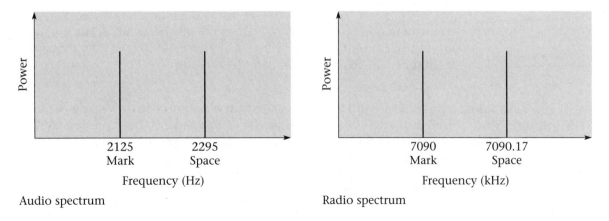

FIGURE 4.5 Spectrum of HF FSK transmission

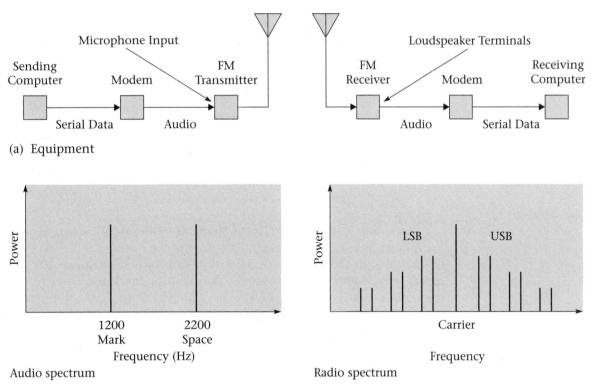

(a) Equipment

(b) Typical spectra (this example is for an amateur packet radio station)

FIGURE 4.6 AFSK using FM equipment

It is possible to build an FSK system with more than two different frequencies in order to increase the number of bits per symbol, but it is usually more efficient to move to a system using phase shifts, or a combination of amplitude and phase shifts, when this is required.

Gaussian Minimum-Shift Keying

A special case of FSK called **Gaussian minimum-shift keying (GMSK)** is used in the GSM cellular radio and PCS systems to be described later. In a minimum-shift system, the mark and space frequencies are separated by half the bit rate, that is:

$$f_m - f_s = 0.5 \ f_b \tag{4.6}$$

where

$$f_m = \text{frequency transmitted for mark (binary 1)}$$
$$f_s = \text{frequency transmitted for space (binary 0)}$$
$$f_b = \text{bit rate}$$

If we use the conventional FM terminology from Chapter 2, we see that GMSK has a deviation each way from the center (carrier) frequency, of

$$\delta = 0.25 \ f_b$$

which corresponds to a modulation index of

$$m_f = \frac{\delta}{f_m}$$

$$= \frac{0.25 f_b}{f_b}$$

$$= 0.25$$

The word *Gaussian* refers to the shape of a filter that is used before the modulator to reduce the transmitted bandwidth of the signal. GMSK uses less bandwidth than conventional FSK, because the filter causes the transmitted frequency to move gradually between the mark and space frequencies. With conventional FSK the frequency transition is theoretically instantaneous, and in practice as rapid as the hardware allows, producing sidebands far from the carrier frequency.

EXAMPLE 4.3 ▼

The GSM cellular radio system uses GMSK in a 200-kHz channel, with a channel data rate of 270.833 kb/s. Calculate:

(a) the frequency shift between mark and space

(b) the transmitted frequencies if the carrier (center) frequency is exactly 880 MHz

(c) the bandwidth efficiency of the scheme in b/s/Hz

SOLUTION

(a) The frequency shift is

$$f_m - f_s = 0.5 \, f_b = 0.5 \times 270.833 \text{ kHz} = 135.4165 \text{ kHz}$$

(b) The shift each way from the carrier frequency is half that found in (a) so the maximum frequency is

$$f_{max} = f_c + 0.25 \, f_b = 880 \text{ MHz} + 0.25 \times 270.833 \text{ kHz} = 880.0677 \text{ MHz}$$

and the minimum frequency is

$$f_{min} = f_c - 0.25 \, f_b = 880 \text{ MHz} - 0.25 \times 270.833 \text{ kHz} = 879.93229 \text{ MHz}$$

(c) The GSM system has a bandwidth efficiency of 270.833/200 = 1.35 b/s/Hz, comfortably under the theoretical maximum of 2 b/s/Hz for a two-level code.

4.3 Phase-Shift Keying

When somewhat higher data rates are required in a band-limited channel than can be achieved with FSK, *phase-shift keying* (PSK) is often used. Measuring phase requires a reference phase, which would be hard to maintain accurately. Usually, the phase of each symbol is compared with that of the previous symbol, rather than with a constant reference. This type of PSK is more completely described as **delta phase-shift keying (DPSK)**. Most DPSK modems use a four-phase system called **quadrature phase-shift keying (QPSK or DQPSK)**. In QPSK, each symbol represents two bits and the bit rate is twice the baud rate. This is called a **dibit system**. Such a system can carry twice as much data in the same bandwidth as can a single-bit system like FSK, provided the signal-to-noise ratio is high enough.

Figure 4.7 is a vector diagram that represents a typical DQPSK system. Phase shifts are given with respect to the phase of the previous symbol. Each of the four possible phase shifts (including no shift at all) is associated with a two-bit sequence, as shown in Table 4.1.

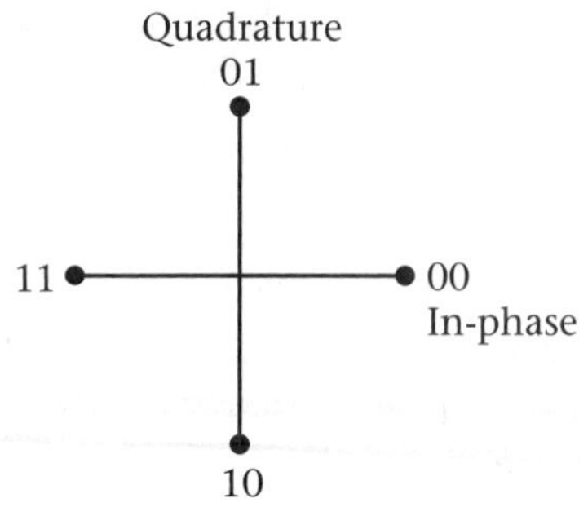

FIGURE 4.7
Delta quadrature
phase-shift keying

TABLE 4.1 DQPSK Coding

Phase Shift (degrees)	Symbol
0	00
+90	01
−90	10
180	11

π/4 Delta Phase-Shift Keying

The system shown in Figure 4.7 and Table 4.1 requires a 180 degree transition for the symbol 11. The transmitted signal has to go to zero amplitude momentarily as it makes this transition. Accurate transmission of this signal therefore requires a linear amplifier, unlike the case for FSK. In fact, the amplifier should be linear all the way down to zero output. This is quite possible, of course, but linear amplifiers are markedly less efficient than nonlinear amplifiers. The need for linearity can be reduced, though not eliminated, by changing to a system called π/4 DQPSK. Here the allowable transitions from the previous phase angle are ±45° and ±135°. Neither of these requires the signal amplitude to go through zero, relaxing the linearity requirements somewhat. A typical π/4 DQPSK system has the state table shown in Table 4.2, and a vector diagram showing the possible transitions

TABLE 4.2 π/4 DQPSK Coding

Phase Shift (degrees)	Symbol
45	00
135	01
−45	10
−135	11

can be found in Figure 4.8. This system is used for the North American TDMA cell phone and PCS systems.

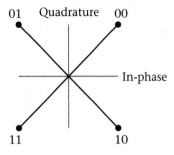

FIGURE 4.8
π/4 DQPSK

EXAMPLE 4.4

The North American TDMA digital cell phone standard transmits at 24.3 kilobaud using DQPSK. What is the channel data rate?

SOLUTION

Since this is a dibit system, the symbol rate, also known as the baud rate, is half the bit rate. Therefore the data rate is 48.6 kb/s.

4.4 Quadrature Amplitude Modulation

The only way to achieve high data rates with a narrowband channel is to increase the number of bits per symbol, and the most reliable way to do that is to use a combination of amplitude and phase modulation known as *quadrature amplitude modulation (QAM)*. For a given system, there is a finite number of allowable amplitude-phase combinations. Figure 4.9(a) is a **constellation diagram** that shows the possibilities for a hypothetical system with sixteen amplitude/phase combinations. Thus each transmitted symbol represents four bits. This diagram is similar to the previous figure except that the vectors are not drawn. Each dot represents a possible amplitude/phase combination or state. With a noiseless channel, the number of combinations could be increased indefinitely, but a practical limit is reached when the difference between adjacent states becomes too small to be detected reliably in the presence of noise and distortion. If a QAM signal is applied to

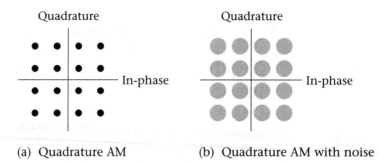

(a) Quadrature AM   (b) Quadrature AM with noise

FIGURE 4.9   Quadrature AM

the oscilloscope, the noise can be seen as a blurring of the points in the constellation, as shown in Figure 4.9(b).

In fixed terrestrial microwave systems QAM is used with quite a large number of states—up to 1024 in some cases. This requires a very high signal-to-noise ratio however, and portable and mobile systems are much more limited.

QAM is more efficient in terms of bandwidth than either FSK or QPSK, but it is also more susceptible to noise. Another disadvantage compared to FSK is that QAM signals, like analog AM signals, vary in amplitude. This means that transmitter amplifiers must be linear.

EXAMPLE 4.5 ▼

A modem uses 16 different phase angles and 4 different amplitudes. How many bits does it transmit for each symbol?

SOLUTION

The number of possible states per symbol is $16 \times 4 = 64$
The number of bits per symbol is $\log_2 64 = 6$

4.5   Multiplexing and Multiple Access

Multiplexing allows several signals to be carried on a single transmission channel or medium. When the signals to be multiplexed originate at different locations, the system is often described by using the term **multiple access**. Both are used in wireless communication.

Frequency-Division Multiplexing and Multiple Access

The simplest multiple access scheme is the one used by radio and television broadcasting stations. Each signal is assigned a portion of the available frequency spectrum on a full-time basis. This is called **frequency-division multiplexing (FDM)** or **frequency-division multiple access (FDMA)** depending on the situation. For instance, over-the-air broadcasts are FDMA while a cable-television system, where all the signals are assigned slots on the same cable by the headend equipment, is an example of FDM. Frequency division can be and is used with both analog and digital signals.

Time-Division Multiplexing and Multiple Access

Time-division multiplexing (TDM) is used mainly for digital communication. In TDM, each information signal is allowed to use all the available bandwidth, but only for part of the time. From Hartley's Law (Equation 4.1), it can be seen that the amount of information transmitted is proportional to both bandwidth and time. Therefore, at least in theory, it is equally possible to divide the bandwidth or the time among the users of a channel. Continuously varying signals, such as analog audio, are not well adapted to TDM, because the signal is present at all times. On the other hand, sampled audio is very suitable for TDM, as it is possible to transmit one sample or bit from each of several sources sequentially, then send the next sample or bit from each source, and so on. As already mentioned, sampling itself does not imply digital transmission, but in practice sampling and digitizing usually go together.

Many signals can be sent on one channel by sending either a sample from each signal, or a fixed number of bits from each signal, in rotation. Time-division multiplexing requires that the total bit rate be multiplied by the number of channels multiplexed. This means that the bandwidth requirement is also multiplied by the number of signals.

TDM in Telephony

TDM is used extensively in digital telephony. The simplest North American standard is known as the DS-1 signal, which consists of 24 PCM voice channels, multiplexed using TDM. Each channel is sampled at 8 kHz, with 8 bits per sample, as previously described. This gives a bit rate of 8 kb/s $\times$ 8 = 64 kb/s for each voice channel.

The DS-1 signal consists of frames, each of which contains the bits representing one sample from each of the 24 channels. One extra bit, called the *framing bit,* is added to each frame to help synchronize the transmitter and receiver. Each frame contains $24 \times 8 + 1 = 193$ bits.

The samples must be transmitted at the same rate as they were obtained in order for the signal to be reconstructed at the receiver without delay. This requires the multiplexed signal to be sent at a rate of 8000 frames per second.

Thus the bit rate is 193×8000 b/s = 1.544 Mb/s. See Figure 4.10 for an illustration of a frame of a DS-1 signal.

FIGURE 4.10
DS-1 signal

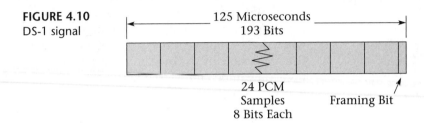

125 Microseconds
193 Bits

24 PCM
Samples
8 Bits Each

Framing Bit

Time-Division Multiple Access

Time-division multiple access (TDMA) is like TDM except that it involves signals originating at different points. The telephone system observed in the previous section uses TDM because all the signals are combined at one point. An example of TDMA is a digital cellular radio system where several signals from mobile units are combined on one channel by assigning each a time slot. TDMA systems are very similar in principle to TDM, but they tend to be more complex to design. One complicating feature in TDMA radio systems is: the propagation time for the signal from a mobile unit to a base station varies with its distance to the base. We will look at the details later when we discuss specific systems.

4.6 Spread-Spectrum Systems

As radio communication systems proliferate and traffic increases, interference problems become more severe. Interference is nothing new, of course, but it has been managed reasonably successfully in the past by careful regulatory control of transmitter locations, frequencies, and power levels. There are some exceptions to this, however. Two examples are CB radio and cordless telephones. In fact, wherever government regulation of frequency use is informal or nonexistent, interference is likely to become a serious problem. Widespread use of such systems as cordless phones, wireless local-area networks, and wireless modems by millions of people obviously precludes the tight regulation associated with services such as broadcasting, making bothersome interference almost inevitable. One approach to the problem, used by cellular radio systems, is to employ a complex system of frequency reuse, with computers choosing the best channel at any given time. However, this system too implies strong central control, if not by government then by one or more service providers, each having exclusive rights to certain radio channels. That is, in fact, the current situation with respect to cellular telephony, but it can cause problems where several widely different services use the same frequency range. The 49-MHz band, for instance, is currently used by cordless phones, baby monitors, remote controlled models, and various other users in an almost completely unregulated way. Similarly, the 2.4-GHz band is shared by wireless LANs, wireless modems, cordless phones—and even microwave ovens!

Another problem with channelized communication, even when tightly controlled, is that the number of channels is strictly limited. If all available channels are in use in a given cell of a cellular phone system, the next attempt to complete a call will be *blocked,* that is, the call will not go through. Service does not degrade gracefully as traffic increases; rather, it continues as normal until the traffic density reaches the limits of the system and then ceases altogether for new calls.

There is a way to reduce interference that does not require strong central control. That technique, known as spread-spectrum communication, has been used for some time in military applications where interference often consists of deliberate jamming of signals. This interference, of course, is not under the control of the communicator, nor is it subject to government regulation.

Military communication systems need to avoid unauthorized eavesdropping on confidential transmissions, a problem alleviated by the use of spread-spectrum techniques. Privacy is also a concern for personal communication systems, but many current analog systems, such as cordless and cellular telephone systems, have nonexistent or very poor protection of privacy.

For these reasons, and because the availability of large-scale integrated circuits has reduced the costs involved, there has recently been a great deal of interest in the use of spread-spectrum technology in personal communication systems for both voice and data.

The basic idea in spread-spectrum systems is, as the name implies, to spread the signal over a much wider portion of the spectrum than usual. A simple audio signal that would normally occupy only a few kilohertz of spectrum can be expanded to cover many megahertz. Thus only a small portion of the signal is likely to be masked by any interfering signal. Of course, the average power density, expressed in watts per hertz of bandwidth, is also reduced, and this often results in a signal-to-noise ratio of less than one (that is, the signal power in any given frequency range is less than the noise power in the same bandwidth).

It may seem at first glance that this would make the signal almost impossible to detect, which is true unless special techniques are used to "de-spread" the signal while at the same time spreading the energy from interfering signals. In fact, the low average power density of spread-spectrum signals is responsible for their relative immunity from both interference and eavesdropping.

EXAMPLE 4.6

A voice transmission occupies a channel 30 kHz wide. Suppose a spread-spectrum system is used to increase its bandwidth to 10 MHz. If the signal has a total signal power of −110 dBm at the receiver input and the system noise temperature referred to the same point is 300 K, calculate the signal-to-noise ratio for both systems.

SOLUTION

Recall from Chapter 1 that thermal noise power is given by

$$P_N = kTB$$

where

P_N = noise power in watts
k = Boltzmann's constant: 1.38×10^{-23} joules/kelvin (J/K)
T = absolute temperature in kelvins
B = noise power bandwidth in hertz

In general, the noise power bandwidth for a system will be approximately equal to the receiver bandwidth. For the signal with a bandwidth of 30 kHz and a noise temperature of 300 K,

$$
\begin{aligned}
P_N \text{ (30 kHz)} &= 1.38 \times 10^{-23} \text{ J/K} \times 300 \text{ K} \times 30 \times 10^3 \text{ Hz} \\
&= 124 \times 10^{-18} \text{ W} \\
&= -129 \text{ dBm}
\end{aligned}
$$

When the signal bandwidth increases to 10 MHz, the signal is spread over a much wider region of the spectrum, and a receiver designed to receive the whole signal bandwidth would need a bandwidth of 10 MHz as well. It would receive a noise power equal to

$$
\begin{aligned}
P_N \text{(10 MHz)} &= 1.38 \times 10^{-23} \text{ J/K} \times 300 \text{ K} \times 10 \times 10^6 \text{ Hz} \\
&= 41.4 \times 10^{-15} \text{ W} \\
&= -104 \text{ dBm}
\end{aligned}
$$

With both signal and noise in dBm, we can subtract to get the signal-to-noise ratio.

For the 30 kHz bandwidth,

$$S/N = -110 \text{ dBm} - (-129 \text{ dBm}) = 19 \text{ dB}$$

For the 10 MHz bandwidth,

$$S/N = -110 \text{ dBm} - (-104 \text{ dBm}) = -6 \text{ dB}$$

Spread-spectrum communication is especially effective in a portable or mobile environment. Cancellation due to the reflection of signals often causes very deep fades, called **Rayleigh fading**, over a narrow frequency range. When one of these fades happens to coincide with the frequency in use, the signal can be lost completely. With spread spectrum, only a small part of the communication will be lost, and this can usually be made up, at least in digital schemes, by the use of error-correcting codes.

There are two important types of spread-spectrum systems. They are known as **frequency hopping** and **direct sequence**.

Frequency-Hopping Systems

Frequency hopping is the simpler of the two spread-spectrum techniques. A frequency synthesizer is used to generate a carrier in the ordinary way. There is one difference, however: instead of operating at a fixed frequency, the synthesizer changes frequency many times per second according to a preprogrammed sequence of channels. This sequence is known as a **pseudo-random noise (PN) sequence** because, to an outside observer who has not been given the sequence, the transmitted frequency appears to hop about in a completely random and unpredictable fashion. In reality, the sequence is not random at all, and a receiver which has been programmed with the same sequence can easily follow the transmitter as it hops and the message can be decoded normally.

Since the frequency-hopping signal typically spends only a few milliseconds or less on each channel, any interference to it from a signal on that frequency will be of short duration. If an analog modulation scheme is used for voice, the interference will appear as a click and may pass unnoticed. If the spread-spectrum signal is modulated using digital techniques, an error-correcting code can be employed that will allow these brief interruptions in the received signal to be ignored, and the user will probably not experience any signal degradation at all. Thus reliable communication can be achieved in spite of interference.

EXAMPLE 4.7

A frequency-hopping spread-spectrum system hops to each of 100 frequencies every ten seconds. How long does it spend on each frequency?

SOLUTION

The amount of time spent on each frequency is

$$t = 10 \text{ seconds}/100 \text{ hops}$$
$$= 0.1 \text{ second per hop}$$

If the frequency band used by the spread-spectrum system contains known sources of interference, such as carriers from other types of service, the frequency-hopping scheme can be designed to avoid these frequencies entirely. Otherwise, the communication system will degrade gracefully as the number of interfering signals increases, since each new signal will simply increase the noise level slightly.

Direct-Sequence Systems

The direct-sequence form of spread-spectrum communication is commonly used with digital modulation schemes. The idea is to modulate the transmitter with a bit stream consisting of pseudo-random noise (PN) that has a much higher rate than the actual data to be communicated. The term *pseudo-random* means that the bit stream appears at first glance to be a random sequence of zeros and ones but is actually generated in such a way as to repeat exactly from time to time. The data to be transmitted is combined with the PN. One common technique is to invert all the bits of the PN stream during the time the "real" data is represented by a one and to leave the PN bit stream unchanged when a data zero is to be transmitted. The extra bits transmitted in this way are called **chips**, and the resulting bit rate is known as the *chipping rate*. Most **direct-sequence spread-spectrum** systems use a chipping rate at least ten times as great as the bit rate of the actual information to be transmitted.

The use of the high-speed PN sequence results in an increase in the bandwidth of the signal, regardless of what modulation scheme is used to encode the bits into the signal. Recall from Hartley's Law (Equation 4.1) that for any given modulation scheme, the bandwidth is proportional to the bit rate. It follows from Hartley's Law that a direct-sequence system transmitting a total of ten bits for each information bit will use ten times as much bandwidth as a narrowband signal with the same type of modulation and the same information rate. That is, the sidebands will extend ten times as far from the carrier, as illustrated in Figure 4.11. Direct-sequence spread-spectrum schemes typically use some form of phase-shift keying (PSK).

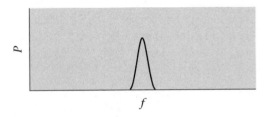

(a) Before spreading   (b) After spreading

FIGURE 4.11 Direct-sequence spread-spectrum signal

EXAMPLE 4.8

A digital communication scheme uses DQPSK. It is to transmit a compressed PCM audio signal which has a bit rate of 16 kb/s. The chipping rate is 10 to 1. Calculate the number of signal changes (symbols) which must be transmitted each second.

SOLUTION

The total bit rate, including the chips, is 10 times the data rate, or 160 kb/s. Since there are four signal states, each state represents two bits. Therefore the symbol rate is

160/2 = 80 kilobaud

Expanding the bandwidth by a factor of ten while keeping the transmitted power constant will decrease the received signal-to-noise ratio by the same factor. As before, the pseudo-random sequence is known to the receiver, which has to separate the information signal from the chips. A **processing gain**, also called **spreading gain**, can be defined equal to the bandwidth expansion (which, for direct-sequence spread spectrum, is also equal to the ratio of chips to information bits):

$$G_p = \frac{B_{RF}}{B_{BB}} \qquad (4.7)$$

where

G_p = processing gain

B_{RF} = RF (transmitted) bandwidth

B_{BB} = baseband (before spreading) bandwidth

The processing gain also describes the amount by which the signal-to-noise ratio of the signal is reduced by the spreading process during transmission. Of course, this reduction is reversed at the receiver. Since signal-to-noise ratio is generally given in decibels, it would make sense to express the processing gain that way too, that is:

G_p (dB) = 10 log G_p

G_p (dB) = $(S/N)_i$ (dB) − $(S/N)_o$ (dB) $\qquad (4.8)$

where

G_p (dB) = processing gain in decibels

$(S/N)_i$ (dB) = signal-to-noise ratio in dB before spreading

$(S/N)_o$ (dB) = signal-to-noise ratio in dB after spreading

EXAMPLE 4.9 ▼

A signal would have a bandwidth of 200 kHz and a signal-to-noise ratio of 20 dB if transmitted without spreading. It is spread using a chipping rate of 50:1. Calculate its bandwidth and signal-to-noise ratio after spreading.

SOLUTION

The bandwidth after spreading can be found from Equation (4.7):

$$G_p = \frac{B_{RF}}{B_{BB}}$$

$$B_{RF} = G_p B_{BB}$$

$$= 50 \times 200 \text{ kHz}$$

$$= 10 \text{ MHz}$$

The processing gain in decibels is

$$G_p \text{ (dB)} = 10 \log G_p$$
$$= 10 \log 50$$
$$= 17 \text{ dB}$$

The signal-to-noise ratio after spreading is given by Equation (4.8):

$$G_p \text{ (dB)} = (S/N)_i \text{ (dB)} - (S/N)_o \text{ (dB)}$$
$$(S/N)_o \text{ (dB)} = (S/N)_i \text{ (dB)} - G_p \text{ (dB)}$$
$$= 20 \text{ dB} - 17 \text{ dB}$$
$$= 3 \text{ dB}$$

▲

Reception of Spread-Spectrum Signals

The type of receiver required for spread-spectrum reception depends on how the signal is generated. For frequency-hopped transmissions, what is needed is a relatively conventional narrowband receiver that hops in the same way as and is synchronized with the transmitter. This requires that the receiver be given the frequency-hopping sequence, and there be some form of synchronizing signal (such as the signal usually sent at the start of a data frame in digital communication) to keep the transmitter and receiver synchronized. Some means must also be provided to allow the receiver to detect the start of a transmission, since, if this is left to chance, the transmitter and receiver will most likely be on different frequencies when a transmission begins.

One way to synchronize the transmitter and receiver is to have the transmitter send a tone on a prearranged channel at the start of each transmission, before it begins hopping. The receiver can synchronize by detecting the end of the tone and then begin hopping according to the prearranged PN sequence. Of course, this method fails if there happens to be an interfering signal on the designated synchronizing channel at the time synchronization is attempted.

A more reliable method of synchronizing frequency-hopping systems is for the transmitter to visit several channels in a prearranged order before beginning a normal transmission. The receiver can monitor all of these channels sequentially, and once it detects the transmission, it can sample the next channel in the sequence for verification and synchronization.

Direct-sequence spread-spectrum transmissions require different reception techniques. Narrowband receivers will not work with these signals, which occupy a wide bandwidth on a continuous basis. A wideband receiver is required, but a conventional wideband receiver would output only noise. In order to distinguish the desired signal from noise and interfering signals, which over the bandwidth of the receiver are much stronger than the desired signal, a technique called autocorrelation is used. Essentially this involves multiplying the received signal by a signal generated at the receiver from the PN code. When the input signal corresponds to the PN code, the output from the autocorrelator will be large; at other times this output will be very small. Of course, once again the transmitter and receiver will probably not be synchronized at the start of a transmission, so the transmitter sends a preamble signal, which is a prearranged sequence of ones and zeros, to let the receiver synchronize with the transmitter.

Code-Division Multiple Access (CDMA)   Traditionally, most analog communication systems have used frequency-division multiplexing, and digital systems have employed time-division multiplexing, in order to combine many information signals into a single transmission channel. When the signals originate from different sources, these two methods (which have already been studied) become frequency-division multiple access (FDMA) and time-division multiple access (TDMA), respectively.

Spread-spectrum communication allows a third method for multiplexing signals from different sources—**code-division multiple access (CDMA)**. All that is required is for each transmitter to be assigned a different pseudo-noise (PN) sequence. If possible, orthogonal sequences should be chosen; that is, the transmitters should never be in the same place at the same time. The PN sequence for the transmitter is given only to the receiver that is to operate with that transmitter. This receiver will then receive only the correct transmissions, and all other receivers will ignore these signals. This technique, which is applicable to both frequency-hopping and direct-sequence transmissions, allows many transmissions to share the same spread-spectrum channel.

It is not necessary for all transmitters to have orthogonal PN sequences in order to use CDMA. If some of the sequences are not orthogonal, there will be some mutual interference between users. However, the system can still operate until there are so many users that the signal-to-noise ratio becomes unacceptably low.

In order that all transmissions other than the desired one blend together at the receiver as noise, it is highly desirable in a CDMA system that all received signals have about the same strength at the receiver. Obviously this is not always possible, such as in an unlicensed environment like that of cordless phones. But where it is possible, the benefits of close control over transmitter power levels are great. We will see this idea applied later when we look at CDMA PCS systems.

 **Summary**

The main points to remember from this chapter are:

- Digital wireless transmission uses frequency, phase, and amplitude variations, just as does analog transmission.

- The maximum data rate for a channel is a function of bandwidth, modulation scheme, and signal-to-noise ratio.

- In general, more complex modulation schemes can achieve higher data rates, but only when the signal-to-noise ratio is high.

- Frequency-shift keying (FSK) uses two (and occasionally more than two) transmitted frequencies to achieve modest data rates with good performance in noisy channels.

- Gaussian minimum-shift keying is a special case of FSK that achieves the minimum bandwidth possible for a two-frequency FSK system at a given data rate.

- Most phase-shift keying (PSK) systems use four phase angles for somewhat higher data rates than are achievable with FSK.

- Quadrature amplitude modulation (QAM) achieves higher data rates than FSK or PSK by using a combination of amplitude and phase modulation. QAM requires a relatively noise-free channel to realize its advantages.

- Frequency-division multiplexing and multiple-access schemes divide available bandwidth among channels, with each operating full-time.

- Time-division multiplexing and multiple-access schemes divide the available time among channels, with each using the full bandwidth.

- Spread-spectrum systems reduce interference by spreading each signal over a large bandwidth.

- Code-division multiple-access (CDMA) schemes assign all channels full-time to the full available bandwidth. Channels are separated by spreading and despreading codes.

Equation List

$$I = ktB \tag{4.1}$$

$$C = 2B \log_2 M \tag{4.2}$$

$$C = B \log_2(1 + S/N) \tag{4.3}$$

$$C = S \log_2 M \tag{4.5}$$

$$f_m - f_s = 0.5 \, f_b \tag{4.6}$$

$$G_p = \frac{B_{RF}}{B_{BB}} \tag{4.7}$$

$$G_p \, (dB) = (S/N)_i \, (dB) - (S/N)_o \, (dB) \tag{4.8}$$

Key Terms

amplitude-shift keying (ASK) data transmission by varying the amplitude of the transmitted signal

audio frequency-shift keying (AFSK) use of an audio tone of two or more different frequencies to modulate a conventional analog transmitter for data transmission

baud rate speed at which symbols are transmitted in a digital communication system

bit rate speed at which data is transmitted in a digital communication system

chips extra bits used to spread the signal in a direct-sequence spread-spectrum system

code-division multiple access (CDMA) system to allow multiple users to use the same frequency using separate PN codes and a spread-spectrum modulation scheme

constellation diagram   in digital communication, a pattern showing all the possible combinations of amplitude and phase for a signal

delta phase-shift keying (DPSK)   digital modulation scheme that represents a bit pattern by a change in phase from the previous state

dibit system   any digital modulation scheme that codes two bits of information per transmitted symbol

direct-sequence spread spectrum   technique for increasing the bandwidth of a transmitted signal by combining it with a pseudo-random noise signal with a higher bit rate

frequency hopping   form of spread-spectrum communication in which the RF carrier continually moves from one frequency to another according to a prearranged pseudo-random pattern

frequency-division multiple access (FDMA)   sharing of a communication channel among multiple users by assigning each a different carrier frequency

frequency-division multiplexing (FDM)   combining of several signals into one communication channel by assigning each a different carrier frequency

frequency-shift keying (FSK)   digital modulation scheme using two or more different output frequencies

Gaussian minimum-shift keying (GMSK)   variant of FSK which uses the minimum possible frequency shift for a given bit rate

mark   in digital communication, a logic one

modem   acronym for modulator-demodulator; device to enable data to be transmitted via an analog channel

multiple access   use of a single channel by more than one transmitter

multiplexing   use of a single channel by more than one signal

phase-shift keying (PSK)   digital modulation scheme in which the phase of the transmitted signal is varied in accordance with the baseband data signal

processing gain   improvement in interference rejection due to spreading in a spread-spectrum system

pseudo-random noise (PN) sequence   a transmitted series of ones and zeros that repeats after a set time, and which appears random if the sequence is not known to the receiver

quadrature AM (QAM)   modulation scheme in which both the amplitude and phase of the transmitted signal are varied by the baseband signal

quadrature phase-shift keying (QPSK or DQPSK)   digital modulation scheme using four different transmitted phase angles

Rayleigh fading   variation in received signal strength due to multipath propagation

space   binary zero

spreading gain   improvement in interference rejection due to spreading in a spread-spectrum system

symbol   in digital communication, the state of the signal at a sampling time

time-division multiple access (TDMA)   system to allow several transmissions to use a single channel by assigning time slots to each

time-division multiplexing (TDM)   system to combine several data streams onto a single channel by assigning time slots to each

● Questions

1. What is the meaning of the term *modem*?
2. What parameters of a sine-wave carrier can be modulated?
3. Name the three most common basic types of digital modulation.
4. Which type of modulation is likely to be used for:
   (a) low data rates
   (b) moderate data rates
   (c) high data rates
5. What signal parameters are varied with *QAM*?
6. What factors limit the maximum data rate for a channel?
7. What is an eye diagram?
8. Explain the difference between the terms *bit rate* and *baud rate*.
9. Explain the origin and meaning of the terms *mark* and *space*.
10. What is the difference between *FSK*, *AFSK*, and *GMSK*?
11. Why is delta phase-shift keying the most common form of PSK?
12. What is the advantage of $\pi/4$ DQPSK?
13. What is represented by the dots in a constellation diagram for a QAM system?
14. Compare the modulation schemes studied in this chapter, listing as many advantages and disadvantages for each as you can.

15. List and describe the three multiple-access systems in common use.

16. What is a DS-1 signal?

17. Compare frequency-hopping and direct-sequence spread-spectrum systems.

18. What happens when a call is *blocked*?

19. How do spread-spectrum systems reduce the effect of fading?

20. Briefly describe what is meant by orthogonal spread-spectrum signals.

● Problems

1. The North American analog cellular radio system uses FM with channels 30 kHz wide. Suppose such a channel were used for digital communication. If the available signal-to-noise ratio is 20 dB, calculate the maximum theoretical bit rate and the corresponding baud rate using:

   (a) a two-level code

   (b) a four-level code

2. How much bandwidth would be required to transmit a DS-1 signal (1.544 Mb/s) using a four-level code:

   (a) assuming a noiseless channel?

   (b) with a signal-to-noise ratio of 15 dB?

3. The AFSK system described in the text operates at 1200 bits per second using an FM signal modulated by tones at 1200 and 2200 Hz, with a frequency deviation of 5 kHz. Calculate the efficiency of this system in bits per second per hertz of bandwidth by using Carson's rule (see Chapter 2) to calculate the approximate radio frequency bandwidth for this system. Is this system bandwidth-efficient?

4. A typical HF radioteletype system uses 170 Hz shift between mark and space frequencies and a bit rate of 45 bits per second. What would be the bit rate if GMSK were used for this system?

5. Consider a QPSK system that will transmit three bits of information per symbol.

   (a) How many phase angles are needed?

   (b) Draw a vector diagram for such a system.

   (c) Would this system have any advantages compared with the dibit systems described in the text? Any disadvantages?

6. A constellation diagram for a modem is shown in Figure 4.12.

FIGURE 4.12

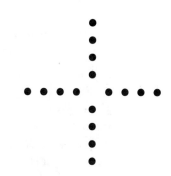

(a) What type of modulation is this?

(b) If the transmitted bit rate is 9600 b/s, what will be the baud rate using this modem?

7. A microwave radio system uses 256-QAM, that is, there are 256 possible amplitude and phase combinations.

(a) How many bits per symbol does it use?

(b) If it has a channel with 40-MHz bandwidth, what is its maximum data rate, ignoring noise?

8. Draw a constellation pattern for a modem that uses eight equally-spaced phase angles and four equally-spaced amplitude levels. If the modem operates at 4800 baud, what is its bit rate?

9. An ordinary broadcast television channel has a bandwidth of 6 MHz. How many FM radio stations (with a bandwidth of 200 kHz each) could be accommodated in one television channel using FDM?

10. Suppose the TV channel described in the previous problem is used for voice-quality transmission using SSBSC AM with a bandwidth of 4 kHz per channel. How many voice channels can be carried?

11. Suppose the TV channel described in the previous problem is used for digital audio communication. Assume the RF system can support 2 b/s/Hz of RF bandwidth.

(a) What is the total available bit rate?

(b) How many telephone voice signals, at a bit rate of 64 kb/s each, could be accommodated on this channel using TDM?

(c) How many high-quality stereo audio signals could be accommodated? Assume 16-bit samples, with a sampling rate of 44.1 kSa/s, and ignore error correction and data compression.

12. Suppose that a voice signal normally occupies 30 kHz of bandwidth and has a signal-to-noise ratio of 20 dB. Spread-spectrum techniques are used to increase its bandwidth to 2 MHz.

   (a) What is the signal-to-noise ratio of the spread signal?

   (b) What is the processing gain, in decibels?

13. Suppose a frequency-hopping system hops among 500 channels. How many orthogonal PN sequences are possible?

14. Suppose there is a narrowband analog signal on one of the channels visited by a frequency-hopping system.

   (a) What is the effect on the narrowband signal of the spread-spectrum signals?

   (b) What is the effect on the spread-spectrum signals of the narrowband signal?

15. A direct-sequence spread-spectrum system uses FSK with a chipping rate of 20 to 1. The signal-to-noise ratio for the spread signal is –5 dB (that is, the signal is 5 dB weaker than the noise in the same bandwidth). If the data is transmitted at 50 kb/s, calculate:

   (a) the chipping rate

   (b) the bandwidth occupied by the spread signal if the modulation scheme used allows 1.5 bits/s/Hz

   (c) the signal-to-noise ratio for the despread signal at the receiver

16. A signal has a bit rate of 20 kb/s. Find the baud rate if the signal is transmitted using:

   (a) FSK with two frequencies

   (b) QPSK with four phase angles

   (c) QAM with four phase angles and four amplitudes

17. Ten voice signals are to be multiplexed and transmitted. The analog signal occupies 4 kHz of bandwidth and can be digitized using a vocoder at 12 kb/s. Calculate the required bandwidth for each of the following possibilities.

   (a) FDMA using analog FM with 12 kHz deviation (use Carson's rule to find the bandwidth for one signal). Ignore guard bands between channels.

   (b) FDMA using SSBSC AM. Ignore guard bands.

   (c) TDM using GMSK. Assume a noiseless channel.

   (d) TDMA using QPSK. Assume a noiseless channel.

   (e) CDMA using frequency-hopping with 10 available channels. Use GMSK and assume a noiseless channel.

   (f) CDMA using direct-sequence, QPSK with a chipping rate of 10:1.

Basic Telephony

5

Objectives

After studying this chapter, you should be able to:

- Describe the topology of the switched telephone network.

- Describe the various signals present on a local-loop telephone line and explain the function of each.

- Describe and compare in-band and out-of-band signaling systems for telephony.

- Explain the advantages of common-channel signaling.

- Describe Signaling System Seven and explain its use in keeping track of calls.

- Explain the use of time-division multiplexing in telephony and perform bit rate calculations with TDM signals.

5.1 Introduction

The **public switched telephone system (PSTN)** is undoubtedly the largest, and probably the most important communication system in the world. The reasons for this are contained in those first two words. It is public in the sense that anyone can connect to it. Because it is switched, it is possible, in theory at least, for anyone to communicate with anyone else. This makes the telephone system very different from broadcasting systems and from private communication networks.

The addition of personal wireless communication to the PSTN has made it even more ubiquitous. Radio systems that were separate entities are increasingly becoming extensions of the telephone system. Consider, for instance, the difference between citizens' band radio and cellular radio. The former is a separate system, useful only when those who wish to communicate have compatible radios and are within radio range. The latter can connect its user to any telephone in the world, regardless of distance.

The telephone network employs many of the most interesting developments in communication practice, such as cellular radio, fiber optics, and digital signal transmission, but it remains in many ways consistent with its origins in the nineteenth century. This will become obvious when we look at telephone signaling systems and the voltage and current levels found on subscriber lines. Compatibility has been maintained in most areas of the system, so simple dial-type telephones can coexist with modern data-communication equipment. Though originally intended only for voice communication, the switched telephone network has been adapted to serve many other needs, including data communication, facsimile, and even video.

This chapter introduces the telephone system and describes the ways in which it can connect with wireless systems. A basic knowledge of ordinary voice telephony (*plain old telephone service* or *POTS* in telephone jargon) will be very useful as we consider more advanced uses.

5.2 Network Topology

Switched networks can be categorized as **circuit-switched** or **packet-switched**. In a circuit-switched network there is a dedicated physical path from transmitter to receiver for the duration of the communication. The PSTN is a circuit-switched network. Packet-switched networks route short bursts of data, called *packets,* from point to point as needed. A *virtual connection* may exist, but it is merely a record of the addresses on the network between which communication takes place. Successive packets may take different paths through the network. The internet is a packet-switched network.

Both types of networks have their advantages. Circuit switching is often more reliable, especially when it is important that messages arrive quickly and in the same order in which they were sent, as is the case with telephony. On the other hand, packet switching can make more efficient use of network resources. In a circuit-switched network the circuit is often idle. For instance, during a phone call circuits are active in both directions, but most of the time only one person is talking. It is possible to use packet-switched networks for telephony (internet phone is an example), and it looks as if this is the direction networks are moving.

Wireless networks can be either circuit-switched (cellular phones) or packet-switched (wireless local-area networks), and we shall examine both types later in this book.

PSTN Structure

Figure 5.1 shows the basic structure, or *topology,* of a local calling area (known as a **Local Access and Transport Area** or **LATA**) in a typical switched telephone system.

Each subscriber is normally connected via a separate twisted-pair line, called a **local loop**, to a **central office**, also called an **end office**, where circuit switching is done. Actually the term *office* can be deceiving: in urban areas, it is quite possible for there to be more than one central office in the same building.

The central office represents one exchange: that is, in a typical seven-digit telephone number, all the lines connected to a single central office begin with the same three digits. Thus there can be ten thousand telephones

FIGURE 5.1
Local access and transport area

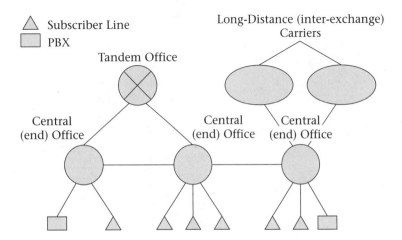

connected to a central office. Subscribers connected to the same central office can communicate with each other by means of the central office switch, which can connect any line to any other line. Modern switches are digital, so the analog local-loop signals are digitized as they enter the central office.

The central offices themselves are connected together by **trunk lines**; any subscriber can contact any other subscriber within a local calling area. There are not enough trunks or enough switching facilities for every subscriber to use the system at once, so there is the possibility of overload. This can make it impossible for a subscriber to place a call, an occurrence that is known as **call blocking**. The likely number of simultaneous conversations is predicted by statistical methods, and should usually be exceeded only during emergencies.

When there are no available trunks between two central offices, sometimes a connection can be made through a **tandem office**, which connects central offices without having any direct connections to individual telephones.

Long-distance calls are completed using a mesh of long-distance switching centers. The network usually lets the system find a direct route from one area of the country to the other, and there is never a need for more than one intermediate switch. Figure 5.2 shows this type of system.

The telephone system in the United States and Canada was formerly a monopoly, and it still is in much of the world. This is changing. Competition for long-distance calls is well established, and competition for the local subscriber loop is just beginning. Each competing long-distance carrier has its own connection to the local access and transport area. To keep the diagram simple, only two long-distance carriers are shown in Figure 5.2. Each carrier has its own connection, called a **point of presence (POP)**, to the local telephone system. Wireless telephone systems are connected to the PSTN in a similar fashion.

FIGURE 5.2
Long-distance
network

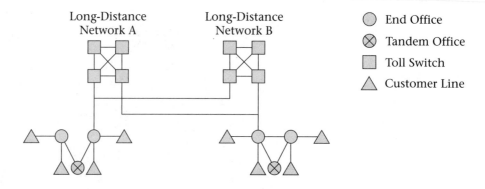

Until now, local-loop competitors have leased their physical connections from the original local monopoly. As local-loop competitors using their own cable (for example, cable television companies) enter the market, they too will require points of presence.

New switching equipment and trunk lines are digital, with time-division multiplexing used to combine many signals on one line. Fiber optics technology is increasingly used for trunk lines; other common media include terrestrial microwave links, geostationary satellites, and coaxial cable. Short distances between central offices may be covered using *multipair* cable, which has many twisted pairs in one protective sheath. So far, most local loops are still analog using twisted-pair copper wire, but that seems likely to change in the future. Eventually the system will be digital from one end to the other, and most of it will employ fiber-optic cable. However, telephone equipment is built for maximum reliability and lasts a long time. It is also expensive, and telephone companies quite naturally expect to use it until it wears out. It will be a few more years before we have a complete digital network.

5.3 The Local Loop and Its Signals

Normally each individual subscriber telephone is connected to the central office by a single twisted pair of wires. The wires are twisted to help cancel their magnetic fields and reduce interference, called **crosstalk**, between circuits in the same cable. It is common practice to run a four-conductor cable to each residence, but only two of these wires (usually red and green) are used for a single line. The others (black and yellow) allow for the installation of a second line without running more cable.

Recently there has been a trend toward running multiplexed digital signals to junction boxes in neighborhoods, switching at that point to analog signals on individual pairs, in an effort to reduce the total amount of copper cable. There has been some installation of optical fiber, though the last section of loop to the customer normally remains copper. In the future, **fiber-in-the-loop (FITL)** may be used. It will cost more because of the necessity of converting back and forth between electrical and optical signals at each subscriber location, but the bandwidth will be vastly increased. This will allow a great number of additional services, such as cable television and high-speed internet access, to be carried on the same fiber.

The local loop performs several functions. It carries voice signals both ways. It must also carry signaling information both ways: dialing pulses or tones to the central office from the customer and dial tones, ringing, busy signals, and prerecorded messages from the network to the subscriber. Wireless telephones must perform all of these functions, although usually not in exactly the same way.

In addition to the previously mentioned functions, the twisted pair must transmit power from the central office to operate the telephone and ring the bell.

When the phone is *on hook* (not in use), the central office maintains a voltage of about 48 V dc across the line. Of the two wires in the twisted pair, one, normally the green, is designated *tip* and the other (red), *ring*. The ring is connected to the negative side of the supply. Most of the time in electronic equipment a red wire is positive but not here! The "tip" and "ring" terminology dates from the days of manual switchboards; it describes the connections to the plugs used in these boards. The positive (tip) side of the supply is grounded.

The central office supply is called the *battery*. The voltage does, in fact, derive from a storage battery that is constantly under charge. This allows the telephone system to function during electrical power outages, whether they occur at the central office or at the customer's premises, and has resulted in a well-deserved reputation for reliability. If wireless systems are to strive for similar performance, they too need emergency power. This is especially important for cellular phone and PCS services, because many people subscribe to these services in order to have emergency communication when needed.

When the phone is on hook, it represents an open circuit to the dc battery voltage. The subscriber signals the central office that he or she wishes to make a call by lifting the receiver, placing the instrument *off hook*. The telephone has a relatively low resistance (about 200 ohms) when off hook, which allows a dc current to flow in the loop. The presence of this current signals the central office to make a line available (the telephone is said to have *seized* the line). When off hook, the voltage across the telephone drops considerably, to about 5 to 10 volts, due to the resistance of the telephone line. Resistance can also be added at the central office, if necessary, to maintain the loop current in the desired range of approximately 20 to 80 mA. Figure 5.3 illustrates this capability.

EXAMPLE 5.1 ▼

A local loop has a resistance of 1 kΩ, and the telephone connected to it has an off-hook resistance of 200 Ω. Calculate the loop current and the voltage across the telephone when the phone is:

(a) on hook

(b) off hook

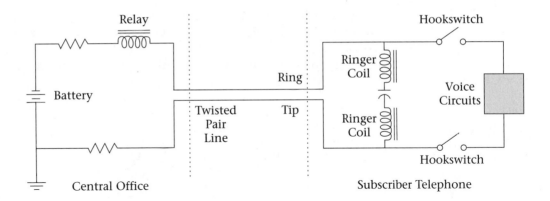

FIGURE 5.3 Local loop

SOLUTION

(a) When the telephone is on hook, its dc resistance is infinite so the current is zero. Since there will be no voltage drop around the loop, except at the phone itself, the full battery voltage will appear across the phone.

(b) When the phone is off hook, the total loop resistance is

$$R_T = 1000 \ \Omega + 200 \ \Omega = 1200 \ \Omega$$

Then the loop current is

$$I = \frac{48 \ \text{V}}{1200 \ \Omega} = 40 \ \text{mA}$$

The voltage across the telephone is

$$V = IR = 40 \ \text{mA} \times 200 \ \Omega = 8 \ \text{V}$$

Once a line has been assigned, the office signals the user to proceed by transmitting a dial tone, which consists of 350 Hz and 440 Hz signals added together.

Dialing can be accomplished in one of two ways. The old-fashioned rotary dial functions by breaking the loop circuit at a 10 Hz rate, with the number of interruptions equal to the number dialed. That is, dialing the number 5 causes five interruptions (pulses) in the loop current. This technique is called *pulse dialing* and can be emulated by some electronic telephones. The second and much more efficient way is for the phone to transmit a combination of two tones for each number. This is officially known as **dual-tone**

multi-frequency (DTMF) dialing and is commonly referred to as *Touch-Tone* or just *tone dialing* (the term Touch-Tone is a registered trade mark of AT&T). Table 5.1 shows the combinations of tones used for each digit. This system improves efficiency because digits can be transmitted in much less time than with pulse dialing.

TABLE 5.1   DTMF Frequencies

Frequencies (Hz)	1209	1336	1477	1633
697	1	2	3	A
770	4	5	6	B
852	7	8	9	C
941	*	0	#	D

The letters A through D are included in the system specifications but are not present on ordinary telephones. Some wireless equipment uses them for special functions.

EXAMPLE 5.2

What frequencies are generated by a telephone using DTMF signaling when the number 9 is pressed?

SOLUTION

Use Table 5.1. Go across from 9 to find 852 Hz; go up to find 1477 Hz. Therefore, the output frequencies are 852 Hz and 1477 Hz.

Assume for now that the called party is connected to the same central office as the calling party; that is, they have the same exchange, and the first three numbers in a typical seven-digit telephone number are the same. When the switch connects to the called party, it must send an intermittent ringing signal to that telephone. The standard for the ringing voltage at the central office is 100 V ac at a frequency of 20 Hz, superimposed on the 48 V dc battery voltage. Of course, the voltage at the telephone will be less than that, due to the resistance of the wire in the local loop. In order to respond to the ac ringing signal when on-hook, without allowing dc current to flow, the telephone ac-couples the ringer to the line. In a conventional telephone with an electromechanical ringer, the ringer consists of two coils and a capacitor in series across the line.

While the called telephone is ringing, the central office switch sends a pulsed ac voltage, called a **ringback signal**, to the calling telephone. The ringback signal consists of 440 and 480 Hz signals added together. When the called phone goes off-hook, the circuit is complete, the ringing voltages are switched off, and conversation can begin. If the circuit corresponding to the called telephone is in use, a busy signal will be returned to the caller.

See Table 5.2 for a summary of the signals described so far.

TABLE 5.2   Local-Loop Voltages and Currents

On-hook voltage	48 V dc
Off-hook voltage (at phone)	5–10 V dc, depending on loop resistance
Off-hook current	23–80 mA dc, depending on loop resistance
Dial tone	350 and 440 Hz
Ringing voltage (at office)	100 V ac, 20 Hz, superimposed on 48 V dc
Ringing voltage (at phone)	Approximately 80 V ac, superimposed on 48 V dc
Ringback voltage	440 and 480 Hz; pulsed 2 s on, 4 s off
Busy signal	480 Hz and 620 Hz; pulsed 0.5 s on, 0.5 s off

The single twisted-pair local loop is required to carry both sides of the conversation simultaneously, providing full-duplex communication. This is called a *two-wire system*. The rest of the network uses separate transmission paths for each direction, a topology called a *four-wire system*. Converting between the two systems is done using a circuit called a **hybrid coil**, shown in Figure 5.4. The same thing can be done electronically. Signals from the transmitter will add at the line and cancel at the receiver. Similarly, signals coming in on the line will cancel at the transmitter and add at the receiver. Deliberately unbalancing the circuit allows a small portion of the transmitter signal to reach the receiver, creating a **sidetone** that lets the user know the line is active and hear what is being transmitted. Hybrid coils are used both in the phone and in the central office line cards.

FIGURE 5.4
Hybrid coil

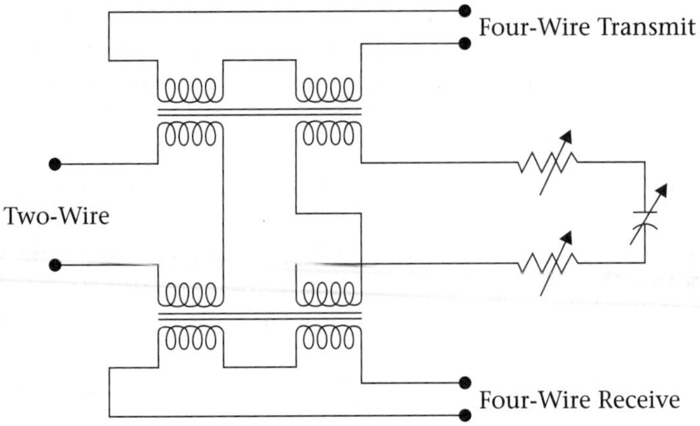

Four-Wire Transmit

Two-Wire

Four-Wire Receive

5.4 Digital Telephony

The telephone system was completely analog at its beginning in the nineteenth century, of course. Over the past 30 years or so, it has gradually been converted to digital technology and the process is not yet complete. The digital techniques were originally designed to work in conjunction with the existing analog system. This, and the fact that many of the standards for digital technology are quite old, should help the reader understand some of the rather peculiar ways things are done in digital telephony.

The basics of digital transmission of analog signals by pulse-code modulation (PCM) were discussed in Chapter 3. We can summarize the results here as they apply to the North American system. The numbers vary slightly in some other countries but the principles are the same.

The analog voice signal is low-pass filtered at about 3.4 kHz and then digitized, using 8-bit samples at a sampling rate of 8 kHz. The signal is compressed, either before or after digitization, to improve its signal-to-noise ratio. The bit rate for one voice signal is then

$$f_b(voice) = 8 \text{ bits/sample} \times 8000 \text{ samples/second} = 64 \text{ kb/s} \qquad (5.1)$$

The sample rate is determined by the maximum frequency to be transmitted, which was chosen for compatibility with existing analog FDM transmission (which uses SSBSC AM with a bandwidth of 4 kHz per channel, including guardbands between channels). An upper frequency limit of about 3.4 kHz has long been considered adequate for voice transmission.

In Chapter 3 we found that a much lower bit rate could be used for telephone-quality voice using data compression and vocoders. These techniques are not employed in the ordinary telephone system, though data compression is used in special situations where bandwidth is limited and expensive, as in intercontinental undersea cables.

The connection of wireless systems to the PSTN requires conversion of standards in both directions. In general, wireless systems use ordinary telephone quality as a guide, though many of these systems fall short of what has been considered *toll quality,* that is, good enough to charge long-distance rates for. Until now, users have been so delighted to have portable telephones that they have been willing to put up with lower quality. This situation is changing quickly now that wireless phones are everywhere.

Time-Division Multiplexing   In Chapter 3 we discussed the simplest form of time-division multiplexed telephone signal. The DS-1 signal frame has one sample (8 bits) from each of 24 telephone channels plus one framing bit. This gives it a bit rate of

$$f_b \ (DS\text{-}1) = 24 \times 64 \text{ kb/s} + 1 \times 8 \text{ kb/s} = 1.544 \text{ Mb/s.} \tag{5.2}$$

When this signal is transmitted over copper wire, the result is known as a *T-1 carrier.* That is, the *signal* includes only the coding into ones and zeros, while the *carrier* also includes the voltage levels used. See Figure 5.5 for a review of the DS-1 frame, which we looked at in Chapter 4.

FIGURE 5.5
DS-1 signal

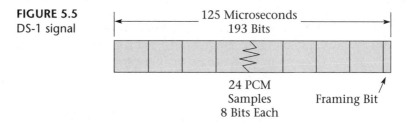

A DS-1 signal can equally well be used for data communication. A single time slot may be used or any number of time slots can be combined for higher bit rates.

The framing bits are used to enable the receiver to determine which bit (and in what sample) is being received at a given time. In addition, the receiver must often distinguish between frames in order to decode signaling information. In one frame out of every six, each of the least significant bits in the 24 samples may be used for signaling information rather than as part of the PCM signal. This information includes on/off hook status, dial tone, dialed digits, ringback, and busy signal. This **bit robbing** for signaling results in a very slight degradation of voice signal quality; for instance, the signal-to-noise ratio is reduced by about two decibels.

The frames are divided into groups of 12 with different signaling information in the sixth and twelfth frames, known as the *A* and *B* frames, in a sequence. A group of twelve frames is called a *superframe*. As a result, the receiver is required to count frames up to 12. To allow the receiver to accomplish this, the framing bit alternates between two sequences, 10001<u>1</u> and 01110<u>0</u>. The underlined bits indicate the *A* and *B* signaling frames, respectively. The "stolen" signaling bits can be used to indicate basic line states such as on-hook and off-hook, ringing, and busy signals.

Unfortunately the effect of bit robbing on data transmission is much greater than it is for voice. Occasional bit errors are acceptable in voice signals, but certainly not in data. To avoid errors when bit robbing is used, one bit from each 8-bit sample is discarded in *every* frame. This reduces the data capacity of one voice channel from 64 kb/s to 56 kb/s. Of course, channels can be combined for higher rates, but the loss in throughput is very substantial. Bit robbing can be eliminated by using common-channel signaling, which is described in the next section.

Digital Signal Hierarchy

The DS-1 signal and T-1 carrier described earlier represent the lowest level in a hierarchy of TDM signals with higher bit rates. All of these signals contain PCM audio signals, each sampled 8,000 times per second. As the number of multiplexed voice signals increases, so does the bit rate. This requires that the channel have a wider frequency response and that variations of time delay with frequency be held to a low level. Twisted-pair lines, when specially conditioned, can be used for the T1 and T2 carriers, but higher data rates require channels with greater bandwidth, such as coaxial cable, microwave radio, or optical fiber. See Table 5.3 for more details.

TABLE 5.3   Digital Signal Hierarchy

Carrier	Signal	Voice Channels	Bit Rate (Mb/s)
T1	DS-1	24	1.544
T1C	DS-1C	48	3.152
T2	DS-2	96	6.312
T3	DS-3	672	44.736
T4	DS-4	4032	274.176
T5	DS-5	8064	560.16

A glance at the table shows that the math does not seem to be exact. For instance, a DS-1C signal carries as many voice channels as two DS-1 signals, but the bit rate is more than twice as great. The difference is:

3.152 Mb/s − 2 × 1.544 Mb/s = 64 kb/s

The extra bits have several uses. They provide synchronization and framing for the demultiplexer. There are also extra bits called *stuff bits* which are added during multiplexing to compensate for differences between the clock rates of the tributaries and the multiplexer. If the tributary clock rate is slow, more stuff bits will be added to build up the bit rate; if it is fast, fewer stuff bits are needed. This **bit stuffing** is more formally called **justification**. Figure 5.6 is an example of the creation of a DS-3 signal by multiplexing other signals.

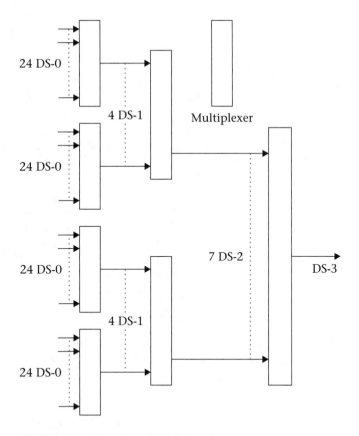

FIGURE 5.6 Creation of DS-3 signal

5.5 Telephone-Network Signaling

We have already looked at some of the control and supervisory signals used with the telephone system. The local-loop signals such as DTMF tones, dial tone, busy signal, and ringback signal are examples. All of these signals use the same channel as the voice, but not at the same time. Since they use the same channel, they are called **in-channel signals**. Their frequencies are also in the same range as voice frequencies; they can be heard by the user, so they are also referred to as **in-band signals**.

We also noted that the telephone instrument communicates its off-hook or on-hook status to the central office by the presence or absence, respectively, of a dc current. This signal is in-channel because it uses the same pair of wires as the voice, but it is **out-of-band** because the dc current is not in the same frequency range as voice signals. Consequently, the central office can receive the off-hook signal continuously, even during the call.

Traditionally, similar methods have been used within the network for such purposes as communicating the number of the calling party to billing equipment, determining which trunk lines are idle (that is, ready for use), and so on. Early systems used dc currents and dial pulses, as with local loops. Later versions used either a switched single-frequency tone at 2600 Hz (called *SF signaling*) or, still later, combinations of tones similar to but not the same as the DTMF system and known as *MF, (multi-frequency)* signaling. For a time, that system was plagued by fraud as people ranging from amateur "phone phreakers" to members of organized crime used so-called "blue boxes" to duplicate network signaling tones and make long-distance calls without paying for them. However, changes to the network soon eliminated or at least greatly reduced this problem.

In-channel but out-of-band signaling is also used. A tone at 3825 Hz can be sent along a long-distance network. It will pass through the allotted 4-kHz channel for each call, but will be filtered out before it reaches the customer telephone. This type of signal replaces the dc loop current as a means of indicating whether the line is in use, since long-distance circuits, unlike local loops, do not have dc continuity.

We have seen that the digital signal hierarchy makes available a few bits, "stolen" from the voice channel, for signaling. These can be used to replace the in-channel tones. However, there are insufficient bits to use for call routing, call display, billing, and other functions of a modern telephone network.

Signaling System Seven

Recently the trend has been to use a completely separate data channel to transmit control information between switches. This **common-channel signaling** reduces fraud, since users have no access to the control channels, and also allows a call to be set up completely before any voice channels are used. The state of the whole network can be known to the control equipment, and the most efficient routes for calls can be planned in advance. Common-channel signaling also makes such services as calling-number identification much more practical.

The current version of common-channel signaling is **signaling system seven (SS7)**. It was introduced to the Bell System in the United States in 1980 and has become, with minor variations, a worldwide system. SS7 is a packet-switched data network linking central offices to each other, to long-distance switching centers, and to centralized databases used for such purposes as call display, credit card validation, voice mail, 800 and 900 number routing and (most interesting for our purposes) cellular and PCS telephone roaming information. SS7 allows much more data to be sent more quickly, and with less interference with voice signals, than older signaling schemes involving in-channel signals.

SS7 uses dedicated 64 kb/s data channels. Usually one digital voice channel in each direction is reassigned for this purpose; the data rate is the same as for a voice channel to accommodate this. If necessary, an analog channel with modems can be used. One 64 kb/s signaling channel can handle the signaling requirements of many voice channels. Figure 5.7 shows how SS7 connects to the rest of the network.

With SS7, calls can be set up with no need to tie up a long-distance voice channel until the connection is made. Since analog local loops do not support common-channel signaling, it is necessary to tie up a voice connection from the subscriber to the central office. If ISDN, described in the next section, is used, voice and control signals can be kept completely separate.

FIGURE 5.7
Signaling system
seven (SS7)

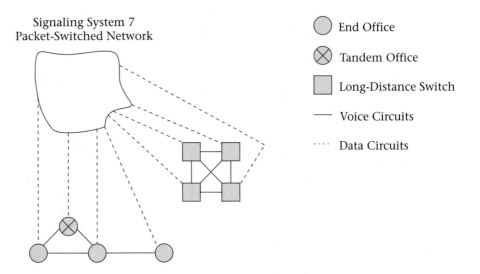

Signaling System 7
Packet-Switched Network

⚪ End Office
⊗ Tandem Office
⬜ Long-Distance Switch
— Voice Circuits
···· Data Circuits

5.6 Digital Local Loops

The analog local loop dates back to the earliest days of telephony, and is certainly outdated. Subscribers now need much greater bandwidth for uses such as high-speed internet access or interactive video than was the case a few years ago. One way to get this bandwidth is to replace the twisted-pair loop with coaxial cable or optical fiber. However, the capacity of a twisted pair is much greater than is required for the single analog signal, with a bandwidth under 4 kHz that it usually carries. Most of the cost of the local loop is in the labor to install it, rather than the cost of the wire itself, so when greater bandwidth is required it makes good sense to redesign the system to increase the capacity of the existing wiring rather than replace the wire. Here is a brief look at two ways of increasing the capacity of twisted-pair local loops.

Integrated Services Digital Network (ISDN)

The **integrated services digital network (ISDN)** concept is designed to allow voice and data to be sent in the same way along the same lines. Currently, most subscribers are connected to the switched telephone network by local loops and interface cards designed for analog signals. This is reasonably well-suited to voice communication, but data can be accommodated only by the use of modems.

As the telephone network gradually goes digital, it seems logical to send data directly over telephone lines without modems. If the local loop could be made digital, with the codec installed in the telephone instrument, there is no reason why the 64 kb/s data rate required for PCM voice could not also be used for data, at the user's discretion.

The integrated services digital network concept provides a way to standardize the above idea. The standard encompasses two types of connections to the network. Large users connect at a primary-access point with a data rate of 1.544 Mb/s. This, you will recall, is the same rate as for the DS-1 signal described earlier. It includes 24 channels with a data rate of 64 kb/s each. One of these channels is the **D (data) channel** and is used for common-channel signaling, that is, for setting up and monitoring calls. The other 23 channels are called **B (bearer) channels** and can be used for voice or data, or combined, to handle high-speed data or digitized video signals, for example.

Individual terminals connect to the network through a basic interface at the basic access rate of 192 kb/s. Individual terminals in a large organization use the basic access rate to communicate with a **private branch exchange (PBX)**, a small switch dedicated to that organization. Residences and small businesses connect directly to the central office by way of a digital local loop. Two twisted pairs can be used for this, though more use of fiber optics is expected in the future.

Basic-interface users have two 64 kb/s B channels for voice or data, one 16 kb/s D channel, and 48 kb/s for network overhead. The D channel is used to set up and monitor calls and can also be employed for low-data-rate applications such as remote meter-reading. All channels are carried on one physical line, using time-division multiplexing. Two pairs are used, one for signals in each direction.

Figure 5.8 shows typical connections to the ISDN. The primary interface is known as a T type interface, and the basic interface has the designation S. Terminal equipment, such as digital telephones and data terminals, designed especially for use with ISDN, is referred to as *TE1* (*terminal equipment type 1*), and connects directly to the network at point S. The network termination equipment designated NT2 could be a PBX, a small computer network called a **local area network**, or a central office. Terminal equipment not especially designed for ISDN is designated *TE2* (*terminal equipment type 2*) and would need a *terminal adapter* (*TA*) to allow it to work with the ISDN. Examples of type two equipment would be ordinary analog telephones, ordinary fax machines, and personal computers with serial ports. Each of these would need a different type of terminal adapter.

FIGURE 5.8
ISDN access

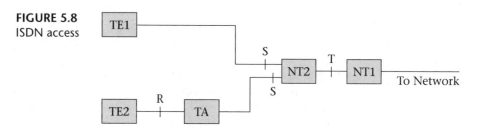

Implementation of the ISDN has been slow, leading some telecommunications people to claim, tongue-in-cheek, that the abbreviation means "it still does nothing." Several reasons can be advanced for this. First, converting local loops to digital technology is expensive, and it is questionable whether the results justify the cost for most small users. Residential telephone users would not notice the difference (except that they would have to replace all their telephones or buy terminal adapters), and analog lines with low-cost modems or fax machines are quite satisfactory for occasional data users. Newer techniques like **Asymmetrical Digital Subscriber Line (ADSL)**, which is described in the next section, and modems using cable-television cable have higher data rates and are more attractive for residential and small-office data communication. Very large data users often need a data rate well in excess of the primary interface rate for ISDN. They are already using other types of networks. It appears possible that the ISDN standard is becoming obsolete before it can be fully implemented.

With this in mind, work has already begun on a revised and improved version of ISDN called *broadband ISDN (B-ISDN)*. The idea is to use much larger bandwidths and higher data rates, so that high-speed data and video can be transmitted. B-ISDN uses data rates of 100 to 600 Mb/s.

Asymmetrical Digital Subscriber Line (ADSL)

The idea behind the asymmetrical digital subscriber line (ADSL) is to use the frequencies above the voice range for high-speed data while leaving the use of the local loop for analog telephony intact. This allows the subscriber to use conventional analog telephones without special adapters, while simultaneously sending and receiving high-speed data. The word *asymmetrical* in the name refers to the fact that the system is designed for faster communication from the network to the subscriber than from the subscriber to the network. Typical uses for ADSL include internet access and interactive television; for both of these the subscriber needs to receive data at a faster rate than it needs to be transmitted.

There are many types of ADSL using different-frequency carriers for downstream (to the subscriber) and upstream (from the subscriber) data. Downstream data rates vary from about 1 to 8 Mb/s, with upstream rates from 160 to 640 kb/s. Most systems use FDM to separate upstream from downstream data, as illustrated in Figure 5.9. Note that the downstream signal has wider bandwidth, as would be expected considering its higher data rate.

Most ADSL systems require the installation of a splitter at the customer premises to separate voice and data signals, but one variety, known as *DSL Lite,* requires no splitter and provides a downstream rate of up to 1.5 Mb/s. ADSL has the advantage over ISDN in that data signals do not have to go through the central office switch. This means that a user can be connected to the internet on a continuous rather than a dial-up basis. It also reduces traffic on the switched network.

FIGURE 5.9
Spectrum of a
typical ADSL system

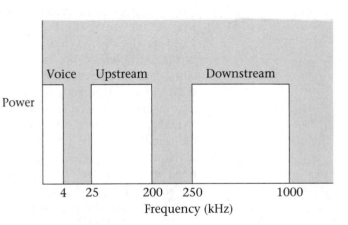

Summary

The main points to remember from this chapter are:

- The public switched telephone network is of great importance for wireless communication because it links wireless users with other users—both wireless and wired.

- Many of the specifications for telephone equipment have their basis in nineteenth century technology and have remained almost unchanged for reasons of compatibility.

- Common-channel signaling allows the telephone network to set up calls economically, without tying up voice lines, and allows the system to provide a considerable amount of data about calls.

- Wireless systems that connect with the telephone system can use its common-channel signaling system to carry information about subscribers roaming out of their local service areas.

- Modern telephone systems are digital except for the subscriber loop, and progress is being made towards digitizing the local loop as well.

- Digital telephone signals are time-division multiplexed with a data rate of 64 kb/s per voice channel.

- Signaling can be done by *robbing* bits from the voice signal but common-channel signaling using Signaling System Seven is more efficient and allows more information to be shared.

- ISDN allows the telephone system to be completely digital from end to end.

- ADSL allows for a conventional analog voice local loop and a high-speed data link to be combined on one twisted pair.

Key Terms

Asymmetrical Digital Subscriber Line (ADSL)   method of providing high-speed data transmission on twisted-pair telephone loops by using high-frequency carriers

B (bearer) channels   in ISDN, channels that carry subscriber communication

bit robbing   use of bits that normally carry payload information for other purposes, such as controlling the communication system

bit stuffing   addition of bits to a bitstream to compensate for timing variations

call blocking   failure to connect a telephone call because of lack of system capacity

central office switch in a telephone system that connects to local subscriber lines

circuit-switched network communication system in which a dedicated channel is set up between parties for the duration of the communication

common-channel signaling use of a separate signaling channel in a telephone system, so that voice channels do not have to carry signaling information

crosstalk interference between two signals multiplexed into the same channel

D (data) channel in ISDN, a communication channel used for setting up calls and not for user communication

dual-tone multi-frequency (DTMF) dialing signaling using combinations of two audio tones transmitted on the voice channel

end office see central office

fiber-in-the-loop (FITL) use of optical fiber for telephone connections to individual customers

hybrid coil a specialized transformer (or its electronic equivalent) that allows telephone voice signals to travel in both directions simultaneously on a single twisted-pair loop

in-band signals control signals sent in a voice channel at voice frequencies

in-channel signals control signals using the same channel as a voice signal

integrated services digital network (ISDN) telephone system using digital local loops for both voice and data, with the codec in the telephone equipment

justification addition of bits to a digital signal to compensate for differences in clock rates; informally known as *bit stuffing*

local access and transport area (LATA) in a telephone system, the area controlled by one central office switch

local area network a small data network, usually confined to a building or cluster of buildings

local loop in a telephone system, the wiring from the central office to an individual customer

out-of-band in telephone signaling, a control signal that is outside the voice frequency range

packet-switched network a communication system that works using data divided into relatively short transmissions called packets; these are routed through the system without requiring a long-term connection between sender and receiver

point of presence (POP)   place where one telephone network connects to another

private branch exchange (PBX)   small telephone switch located on customer premises

public switched telephone system (PSTN)   the ordinary public wireline phone system

ringback signal   in telephony, a signal generated at the central office and sent to the originating telephone to indicate that the destination telephone is ringing

sidetone   in telephony, the presence in the receiver of sounds picked up by the transmitter of the same telephone

signaling system seven (SS7)   system used in telephony which transmits all call setup information on a packet-data network that is separate from the voice channels used for telephone conversations

tandem office   telephone switch that connects only to other switches, and not to individual customers

trunk lines   transmission line carrying many signals, either on multiple pairs or multiplexed together on a single twisted-pair, coaxial cable, or optical fiber

● Questions

1. Explain briefly how the telephone network differs from a broadcasting network.
2. Explain the difference between circuit-switched and packet-switched networks. Is the PSTN mainly circuit-switched or packet-switched?
3. What is a central office?
4. What is the difference between a tandem office and an end office?
5. What is a trunk line?
6. What is meant by a LATA?
7. How do wireless telephone providers connect to the wired telephone network?
8. How has the breakup of the Bell monopoly changed the North American telephone network?
9. What is meant by call blocking, and why does it happen?

10. How many wires are needed (for a single line) from the individual telephone set to the central office?

11. How many wires are normally contained in the cable from an individual residence subscriber to the network? Why is this number different from the answer to Question 10 above?

12. Explain the meaning of the terms *tip* and *ring*. Which has negative polarity?

13. Explain how pulse dialing works.

14. What is meant by DTMF dialing, and why is it better than pulse dialing?

15. What is the function of the hybrid coil in a telephone instrument?

16. What is sidetone and why is it used in a telephone instrument?

17. Approximately how much bandwidth, at baseband, is needed for one channel of telephone-quality audio?

18. List the steps required in originating a local call. Include the appropriate voltages and frequencies that appear at the telephone instrument.

19. Describe the difference between in-band and out-of-band signaling, and give an example of each.

20. Describe the difference between in-channel and common-channel signaling. Which is the more modern system?

21. How does common-channel signaling reduce the vulnerability of the telephone system to fraudulent use?

22. Name some types of information that are carried by Signaling System Seven.

23. What type of data channel is used by SS7?

24. What type of modulation is used in FDM telephony?

25. What type of modulation is used in TDM telephony?

26. What is meant by *bit robbing*? What is its function and why is it undesirable for data connections?

27. What is meant by *bit stuffing*? When and why is it necessary?

28. Compare basic-rate ISDN and ADSL as technologies for voice telephony. Compare number of lines and type of equipment needed.

29. Compare basic-rate ISDN and ADSL as technologies for data communication. Compare data rates and connection type.

Problems

1. Suppose the voltage across a telephone line, at the subscriber, drops from 48 V to 10 V when the phone goes off hook. If the telephone instrument has a resistance of 200 ohms when off hook and represents an open circuit when on hook, calculate:

   (a) the current that flows when the phone is off hook

   (b) the combined resistance of the local loop and the power source at the central office

2. The local loop has a resistance of 650 ohms and the telephone instrument has a ringer voltage of 80 volts when the voltage at the central office is 100 V. Calculate the impedance of the ringer in the telephone.

3. Find the DTMF frequencies for the number 8.

4. What number is represented by tones of 770 and 1209 Hz, in the DTMF system?

5. Calculate the overhead of a DS-4 signal:

   (a) in bits per second

   (b) as a percentage of the total bit rate

6. By what percentage does the use of bit robbing reduce the data capacity of a DS-1 signal?

7. What is the proportion of overhead in a basic rate ISDN signal? (Assume the D channel is part of the overhead.) Compare with the overhead in a DS-1 signal.

8. The fastest modem for use on an analog telephone line operates at 56 kb/s in both directions. If an ADSL system has an upstream rate of 640 kb/s and a downstream rate of 1.5 Mb/s, by what factor does it exceed the modem data rate in each direction?

9. Compare the data rate available with basic-rate ISDN and that available with the ADSL system described in the previous problem. By what factor does ADSL exceed basic-rate ISDN in each direction:

   (a) if a voice call is being made simultaneously with data?

   (b) if no voice call is being made so that the whole available ISDN rate can be used for data?

10. Telephone signals can be carried by radio using either analog or digital modulation schemes. Compare the bandwidth required to carry each of the following signals:

    (a) an analog voice signal with a baseband bandwidth of 4 kHz using SSBSC AM (often used for terrestrial microwave links where many voice signals are transmitted together)

(b) the same analog voice signal using FM with a frequency deviation of 12 kHz (used for cellular phones)

(c) a standard digital voice signal using QPSK and assuming a channel with a signal-to-noise ratio of 20 dB. (Your answer to this question may suggest why data compression and vocoders are often used when digital signals are to be transmitted by radio.)

Transmission Lines and Waveguides

6

Objectives

After studying this chapter, you should be able to:

- Give several examples of transmission lines and explain what parameters of a transmission line must be considered as the frequency increases.

- Define characteristic impedance and calculate the impedance of a coaxial or open-wire transmission line.

- Define reflection coefficient and standing-wave ratio and calculate them in practical situations.

- Calculate losses on transmission lines.

- Name the dominant mode in a rectangular waveguide and describe its characteristics.

- Calculate the cutoff frequency, phase and group velocity, guide wavelength, and characteristic impedance, for the TE_{10} mode in a rectangular waveguide.

- Describe several methods for coupling power into and out of waveguides.

- Explain the operation of passive microwave components, including waveguide bends and tees, resonant cavities, attenuators, and loads.

- Describe the operation and use of circulators and isolators.

- Explain the importance of impedance matching with respect to transmission lines.

6.1 Introduction

Radio communication systems rely on the propagation of electromagnetic waves through air and space. The waves also travel along transmission lines and waveguides en route to and from antennas. In order to understand these systems, we must first learn something of the nature of electromagnetic radiation and then consider the propagation of these waves along conductors and in space. In this chapter we will study the propagation of electromagnetic waves on transmission lines and in waveguides. The next chapter deals with propagation in space.

6.2 Electromagnetic Waves

Radio waves are one form of electromagnetic radiation. Other forms include infrared, visible light, ultraviolet, x-radiation, and gamma rays. Figure 6.1 shows the place of radio waves at the low-frequency end of the frequency spectrum.

Electromagnetic radiation, as the name implies, involves the creation of electric and magnetic fields in free space or in some physical medium. When the waves propagate in a transmission line consisting of a pair of conductors, the fields can be represented as voltages and currents. When the propagation is in a dielectric medium or in free space, we need to work with the fields directly. Both types of wave propagation are essential to wireless communication.

Sinusoidal waves of voltage and current should be familiar from basic electricity and from our studies of various signals earlier in this book. When a sine wave propagates along a line or through space, it varies sinusoidally in both time and space. Just as the time taken for one repetition of a wave is called the period, the distance between identical points on the wave, at the

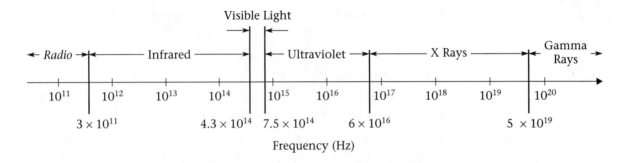

FIGURE 6.1 The electromagnetic spectrum

same instant of time, is known as the **wavelength** of the signal. Another way of putting this is to say that the wavelength is the distance a wave travels in one period.

To arrive at an expression for wavelength on a line, we start with the definition of velocity:

$$v = \frac{d}{t}$$

where

v = velocity in meters per second

d = distance in meters

t = time in seconds

Since we are trying to find wavelength, we substitute the period T of the signal for t, the wavelength λ for d, and v_p, the propagation velocity, for v. This gives us

$$v_p = \frac{\lambda}{T}$$

More commonly, frequency is used rather than period. Since

$$f = \frac{1}{T}$$

the above equation becomes

$$v_p = f\lambda \tag{6.1}$$

Equation (6.1) applies to any kind of wave, not just radio waves. The propagation velocity varies greatly for different types of waves. For electromagnetic waves in free space, v_p is usually represented as c, where c is approximately equal to 300×10^6 meters per second.

EXAMPLE 6.1 ▼

Find the wavelength of a radio wave in free space, with a frequency of 100 MHz.

SOLUTION

From Equation (1.1), and noting that here $v = c$:

$$c = f\lambda \qquad \lambda = \frac{c}{f}$$

$$= \frac{300 \times 10^6 \text{ m/s}}{100 \times 10^6 \text{ Hz}} = 3\,\text{m}$$

6.3 Propagation on Transmission Lines

When dc or low-frequency signals are considered, conducting paths such as wire cables are usually ignored. At most, the resistance of the conductors may be taken into consideration. When the path length is an appreciable fraction of the signal wavelength, however, the techniques described in this section must be used to analyze the path as a transmission line. Thus, transmission line techniques must be used for long lines, high frequencies, or a combination of the two. In general, if a line is longer than about one-sixteenth of the signal wavelength, it should be considered a transmission line.

Almost any configuration of two or more conductors can operate as a transmission line, but there are several types that have the virtues of being relatively easy to study as well as very commonly used in practice.

Figure 6.2 shows some coaxial cables, in which the two conductors are concentric, separated by an insulating dielectric. Figure 6.2(a) shows a solid-dielectric cable; the one in Figure 6.2(b) uses air (and occasional plastic

FIGURE 6.2
Coaxial cables

(a) Solid dielectric

Air dielectric.

(b) Air dielectric

spacers) for the dielectric. When high power is used it is important to keep the interior of the line dry, so these lines are sometimes pressurized with nitrogen to keep out moisture. Coaxial cables are referred to as unbalanced lines because of their lack of symmetry with respect to ground (usually the outer conductor is grounded).

Parallel-line transmission lines are also common. They are not used for signal transmission as frequently as coaxial cable because they tend to radiate energy into space and also to absorb interfering signals. However, parallel lines are often used as part of antenna arrays. Ordinary television twin-lead, as shown in Figure 6.3(a), is an example of a parallel transmission line. Parallel lines can be shielded to reduce interference problems, as illustrated in Figure 6.3(b).

Parallel lines are usually operated as balanced lines; that is, the impedance to ground from each of the two wires is equal. This ensures that the currents in the two wires will be equal in magnitude and opposite in sign, reducing radiation from the cable and its susceptibility to outside interference. Coaxial cable, on the other hand, is inherently unbalanced and should be used with the outer shield grounded. When properly used, there is no connection between the signal currents in the cable, which travel on the outside of the inner conductor and the inside of the outer conductor, and any interference currents, which are only on the outside of the outer conductor.

Twisted pairs of wires are often used as transmission lines for relatively low frequencies, because of their low cost and because of the large amount of twisted-pair line that is already installed as part of the telephone system. Transmission-line techniques do not need to be used at audio frequencies

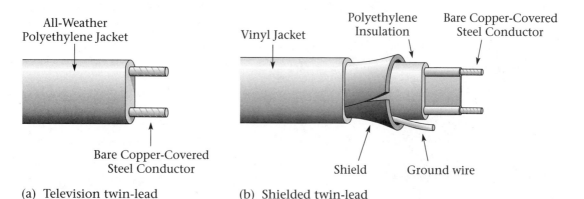

(a) Television twin-lead (b) Shielded twin-lead

FIGURE 6.3 Parallel-line cables

unless the distance is great. When higher frequencies are used, as when tele-phone signals are multiplexed using TDM or twisted-pair line is used for local-area computer networks, these techniques are required. Twisted-pair line is illustrated in Figure 6.3(c).

FIGURE 6.3
(*continued*)

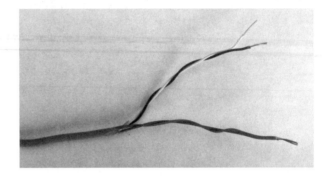

(c) Twisted-pair cable

In the microwave region, wavelengths are short (millimeters to centi-meters) and even relatively short printed-circuit board traces exhibit trans-mission-line effects. Often these traces are carefully designed to have known characteristics. There are two main types of such lines. **Microstrip** lines consist of a circuit-board trace on one side of a dielectric substrate (the circuit board) and a ground plane on the other side. **Striplines** have a ground plane on each side of the board with a conducting trace in the center. Both of these lines are illustrated in Figure 6.4. Of the two, microstrips are easier to build and to connect to components, but striplines radiate less energy.

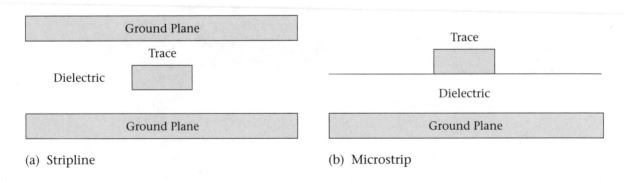

(a) Stripline   (b) Microstrip

FIGURE 6.4   Stripline and microstrip lines

Electrical Model of a Transmission Line

In the analysis of transmission lines it is necessary to consider the resistance of the conductors, the conductance of the dielectric separating them, and the inductance and capacitance of the line. However, it is not sufficient to measure the total value of each quantity for the length of line in use. Rather, the line must be divided into small (ideally, infinitesimally small) sections and the factors listed above must be considered for each section. Another way of saying this is that we must use *distributed* rather than *lumped* constants in our analysis.

Models for infinitesimal sections of both balanced and unbalanced lines are shown in Figure 6.5. The figure allows for R, the resistance of the wire; G, the conductance of the dielectric; L, the series inductance; and C, the shunt capacitance. The values for all of these must be given per unit length; for example, the resistance can be expressed in ohms per meter. However, the length of line we are considering is certainly much smaller than a meter. In fact, the idea is to allow the section to shrink until its length is infinitesimal.

At dc and low frequencies, the inductance has no effect because its reactance is very small compared with the resistance of the line. Similarly, the reactance of the shunt capacitance is very large, so the effect of the capacitance is negligible as well. The line is characterized by its resistance and possibly by the conductance of the dielectric, though this can usually be neglected.

FIGURE 6.5
Transmission-line models

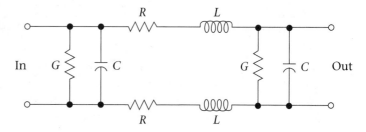

(a) Balanced line

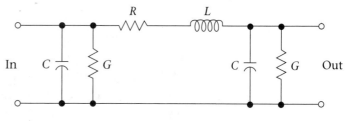

(b) Unbalanced line

As the frequency increases, the inductance and capacitance have more effect. The higher the frequency, the larger the series inductive reactance and the lower the parallel capacitive reactance. In fact, it is often possible at high frequencies to simplify calculations by neglecting the resistive elements and considering only the inductance and capacitance of the line. Such a line is called *lossless,* since the inductive and capacitive reactances store energy but do not dissipate it.

Characteristic Impedance

Consider a lossless line as described in the previous section. Suppose that a voltage step is applied to one end of an infinite length of such a line, as shown in Figure 6.6. Energy from the source will move outward along the line as the capacitors and inductors gradually become charged with energy. Since the line is infinite in length, the charging process will never be completed, and a constant current and voltage will appear at the source.

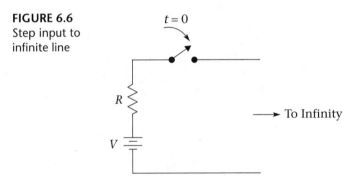

FIGURE 6.6
Step input to infinite line

The ratio between voltage and current has the dimensions of impedance, expressed in ohms. For a lossless line, this ratio is a real number representing a resistance, even though the lossless line has no actual resistance. Instead of being dissipated as heat, as in a resistor, the energy from the source continues to move down the line forever, since its length is infinite.

The ratio just described is called the **characteristic impedance** of the line. It can be shown that for any transmission line,

$$Z_0 = \sqrt{\frac{R + j\omega L}{G + j\omega C}} \tag{6.2}$$

where

Z_0 = characteristic impedance in ohms
R = conductor resistance in ohms per unit length
j = $\sqrt{-1}$
L = inductance in henrys per unit length

G = dielectric conductance in siemens per unit length

C = capacitance in farads per unit length

ω = frequency in radians per second

In general the impedance is complex and is a function of frequency as well as the physical characteristics of the line. For a lossless line, however, R and G are zero and Equation (6.2) simplifies to

$$Z_0 = \sqrt{\frac{L}{C}} \tag{6.3}$$

Of course, there is no such thing as a completely lossless line, but many practical lines approach the ideal closely enough that the characteristic impedance can be approximated by Equation (6.3). This is especially true at high frequencies: as ω gets larger, the values of R and G become less significant in comparison with ωL and ωC. For this reason, Equation (6.3) is often referred to as the *high-frequency model* of a transmission line. Equation (6.3) gives a characteristic impedance that is a real number and does not depend on frequency or the length of the line, but only on such characteristics as the geometry of the line and the permittivity of the dielectric. For coaxial cable, the characteristic impedance is given by:

$$Z_0 = \frac{138}{\sqrt{\epsilon_r}} \log \frac{D}{d} \tag{6.4}$$

where

Z_0 = characteristic impedance of the line

D = inside diameter of the outer conductor

d = diameter of the inner conductor

ϵ_r = relative permittivity of the dielectric, compared with that of free space. ϵ_r is also called the dielectric constant.

See Figure 6.7(a) for an illustration of the geometry. Here it is quite obvious that increasing the dielectric constant, or reducing the diameter of the cable, reduces the characteristic impedance.

FIGURE 6.7
Cable cross sections

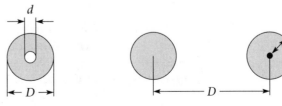

(a) Coaxial cable   (b) Air-dielectric parallel line

EXAMPLE 6.2 ▼

Find the characteristic impedance of a coaxial cable using a solid polyethylene dielectric having $\epsilon_r = 2.3$, with an inner conductor 2 mm in diameter and an outer conductor 8 mm in inside diameter.

SOLUTION

From Equation (6.4),

$$Z_0 = \frac{138}{\sqrt{\epsilon_r}} \log \frac{D}{d}$$

$$= \frac{138}{\sqrt{2.3}} \log \frac{8}{2}$$

$$= 54.8 \ \Omega$$

In practice, it is usually not necessary to use Equation (6.4) to find the impedance of a coaxial cable, since this is part of the cable specifications. In fact, the impedance is often marked right on the insulating jacket. For coaxial cable, there are a few standard impedances that fulfill most requirements. Some examples are shown in Table 6.1.

TABLE 6.1   Coaxial Cable Applications

Impedance (ohms)	Application	Typical Type Numbers
50	radio transmitters	RG-8/U
50	communication receivers	RG-58/U
75	cable television	RG-59/U
93	computer networks	RG-62/U

Parallel line normally has a dielectric consisting mostly of air. Spacers used with open-wire line or the plastic separating the conductors in twinlead will raise the effective dielectric constant slightly, but a reasonable approximation to the impedance can be derived by assuming an air dielectric:

$$Z_0 = 276 \log \frac{D}{r} \tag{6.5}$$

where

Z_0 = characteristic impedance of the line

D = separation between conductors (center to center)

r = conductor radius

See Figure 6.7(b).

EXAMPLE 6.3

An open-wire line uses wire with a diameter of 2 mm. What should the wire spacing be for an impedance of 150 Ω?

SOLUTION

First convert the diameter to a radius of 1 mm. Next rearrange Equation (6.5) to express D as the unknown.

$$Z_0 = 276 \log \frac{D}{r}$$

$$\log \frac{D}{r} = \frac{Z_0}{276}$$

$$\frac{D}{r} = \log^{-1} \frac{Z_0}{276}$$

$$D = r \log^{-1} \frac{Z_0}{276}$$

$$= 1 \text{ mm} \times \log^{-1} \frac{150}{276}$$

$$= 3.5 \text{ mm}$$

The impedance equations for stripline and microstrip transmission lines are somewhat more complicated. A number of approximations can be used, but practically all design work with these lines is done with the aid of specialized computer programs, which have the equations built in. The approximations will not be used here.

Propagation Velocity and Velocity Factor   The speed at which energy is propagated along a transmission line is always less than the speed of light. It varies from about 66% of this velocity on coaxial cable with solid polyethylene dielectric, through 78% for polyethylene foam dielectric, to about 95% for air-dielectric cable. Rather than specify the actual velocity of propagation, it is normal for manufacturers to specify the velocity factor, which is given simply by

$$v_f = \frac{v_p}{c} = \frac{1}{\sqrt{\epsilon_r}}$$ (6.6)

where

v_f = velocity factor, as a decimal fraction
v_p = propagation velocity on the line
c = speed of light in free space
ϵ_r = relative permittivity of the dielectric

The velocity factor for a transmission line depends entirely on the dielectric used. It is commonly expressed as a percentage found by multiplying the value from Equation (6.6) by 100.

EXAMPLE 6.4

Find the velocity factor and propagation velocity for a coaxial cable with a Teflon dielectric ($\epsilon_r = 2.1$).

SOLUTION

From Equation (6.6),

$$v_f = \frac{1}{\sqrt{\epsilon_r}} = \frac{1}{\sqrt{2.1}} = 0.69$$

$$v_p = v_f c = 0.69 \times 300 \times 10^6 \text{ m/s} = 207 \times 10^6 \text{ m/s}$$

Traveling Waves on a Transmission Line

Suppose that a sinusoidal wave is applied to one end of a transmission line. The line can be infinite in length, or it can be terminated with a resistance equal to its characteristic impedance. This will ensure that any signal reaching the end of the line will disappear into the resistance, so that, from the source the line will look as though it were of infinite length. Let the source resistance also be equal to Z_0; the reason for this will soon become apparent. The voltage applied to the line at a given instant moves down the line and appears farther away as time goes on. See Figure 6.8 for the setup.

If the input is sinusoidal, any point on the line will also show a sinusoidal voltage, delayed more in time as we move farther down the line. The

FIGURE 6.8
AC applied to a
matched line

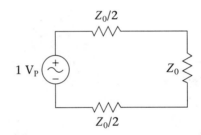

signal appears as a wave that travels down the line. In fact, such a signal is called a *traveling wave*. The signal at any point along the line is the same as that at the source except for a time delay.

With a sine wave, of course, a time delay is equivalent to a phase shift. A time delay of one period will cause a phase shift of 360 degrees, or one complete cycle; a wave that has been delayed that much will be indistinguishable from one that has not been delayed at all. As mentioned earlier, the length of line that causes a delay of one period is known as a wavelength, for which the usual symbol is λ. If we could look at the voltage along the line at one instant of time, the resulting "snapshot" would look like the input sine wave, except that the horizontal axis would be distance rather than time, and one complete cycle of the wave would occupy one wavelength instead of one period.

Figure 6.9 shows how the voltage varies along the line for four instants of time. Note how the crest of the wave moves down the line from one frame to the next.

Lengths of line that are not equal to a wavelength provide a phase delay proportional to their length. Since a length λ produces a phase shift of 360 degrees, the phase delay produced by a given line is simply

$$\phi = \frac{360L}{\lambda} \tag{6.7}$$

where

> ϕ = the phase shift in degrees
> L = the length of the line
> λ = the wavelength on the line

In many applications the time delay and phase shift due to a length of transmission line are of no concern. However, there are times (for example, when two signals must arrive at a given point in phase with each other) when it is important to consider phase shift. Transmission lines can also be used to deliberately introduce phase shifts and time delays where they are required.

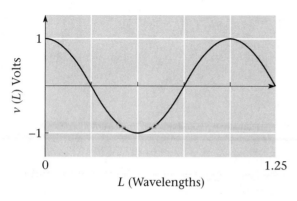

(a) $t = 0$

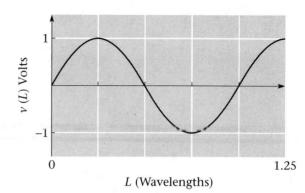

(b) $t = T/4$

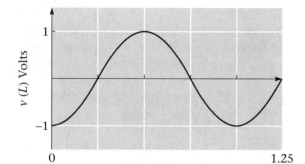

(c) $t = T/2$

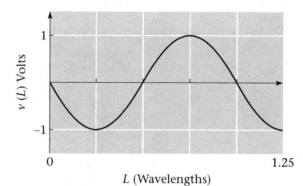

(d) $t = 3T/4$

FIGURE 6.9 Traveling waves

EXAMPLE 6.5

What length of standard RG-8/U coaxial cable would be required to obtain a 45° phase shift at 200 MHz?

SOLUTION

The velocity factor for this line is 0.66 so

$$
\begin{aligned}
v_p &= v_f c \\
&= 0.66 \times 300 \times 10^6 \text{ m/s} \\
&= 198 \times 10^6 \text{ m/s}
\end{aligned}
$$

The wavelength on the line of a 200 MHz signal is found from Equation (6.1):

$$v_p = f\lambda$$

$$\lambda = \frac{v_p}{f}$$

$$= \frac{198 \times 10^6 \text{ m/s}}{200 \times 10^6 \text{ Hz}}$$

$$= 0.99 \text{ m}$$

The required length for a phase shift of 45° would be

$$L = \frac{0.99 \text{ m} \times 45}{360}$$

$$= 124 \text{ mm}$$

6.4 Reflections and Standing Waves

It has been shown that a sine wave applied to a matched line results in an identical sine wave, except for phase, appearing at every point on the line as the incident wave travels down it. If the line is unmatched, a reflected wave from the load adds to the incident wave from the source.

First, let us consider a transmission line terminated in an open circuit. Assume that the line is reasonably long, say one wavelength. The situation is sketched in Figure 6.10. This figure shows the situation at several points in the cycle. Since the open circuit cannot dissipate any power, all the energy in the incident wave must be reflected. The net current at the open circuit must be zero, and the reflected current must have the same amplitude as, and opposite polarity from, the incident voltage at the load. Therefore, the reflected voltage has the same amplitude and polarity as the incident voltage. The reflected wave propagates down the line until it is dissipated in the source impedance, which has been chosen to match the characteristic impedance of the line. (If there is a mismatch at both ends, the situation will be complicated by multiple reflections.)

At every point on the line, the instantaneous values of incident and reflected voltage add algebraically to give the total voltage. From the figure, it can be inferred that the voltage at every point on the line varies sinusoidally, but due to constructive and destructive interference between the incident and reflected waves, the peak amplitude of the voltage varies greatly. At the

FIGURE 6.10
Waves on an
open-circuited line

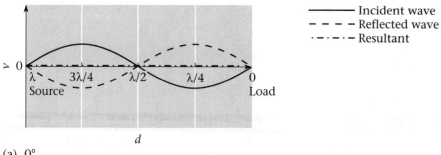

(a) 0°

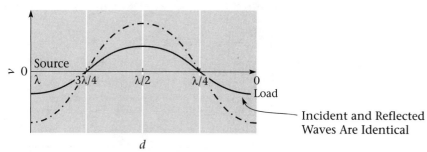

(b) 90°

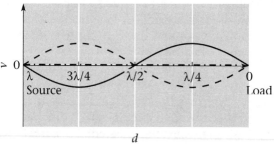

(c) 180°

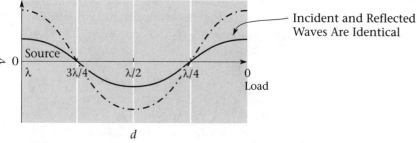

(d) 270°

open-circuited end of the line the peak voltage is at maximum. One-quarter wavelength away from that end, the incident and reflected voltages cancel exactly, because the two signals have equal amplitude and opposite phase. At a distance of one-half wavelength from the open-circuited end, there is another voltage maximum, and the process continues for every half-wavelength segment of line.

A sketch showing the variation of peak (or RMS) voltage along the line is shown in Figure 6.11. It is important to realize that this figure does not represent either instantaneous or dc voltages. There is no dc on this line at all. The figure shows how the amplitude of a sinusoidal voltage varies along the line.

FIGURE 6.11
Standing waves on an
open-circuited line

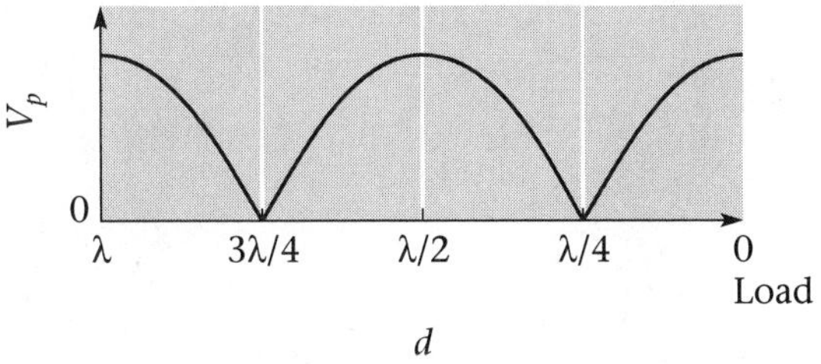

The interaction between the incident and reflected waves, which are both traveling waves, causes what appears to be a stationary pattern of waves on the line. It is customary to call these standing waves because of this appearance. Of course, waves do not really stand still on a line; it is only the interference pattern that stands still.

For comparison, Figure 6.12 shows the standing waves of voltage on a line with a shorted end. Naturally, there is no voltage at the shorted end, since there can never be a voltage at a short circuit. The reflected wave must therefore have a voltage equal in magnitude and opposite in sign to the incident voltage. A voltage maximum occurs one-quarter wavelength from the end, another null occurs at one-half wavelength, and so on.

The current responds in just the opposite way from the voltage. For the open line, the current must be zero at the open end. It is a maximum

FIGURE 6.12
Standing waves on
a short-circuited line

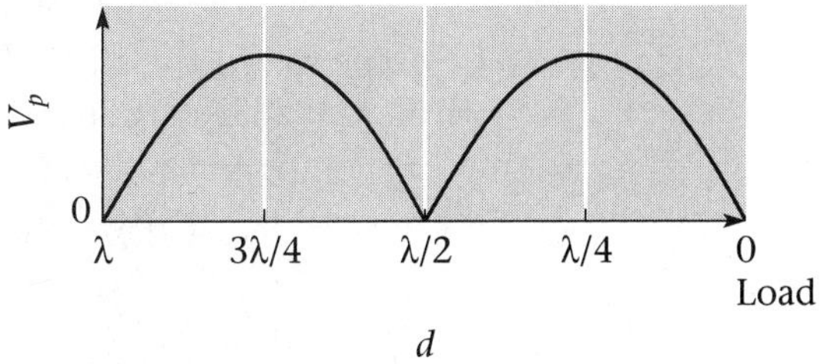

one-quarter wavelength away, zero again at a distance of one-half wavelength, and so on. In other words, the graph of current as a function of position for an open-circuited line looks just like the graph of voltage versus position for a short-circuited line. Similarly, the current on a short-circuited line has a maximum at the termination, a minimum one-quarter wavelength away, and another maximum at a distance of a half wavelength, just like the voltage curve for the open-circuited line.

What happens when the line is mismatched, but not so drastically as discussed above? There will be a reflected wave, but it will not have as large an amplitude as the incident wave since some of the incident signal will be dissipated in the load. The amplitude and phase angle will depend on the load impedance compared to that at the line. The incident and reflected voltages are related by the coefficient of reflection:

$$\Gamma = \frac{V_r}{V_i} \tag{6.8}$$

where

Γ = coefficient of reflection

V_r = reflected voltage

V_i = incident voltage

It can be shown that

$$\Gamma = \frac{Z_L - Z_0}{Z_L + Z_0} \tag{6.9}$$

where

Z_L = load impedance

Z_0 = characteristic impedance of the line

In general, Γ is complex, but for a lossless line it is a real number if the load is resistive. It is positive for $Z_L > Z_0$ and negative for $Z_L < Z_0$. For $Z_L = Z_0$, the reflection coefficient is of course zero. A positive real coefficient means that the incident and reflected voltages are in phase at the load.

When a reflected signal is present but of lower amplitude than the incident wave, there will be standing waves of voltage and current, but there will be no point on the line where the voltage or current remains zero over the whole cycle. See Figure 6.13 for an example.

It is possible to define the **voltage standing-wave ratio** (*VSWR* or just *SWR*) as follows:

$$SWR = \frac{V_{max}}{V_{min}} \tag{6.10}$$

FIGURE 6.13
Standing waves on a
mismatched line

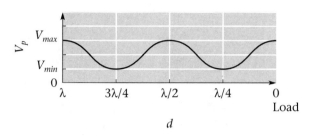

where

SWR = voltage standing-wave ratio
V_{max} = maximum rms voltage on the line
V_{min} = minimum rms voltage on the line

The SWR concerns magnitudes only and is thus a real number. It must be positive and greater than or equal to 1. For a matched line the SWR is 1 (sometimes expressed as 1:1 to emphasize that it is a ratio), and the closer the line is to being matched, the lower the SWR. The SWR has the advantage of being easier to measure than the reflection coefficient, but the latter is more useful in many calculations. Since both are essentially measures of the amount of reflection on a line, it is possible to find a relationship between them.

The maximum voltage on the line occurs where the incident and reflected signals are in phase, and the minimum voltage is found where they are out of phase. Therefore, using absolute value signs to emphasize the lack of a need for phase information, we can write

$$V_{max} = |V_i| + |V_r| \tag{6.11}$$

and

$$V_{min} = |V_i| - |V_r| \tag{6.12}$$

Combining Equations (6.10), (6.11), and (6.12), we get

$$SWR = \frac{V_{max}}{V_{min}} \tag{6.13}$$

$$= \frac{|V_i| + |V_r|}{|V_i| - |V_r|}$$

$$= \frac{|V_i|\left(1 + \dfrac{|V_r|}{|V_i|}\right)}{|V_i|\left(1 - \dfrac{|V_r|}{|V_i|}\right)}$$

$$= \frac{1 + \dfrac{|V_r|}{|V_i|}}{1 - \dfrac{|V_r|}{|V_i|}}$$

$$= \frac{1 + |\Gamma|}{1 - |\Gamma|}$$

A little algebra will show that $|\Gamma|$ can also be expressed in terms of SWR:

$$|\Gamma| = \frac{SWR - 1}{SWR + 1} \tag{6.14}$$

For the special, but important, case of a lossless line terminated in a resistive impedance, it is possible to find a simple relationship between standing-wave ratio and the load and line impedances. First, suppose that $Z_L > Z_0$. Then from Equation (6.13),

$$SWR = \frac{1 + |\Gamma|}{1 - |\Gamma|} \tag{6.15}$$

$$= \frac{1 + \dfrac{Z_L - Z_0}{Z_L + Z_0}}{1 - \dfrac{Z_L - Z_0}{Z_L + Z_0}}$$

$$= \frac{Z_L + Z_0 + Z_L - Z_0}{Z_L + Z_0 - Z_L + Z_0}$$

$$= \frac{Z_L}{Z_0}$$

It is easy to show, in a similar way, that if $Z_0 > Z_L$, then

$$SWR = \frac{Z_0}{Z_L} \tag{6.16}$$

Use of the appropriate equation will always give an SWR that is greater than or equal to one, and positive.

EXAMPLE 6.6 ▼

A 50-Ω line is terminated in a 25-Ω resistance. Find the SWR.

SOLUTION

In this case, $Z_0 > Z_L$ so the solution is given by Equation (6.16).

$$SWR = \frac{Z_0}{Z_L}$$

$$= \frac{50}{25}$$

$$= 2$$

The presence of standing waves causes the voltage at some points on the line to be higher than it would be with a matched line, while at other points the voltage is low but the current is higher than with a matched line. This situation results in increased losses. In a transmitting application, standing waves put additional stress on the line and can result in failure of the line or of equipment connected to it. For instance, if the transmitter happens to be connected at or near a voltage maximum, the output circuit of the transmitter may be subjected to a dangerous overvoltage condition. This is especially likely to damage solid-state transmitters, which for this reason are often equipped with circuits to reduce the output power in the presence of an SWR greater than about 2:1.

Reflections can cause the power delivered to the load to be less than it would be with a matched line for the same source, because some of the power is reflected back to the source. Since power is proportional to the square of voltage, the fraction of the power that is reflected is Γ^2, that is,

$$P_r = \Gamma^2 P_i \tag{6.17}$$

where

P_r = power reflected from the load
P_i = incident power at the load
Γ = voltage reflection coefficient

Sometimes Γ^2 is referred to as the *power reflection coefficient.*

The amount of power absorbed by the load is the difference between the incident power and the reflected power, that is,

$$P_L = P_i - \Gamma^2 P_i \tag{6.18}$$

$$= P_i (1 - \Gamma^2)$$

EXAMPLE 6.7

A generator sends 50 mW down a 50-Ω line. The generator is matched to the line but the load is not. If the coefficient of reflection is 0.5, how much power is reflected and how much is dissipated in the load?

SOLUTION

The amount of power that is reflected is, from Equation (6.17):

$$P_r = \Gamma^2 P_i$$
$$= 0.5^2 \times 50 \text{ mW}$$
$$= 12.5 \text{ mW}$$

The remainder of the power reaches the load. This amount is

$$P_L = P_i - P_r$$
$$= 50 \text{ mW} - 12.5 \text{ mW}$$
$$= 37.5 \text{ mW}$$

Alternatively, the load power can be calculated directly from Equation (6.18):

$$P_L = P_i (1 - \Gamma^2)$$
$$= 50 \text{ mW}(1 - 0.5^2)$$
$$= 37.5 \text{ mW}$$

Since SWR is easier to measure than the reflection coefficient, an expression for the power absorbed by the load in terms of the SWR would be useful. It is easy to derive such an expression by using the relationship between Γ and SWR given in Equation (6.13). The derivation is left as an exercise; the result is

$$P_L = \frac{4SWR}{(1 + SWR)^2} P_i \tag{6.19}$$

EXAMPLE 6.8

A transmitter supplies 50 W to a load through a line with an SWR of 2. Find the power absorbed by the load.

SOLUTION

From Equation (6.19),

$$P_L = \frac{4SWR}{(1 + SWR)^2}\, P_i$$

$$= \frac{4 \times 2}{(1 + 2)^2} \times 50 \text{ W}$$

$$= 44.4 \text{ W}$$

Reflections on transmission lines can cause problems in receiving applications as well. For instance, reflections on a television antenna feedline can cause a double image or "ghost" to appear. In data transmission, reflections can distort pulses, causing errors.

Variation of Impedance Along a Line

A matched line presents its characteristic impedance to a source located any distance from the load. If the line is not matched, however, the impedance seen by the source can vary greatly with its distance from the load. At those points where the voltage is high and the current low, the impedance is higher than at points with the opposite current and voltage characteristics. In addition, the phase angle of the impedance can vary. At some points, a mismatched line may look inductive, at others capacitive, and at a few points, resistive. Very near the load, the impedance looking into the line is close to Z_L. This is one reason why transmission-line techniques need to be used only with relatively high frequencies and/or long lines: a line shorter than about one-sixteenth of a wavelength can usually be ignored.

The impedance that a lossless transmission line presents to a source varies in a periodic way. We have already noticed that the standing-wave pattern repeats itself every one-half wavelength along the line; the impedance varies in the same fashion. At the load and at distances from the load that are multiples of one-half wavelength, the impedance looking into the line is that of the load.

The impedance at any point on a lossless transmission line is given by the equation

$$Z = Z_0 \frac{Z_L \cos \theta + jZ_0 \sin \theta}{Z_0 \cos \theta + jZ_L \sin \theta} \tag{6.20}$$

where

Z = impedance looking from the source toward the load
Z_L = load impedance
Z_0 = characteristic impedance of the line
θ = distance to the load in degrees (for example a quarter wavelength would be 90°)

Provided that $\cos \theta \neq 0$, this simplifies to

$$Z = Z_0 \frac{Z_L + jZ_0 \tan \theta}{Z_0 + jZ_L \tan \theta} \tag{6.21}$$

These equations can be fairly tedious to work with, especially when Z_L is complex. Most transmission-line impedance calculations are done using computers. Many of the computer programs give their results in the form of a Smith chart, which will be described in Appendix A.

Characteristics of Shorted and Open Lines

Though a section of transmission line that is terminated in an open or short circuit is useless for transmitting power, it can serve other purposes. Such a line can be used as an inductive or capacitive reactance or even as a resonant circuit. In practice, short-circuited sections are much more useful, because open-circuited lines tend to radiate energy from the open end.

The impedance of a short-circuited line can be found from Equation (6.21), by setting Z_L equal to zero. The impedance looking toward the short circuit is

$$Z = jZ_0 \tan \theta \tag{6.22}$$

Note that the impedance has no resistive component, since no power can be dissipated in this line. For short lengths (less than one-quarter wavelength or 90°), the impedance is inductive. At one-quarter wavelength, the line looks like a parallel-resonant circuit, with infinite impedance. The line is capacitive for lengths between one-quarter and one-half wavelength, since the tangent of the corresponding angle is negative. At a length of one-half wavelength, or 180°, $\tan \theta$ is zero, and the line behaves like a series-resonant circuit having zero impedance. For longer lines the cycle repeats.

Figure 6.14 shows graphically how the impedance varies with length. It is easy to remember that a short length of shorted line is inductive by visualizing the shorted end as a loop or coil hence an inductance.

The open-circuited line, on the other hand, is capacitive in short lengths. You can remember this by thinking of the two parallel lines at the

FIGURE 6.14
Impedance of a
shorted line

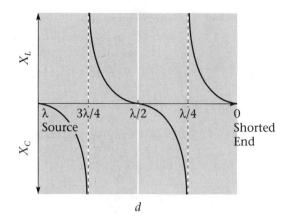

end of the transmission line as a capacitor. It is series-resonant at a length of one-quarter wavelength, inductive between one-quarter and one-half wavelength, and parallel-resonant at a length of one-half wavelength. For longer lines the cycle repeats. Figure 6.15 shows graphically how the impedance varies with length for this line.

FIGURE 6.15
Impedance of an
open-circuited line

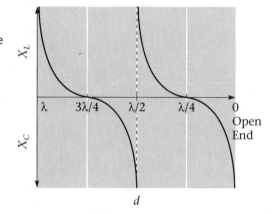

Short transmission-line sections, called stubs, can be substituted for capacitors, inductors or tuned circuits in applications where lumped-constant components (conventional inductors or capacitors) would be inconvenient or impractical. At VHF and UHF frequencies the required values of inductance and capacitance are often very small. It would be difficult to build a 1 pF capacitor, for instance; the capacitance between its leads might well be more than that. In addition, physically small components are difficult to use where large amounts of power are involved, as in transmitters, because of the large voltages and currents that must be handled. A short section of transmission line can avoid all these problems. Also, an air-dielectric transmission line can have much higher Q than a typical lumped-constant resonant circuit.

EXAMPLE 6.9 ▼

A series-tuned circuit operating at a frequency of 1 GHz is to be constructed from a shorted section of air-dielectric coaxial cable. What length should be used?

SOLUTION

The velocity factor of an air-dielectric line is about 0.95 so the propagation velocity is, from Equation (6.6),

$$
\begin{aligned}
v_p &= v_f c \\
&= 0.95 \times 300 \times 10^6 \text{ m/s} \\
&= 285 \times 10^6 \text{ m/s}
\end{aligned}
$$

The wavelength on the line is given by Equation (6.1):

$$ v_p = f\lambda $$

$$ \lambda = \frac{v_p}{f} $$

$$ = \frac{285 \times 10^6 \text{ m/s}}{1000 \times 10^6 \text{ Hz}} $$

$$ = 285 \text{ mm} $$

Since this is a shorted stub, a half wavelength section will be series resonant. Therefore the length will be

$$ L = \frac{\lambda}{2} $$

$$ = \frac{285 \text{ mm}}{2} $$

$$ = 143 \text{ mm} $$

▲

Shorted stubs can be used anywhere a capacitance, inductance, or tuned circuit is required. Of course, this technique is more practical at VHF frequencies and higher where the length of the stub is reasonable and it is often hard to find discrete components with sufficiently small values.

6.5 Transmission Line Losses

Most of the losses in a properly constructed and used transmission line are due to conductor resistance and, at higher frequencies, dielectric conductance. Both of these losses increase with frequency. Some transmission lines also radiate energy. This is particularly true of lines that are improperly connected; for instance, a coaxial line that is used in a balanced circuit will radiate energy from its shield. When parallel lines are used at very high frequencies, such that the distance between the wires is a substantial portion of the wavelength, they too can radiate.

Transmission line loss is usually specified in decibels for a given length, for instance, 100 m or sometimes 100 feet. The decibel loss is proportional to distance, so it is easy to find the loss for any other length of line.

EXAMPLE 6.10 ▾

A transmission line is specified to have a loss of 4 dB/100 m at 800 MHz.

(a) Find the loss in decibels of 130 m of this cable.

(b) Suppose that a transmitter puts 100 W of power into this cable. How much power reaches the load?

SOLUTION

(a) $L(\text{dB}) = \dfrac{4\text{dB}}{100\text{ m}} \times 130\text{ m}$

$\quad\quad = 5.2\text{ dB}$

(b) For this part we need the loss as a power ratio.

$$L = \text{antilog}\left(\frac{L(\text{dB})}{10}\right)$$

$$= \text{antilog}\left(\frac{5.2}{10}\right)$$

$$= 3.31$$

$$L = \frac{P_i}{P_o}$$

$$P_o = \frac{P_i}{L}$$

$$= \frac{100 \text{ W}}{3.31}$$

$$= 30.2 \text{ W}$$

6.6 Waveguides

Transmission line losses increase rapidly with frequency. At frequencies of several gigahertz or more, the losses in conventional transmission lines are such as to make long cable runs impractical. Luckily, there is a different form of transmission line that is impractical at low frequencies but very useful in the microwave region. This type of line, called a **waveguide**, generally consists of a hollow, air-filled tube made of conducting material. Rectangular waveguides of brass or aluminum, sometimes silver-plated on the inside, are most common, but elliptical and circular cross-sections are also used. Elliptical waveguides can be semi-flexible and are used extensively for connecting equipment at the base of a tower to antennas on the tower. Circular guides are useful when rotating antennas are used, as in radar antennas, because of their symmetry. Examples are shown in Figure 6.16.

Electromagnetic waves are launched into one end and propagate down the guide. The waves fill the entire space within the guide, and can be visualized as rays reflecting from the walls of the guide. Since the waves spend most of their time in the air within the guide, losses due to currents in the walls are relatively small. The waves are completely contained within the conducting walls, so there is no loss due to radiation from the guide. Waveguides, then, solve the problem of high transmission-line losses for microwave signals.

It is possible to build a waveguide for any frequency, but waveguides operate essentially as high-pass filters, that is, for a given waveguide cross-section there is a cutoff frequency, below which waves will not propagate. At frequencies below the gigahertz range, waveguides are too large to be practical for most applications.

FIGURE 6.16
Waveguides

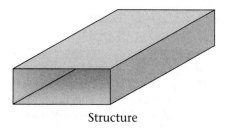

Structure

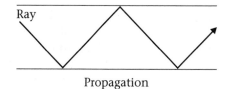

Propagation

(a) Rectangular waveguide

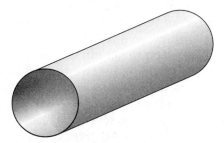

(b) Circular waveguide

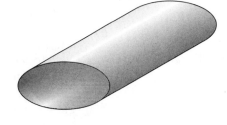

(c) Elliptical waveguide

Modes and Cutoff Frequency There are a number of ways (called *modes*) in which electrical energy can propagate along a waveguide. All of these modes must satisfy certain boundary conditions. For instance, assuming an ideal conductor for the guide, there cannot be any electric field along the wall of the waveguide. If there were such a field, there would have to be a voltage gradient along the wall, and that is impossible since there cannot be any voltage across a short circuit.

Modes are most easily understood by thinking of a wave moving through the guide as if it were a ray of light. Figure 6.17 shows the idea. For each different mode, the ray will strike the walls of the waveguide at a different angle. As the angle a ray makes with the wall of the guide becomes larger, the distance the ray must travel to reach the far end of the guide becomes greater. Though the propagation in the guide is at the speed of light,

FIGURE 6.17
Multimode propagation

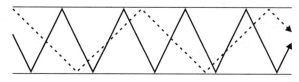

- - - - - - - Low-order mode: faster propagation
———— High-order mode: slower propagation

the greater distance traveled causes the effective velocity down the guide to be reduced.

It is desirable to have only one mode propagating in a waveguide. To see the effect of *multimode* propagation (more than one mode propagating at a time), consider a brief pulse of microwave energy applied to one end of a waveguide. The pulse will arrive at the far end at several different times, one for each mode. Thus a brief pulse will be spread out over time, an effect called **dispersion**. If another pulse follows close behind, there may be interference between the two. For this reason it is undesirable to have more than one mode propagating.

Each mode has a cutoff frequency below which it will not propagate. Single-mode propagation can be achieved by using only the mode with the lowest cutoff frequency, which is called the *dominant mode*. The waveguide is used at frequencies between the cutoff frequency for the dominant mode and that of the mode with the next lowest cutoff frequency.

Modes are designated as *transverse electric (TE)* or *transverse magnetic (TM)* modes according to the pattern of electric and magnetic fields within the waveguide. Figure 6.18 shows several examples of TE modes in a rectangular waveguide. The electric field strength is represented by the arrows, with the length of the arrows proportional to the field strength. Note that in all cases the field strength is zero along the walls of the guide. The field strength varies sinusoidally across the guide cross section. The first number

FIGURE 6.18
TE modes in rectangular waveguide

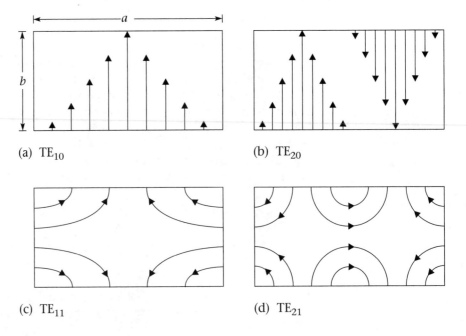

(a) TE_{10}

(b) TE_{20}

(c) TE_{11}

(d) TE_{21}

following the TE designation represents the number of half-cycles of the wave along the long dimension (*a*) of the rectangular guide, and the second represents the number of variations along the short dimension (*b*). Of course, *a* and *b* refer to inside dimensions.

In a rectangular waveguide, TE_{10} is the dominant mode. In a typical rectangular waveguide with $a = 2b$, the TE_{10} and TE_{20} modes each have a cutoff frequency twice that of the TE_{10} mode, giving an approximate 2:1 frequency range for the waveguide in its dominant mode. In what follows, a rectangular waveguide operating in the TE_{10} mode will be assumed unless otherwise stated.

The cutoff frequency for the TE_{10} mode can easily be found. For this mode to propagate, there must be at least one-half wavelength along the wall. Therefore, at the cutoff frequency,

$$a = \frac{\lambda_c}{2} \tag{6.23}$$

where

$$a = \text{the longer dimension of the waveguide cross section}$$
$$\lambda_c = \text{cutoff wavelength}$$

Assuming that the waveguide has an air dielectric, the cutoff frequency can be found as follows. From Equation (6.23),

$$\lambda_c = 2a$$

From earlier work we know that for propagation in free space,

$$\lambda = \frac{c}{f}$$

where

$$\lambda = \text{wavelength in meters}$$
$$c = 300 \times 10^6 \text{ meters per second}$$
$$f = \text{frequency in hertz}$$

Therefore, at the cutoff frequency f_c,

$$\lambda_c = \frac{c}{f_c} \tag{6.24}$$

$$2a = \frac{c}{f_c}$$

$$f_c = \frac{c}{2a}$$

EXAMPLE 6.11

Find the cutoff frequency for the TE_{10} mode in an air-dielectric waveguide with an inside cross section of 2 cm by 4 cm. Over what frequency range is the dominant mode the only one that will propagate?

SOLUTION

The larger dimension, 4 cm, is the one to use in calculating the cutoff frequency. From Equation (6.24), the cutoff frequency is

$$f_c = \frac{c}{2a}$$

$$= \frac{300 \times 10^6 \text{ m/s}}{2 \times 4 \times 10^{-2} \text{ m}}$$

$$= 3.75 \text{ GHz}$$

The dominant mode is the only mode of propagation over a two-to-one frequency range, so the waveguide will be usable to a maximum frequency of $3.75 \times 2 = 7.5$ GHz.

Group and Phase Velocity

Assuming an air dielectric, the wave travels inside the waveguide at the speed of light. However, it does not travel straight down the guide but reflects back and forth from the walls. The actual speed at which a signal travels along the guide is called the **group velocity**, and it is considerably less than the speed of light. The group velocity in a rectangular waveguide is given by the equation

$$v_g = c\sqrt{1 - \left(\frac{f_c}{f}\right)^2} \tag{6.25}$$

where

f_c = cutoff frequency

f = operating frequency

From the above equation, it can be seen that the group velocity is a function of frequency, becoming zero at the cutoff frequency. At frequencies below cutoff, of course, there is no propagation, so the equation does not apply. The physical explanation of the variation of group velocity is that the angle the wave makes with the wall of the guide varies with frequency. At frequencies near the cutoff value, the wave moves back and forth across the guide more often while traveling a given distance down the guide than it does at higher frequencies. Figure 6.19 gives a qualitative idea of the effect.

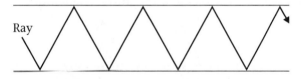

(a) Frequency just above cutoff

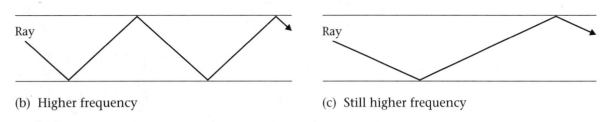

(b) Higher frequency   (c) Still higher frequency

FIGURE 6.19   Variation of group velocity with frequency

EXAMPLE 6.12

Find the group velocity for the waveguide of Example 6.10, at a frequency of 5 GHz.

SOLUTION

From Example 6.10, the cutoff frequency is 3.75 GHz. Therefore, from Equation (6.25), the group velocity is

$$v_g = c\sqrt{1 - \left(\frac{f_c}{f}\right)^2}$$

$$= 300 \times 10^6 \text{ m/s}\sqrt{1 - \left(\frac{3.75}{5}\right)^2}$$

$$= 198 \times 10^6 \text{ m/s}$$

Since the group velocity varies with frequency, dispersion exists even for single-mode propagation. If a signal has components of different frequencies, the higher-frequency components will travel faster than the lower-frequency components. This can be a real problem for pulsed and other wideband signals. For instance, the upper sideband of an FM or AM signal travels faster than the lower sideband.

EXAMPLE 6.13

A waveguide has a cutoff frequency for the dominant mode of 10 GHz. Two signals with frequencies of 12 and 17 GHz respectively propagate down a 50 m length of the guide. Calculate the group velocity for each and the difference in arrival time for the two.

SOLUTION

The group velocities can be calculated from Equation (6.25):

$$v_g = c\sqrt{1 - \left(\frac{f_c}{f}\right)^2}$$

For the 12 GHz signal,

$$v_g = 300 \times 10^6 \text{ m/s}\sqrt{1 - \left(\frac{10}{12}\right)^2}$$

$$= 165.8 \times 10^6 \text{ m/s}$$

Similarly, the 17-GHz signal has $v_g = 242.6 \times 10^6$ m/s. The 12-GHz signal will travel the 50 m in:

$$t_1 = \frac{50 \text{ m}}{165.8 \times 10^6 \text{ m/s}} = 301.6 \text{ ns}$$

Similarly, it can be shown that the travel time for the 17-GHz signal is:

$$t_2 = 206.1 \text{ ns}$$

The difference in travel times for the two signals is:

$$\Delta_t = 301.6 \text{ ns} - 206.1 \text{ ns} = 95.5 \text{ ns}$$

It is often necessary to calculate the wavelength of a signal in a waveguide. For instance, it may be required for impedance matching. It might seem that the wavelength along the guide could be found using the group velocity, in much the same way that the velocity factor of a transmission line is used. However, this common-sense approach does not work because what is really important in impedance-matching calculations is the change in phase angle along the line.

Figure 6.20 shows how the angle varies along the guide. The guide wavelength shown represents 360 degrees of phase variation. The guide wavelength is always larger than the free-space wavelength. Interestingly, the more slowly the waves in the guide propagate along it, the more quickly the phase angle varies along the guide.

FIGURE 6.20
Variation of phase angle along a waveguide

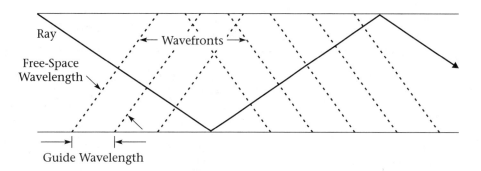

The term **phase velocity** is used to describe the rate at which the wave *appears* to move along the wall of the guide, based on the way the phase angle varies along the walls. Surprisingly, the phase velocity in a waveguide is always greater than the speed of light. Of course, the laws of physics prevent anything from actually moving that fast (except in science fiction stories). A phase velocity greater than the speed of light is possible because phase velocity is not really the velocity of anything.

A similar effect can be seen with water waves at a beach. If the waves approach the shore at an angle, as in Figure 6.21, the crest of the wave will appear to run along the shore at a faster rate than that at which the waves approach the beach. Once again, there is nothing physical moving at the higher velocity.

FIGURE 6.21
Water waves at a shoreline

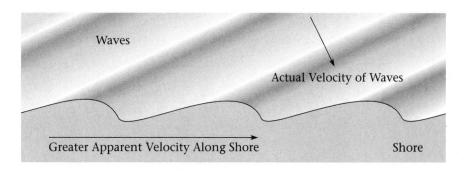

The relationship between phase velocity and group velocity is very simple. The speed of light is the geometric mean of the two, that is,

$$v_g v_p = c^2 \tag{6.26}$$

It is easy to modify the equations given earlier for group velocity, so that the phase velocity can be calculated without first finding the group velocity. The derivation is left as an exercise; the result is given below.

$$v_p = \frac{c}{\sqrt{1 - \left(\dfrac{f_c}{f}\right)^2}} \tag{6.27}$$

EXAMPLE 6.14 ▼

Find the phase velocity for the waveguide used in Example 6.10, at a frequency of 5 GHz.

SOLUTION

The answer can be found directly from Equation (6.27):

$$
\begin{aligned}
v_p &= \frac{c}{\sqrt{1 - \left(\dfrac{f_c}{f}\right)^2}} \\[2mm]
&= \frac{300 \times 10^6 \text{ m/s}}{\sqrt{1 - \left(\dfrac{3.75}{5}\right)^2}} \\[2mm]
&= 454 \times 10^6 \text{ m/s}
\end{aligned}
$$

Alternatively, it can be found from Equation (6.26), since the group velocity is already known:

$$
\begin{aligned}
v_g v_p &= c^2 \\[2mm]
v_p &= \frac{c^2}{v_g} \\[2mm]
&= \frac{(300 \times 10^6 \text{ m/s})^2}{198 \times 10^6 \text{ m/s}} \\[2mm]
&= 454 \times 10^6 \text{ m/s}
\end{aligned}
$$

Waveguide Impedance Like any transmission line, a waveguide has a characteristic impedance. Unlike wire lines, however, the waveguide impedance is a function of frequency. You might expect that the impedance of a waveguide with an air dielectric would have some relationship to the impedance of free space, which is 377 Ω, and that is true. The characteristic impedance of a waveguide, in ohms, is given by

$$Z_o = \frac{377}{\sqrt{1 - \left(\dfrac{f_c}{f}\right)^2}} \tag{6.28}$$

EXAMPLE 6.15

Find the characteristic impedance of the waveguide used in the previous examples, at a frequency of 5 GHz.

SOLUTION

From Equation (6.28),

$$Z_0 = \frac{377}{\sqrt{1 - \left(\dfrac{f_c}{f}\right)^2}}$$

$$= \frac{377}{\sqrt{1 - \left(\dfrac{3.75}{5}\right)^2}}$$

$$= 570 \ \Omega$$

Impedance matching with waveguides can be accomplished as for other types of transmission lines. The only difference in principle is that the phase velocity v_p must be used in calculating λ_g, the wavelength in the guide. That is, the wavelength is given by

$$\lambda_g = \frac{v_p}{f} \tag{6.29}$$

Another way to find the guide wavelength, given the free-space wavelength, is to use the equation

$$\lambda_g = \frac{\lambda}{\sqrt{1 - \left(\dfrac{f_c}{f}\right)^2}} \tag{6.30}$$

where

$$\lambda = \text{the free-space wavelength}$$
$$\lambda_g = \text{the wavelength in the guide}$$

EXAMPLE 6.16

Find the guide wavelength for the waveguide used in the previous examples.

SOLUTION

Since the phase velocity is already known, the easiest way to solve this problem is to use Equation (6.29).

$$\lambda_g = \frac{v_p}{f}$$

$$= \frac{454 \times 10^6 \text{ m/s}}{5 \times 10^9 \text{ Hz}}$$

$$= 9.08 \text{ cm}$$

Techniques for matching impedances using waveguides also differ from those used with conventional transmission lines. Shorted stubs of adjustable length can be used, but a simpler method is to add capacitance or inductance by inserting a tuning screw into the guide, as shown in Figure 6.22. As the

FIGURE 6.22
Tuning screw

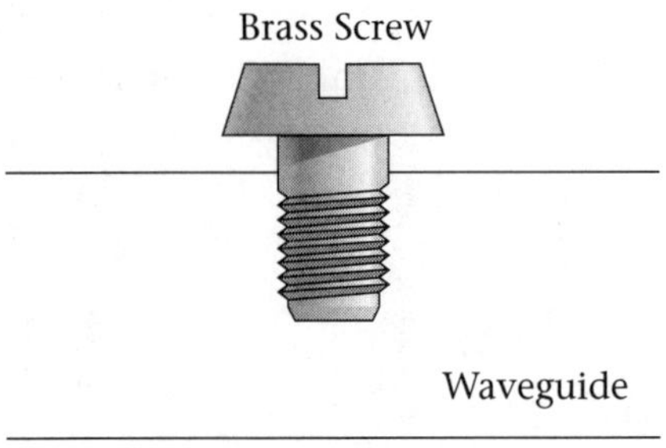

Brass Screw

Waveguide

screw is inserted farther into the guide, the effect is first capacitive, then series-resonant, and finally inductive.

Coupling Power into and out of Waveguides

There are three basic ways to launch a wave down a guide. Figure 6.23 shows all three. Figure 6.23(a) shows the use of a probe. The probe couples to the electric field in the guide and should be located at an electric-field maximum. For the TE_{10} mode, it should be in the center of the a (wide) dimension. The probe launches a wave along the guide in both directions. Assuming that propagation in only one direction is desired, it is only necessary to place the probe a quarter-wavelength from the closed end of the guide. This closed end represents a short circuit, and following the same logic as for transmission lines, there will be an electric-field maximum at a distance of one-quarter wavelength from the guide. This is the waveguide equivalent of the voltage maximum one-quarter wavelength from the shorted end of a transmission line.

The wave emitted by the probe reflects from the shorted end of the guide. Just as with conventional lines, there is a 180° phase shift at the reflecting surface and this, combined with the 180° shift due to the total path length of one-half wavelength, results in the reflected wave being in phase with (and adding to) the direct wave in the direction along the length of the guide. The distance to the end of the guide will have to be calculated using the guide wavelength described earlier.

Figure 6.23(b) shows another way to couple power to a guide. A loop is used to couple to the magnetic field in the guide. It is placed in a location of maximum magnetic field, which for the TE_{10} mode occurs close to the end wall of the guide. It may help you understand this if you think of the magnetic field in a waveguide as equivalent to current in a conventional transmission line and consider the electric field as the equivalent of voltage.

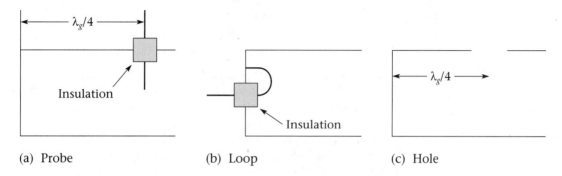

(a) Probe (b) Loop (c) Hole

FIGURE 6.23 Coupling power to a waveguide

There is, of course, a current maximum at the short-circuited end of a transmission line.

A third way of coupling energy, shown in Figure 6.23(c), is simply to put a hole in the waveguide, so that electromagnetic energy can propagate into or out of the guide from the region exterior to it.

Waveguides are reciprocal devices, just like transmission lines. Therefore, the same means can be used to couple power into and out of a waveguide.

Variations on the above methods are also possible. For instance, it is possible to use two holes spaced one-quarter wavelength apart to launch a wave in only one direction. Figure 6.24 shows how this works. A signal moving down the main guide in the direction of the arrow will be coupled to the secondary guide where it will again move in the direction indicated by the arrow. It will not propagate in the other direction because the signals from the two holes are 180 degrees out of phase with each other in that direction. The signal in the main guide reaches hole B 90 degrees after it gets to hole A. The wave propagating to the left from hole B is delayed another 90 degrees before joining the wave propagated from hole A. A wave propagating down the main guide in the opposite direction will also be coupled to the secondary guide, but it will propagate in the opposite direction and will be absorbed by the resistive material in the end of the secondary guide. Though only two holes are shown, in practice there is often a greater number to provide more coupling between the guides. Devices of this type are called **directional couplers**.

FIGURE 6.24
Two-hole directional coupler

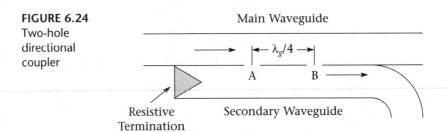

Directional couplers are characterized by their insertion loss, coupling, and directivity. All three are normally specified in decibels. The *insertion loss* is the amount by which a signal in the main guide will be attenuated. The *coupling* specification gives the amount by which the signal in the main guide is greater than that coupled to the secondary waveguide. The *directivity* refers to the ratio between the power coupled to the secondary guide, for signals travelling in the two possible directions along the main guide.

EXAMPLE 6.17

A signal with a level of 20 dBm enters the main waveguide of a directional coupler in the direction of the arrow. The coupler has an insertion loss of 1 dB, coupling of 20 dB, and directivity of 40 dB. Find the strength of the signal emerging from each guide. Also find the strength of the signal that would emerge from the secondary guide, if the signal in the main guide were propagating in the other direction.

SOLUTION

The signal level in the main guide is

$$20 \text{ dBm} - 1 \text{ dB} = 19 \text{ dBm}$$

The signal level in the secondary guide is

$$20 \text{ dBm} - 20 \text{ dB} = 0 \text{ dBm}$$

If the signal direction in the main guide were reversed, the signal level in the secondary guide would be reduced by 40 dB to

$$0 \text{ dBm} - 40 \text{ dB} = -40 \text{ dBm}$$

Waveguide Components

The use of waveguides requires redesign of some of the ordinary components that are used with feedlines. The lowly tee connector is an example. In addition, other components, such as resonant cavities, are too large to be practical at lower frequencies. Several other varieties of microwave passive components will also be described.

Bends and Tees

Anything that changes the shape or size of a waveguide has an effect on the electric and magnetic fields inside. If the disturbance is sufficiently large, there will be a change in the characteristic impedance of the guide. However, as long as any bend or twist is gradual, the effect will be minimal. Figure 6.25 shows examples of bends and twists. The bends are designated as *E-plane bends,* Figure 6.25(a), or *H-plane bends* shown in Figure 6.25(b). Since the rectangular guide shown normally operates in the TE_{10} mode, the electric field lines are perpendicular to the long direction. Therefore the E-plane bend changes the direction of the electric field lines. Similar logic holds for the H-plane bend.

FIGURE 6.25
Waveguide bends

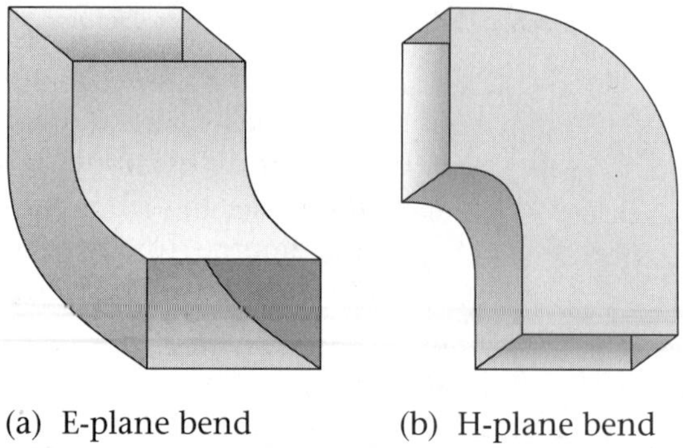

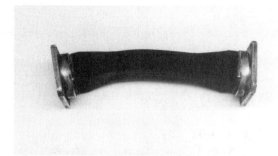

(a) E-plane bend   (b) H-plane bend

Rigid waveguide, with its carefully designed gradual bends, resembles plumbing and is just as tricky to install. Flexible waveguide, as shown in Figure 6.26, is used for awkward installations.

FIGURE 6.26
Flexible waveguide

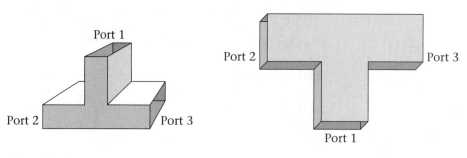

One of the more common components used with ordinary transmission line is the tee, which allows one line to branch into two. Tees can also be built for waveguides. Figure 6.27 shows E-plane and H-plane tees, named in the same way as the bends described above. A signal applied to port 1

FIGURE 6.27
Waveguide tees

Port 1

Port 2   Port 3

(a) E-plane tee

Port 2   Port 3

Port 1

(b) H-plane tee

appears at each of the other ports. For the H-plane tee, the signal is in phase at the two outputs, while the E-plane tee produces two out-of-phase signals. Sometimes the H-plane tee is referred to as a shunt tee, and the E-plane tee is called a series tee.

The *hybrid* or *"magic" tee* shown in Figure 6.28 is a combination of E-plane and H-plane tees. It has some interesting features. In particular, it can provide isolation between signals. An input at port 3 will result in equal and in-phase outputs at ports 1 and 2 but no output at port 4. On the other hand, a signal entering via port 4 will produce equal and out-of-phase outputs at ports 1 and 2 but no output at port 3.

FIGURE 6.28
Hybrid tee

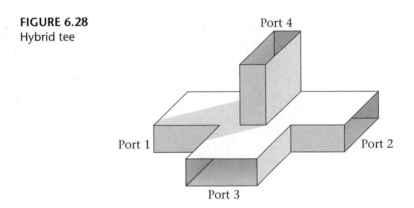

Port 4

Port 1

Port 2

Port 3

Cavity Resonators

In our discussion of waveguides it was noted that the waves reflect from the walls as they proceed down the guide. Suppose that instead of using a continuous waveguide, we were to launch a series of waves in a short section of guide called a cavity, as illustrated in Figure 6.29. The waves would, of course, reflect back and forth from one end to the other. A cavity of random size and shape would have random reflections with a variety of phase angles and a good deal of cancellation. However, suppose that the cavity had a

FIGURE 6.29
Rectangular cavity

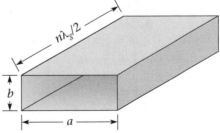

$n\lambda_g/2$

b

a

length of exactly one-half wavelength. Waves would reflect from one end to the other and would be in phase with the incident signal. There would be a buildup of field strength within the cavity. In a perfectly lossless cavity, this could continue forever. Of course, losses in the walls of a real cavity would result in the signal eventually dying out unless sustained by new energy input.

This description should sound familiar: it is a description of resonance. Like any other resonant device, a waveguide cavity has a Q; the Q for resonant cavities is very high, on the order of several thousand. Cavities can be tuned by changing their size (for instance, by moving a short-circuiting plate at one end).

The rectangular cavity described above is not the only possible type. Figure 6.30 shows several other types. Resonant cavities of various sorts are found in many types of microwave devices and are even used in the VHF region when very high Q is required and the considerable size of the cavities is acceptable.

FIGURE 6.30
Resonant cavities

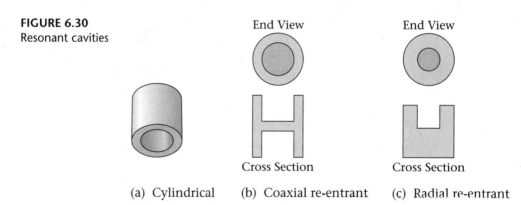

(a) Cylindrical   (b) Coaxial re-entrant   (c) Radial re-entrant

Waveguide Attenuators and Loads

At lower frequencies a load is simply a resistor, and an attenuator is a combination of resistors designed to preserve the characteristic impedance of a system while reducing the amount of power applied to a load. Once again, this lumped-constant approach becomes less practical as the frequency increases. At microwave frequencies, an ordinary carbon or metal-film resistor would have a complex equivalent circuit involving a good deal of distributed inductance and capacitance. Nonetheless, the basic idea of using resistive material to absorb energy is still valid.

Figure 6.31 shows waveguide versions of attenuators. The flap attenuator shown in Figure 6.31(a) uses a carbon flap that can be inserted to a greater or lesser extent into the waveguide. The fields inside the guide are, of course, present in the carbon vane as well. A current flows in the flap, causing power loss.

FIGURE 6.31
Waveguide
attenuators

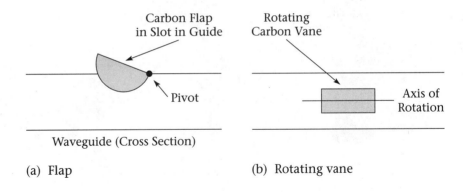

(a) Flap   (b) Rotating vane

The attenuator shown in Figure 6.31(b) uses a rotating vane. When the vane is rotated so that the electric field is perpendicular to its surface, little loss occurs, but when the field runs along the surface of the vane, a much larger current is induced, causing greater loss. Both fixed and variable vane attenuators are available with a variety of attenuation values.

Figure 6.32 illustrates a terminating load for a waveguide. The carbon insert is designed to dissipate the energy in the guide without reflecting it.

FIGURE 6.32
Waveguide
termination

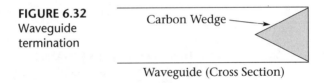

Circulators and Isolators

Isolators and circulators are useful microwave components that generally use ferrites in their operation. The theory of operation is beyond the scope of this text, but a brief description of their characteristics is in order.

An isolator is a device that allows a signal to pass in only one direction. In the other direction, it is greatly attenuated. An isolator can be used to shield a source from a mismatched load. Energy will still be reflected from the load, but instead of reaching the source, the reflected power is dissipated in the isolator. Figure 6.33 illustrates this application of an isolator.

The circulator, shown schematically in Figure 6.34, is a very useful device that allows the separation of signals. The three-port circulator shown allows a signal introduced at any port to appear at, and only at, the next port in counterclockwise rotation. For instance, a signal entering at port 1 appears at port 2 but not at port 3. Circulators can have any number of ports from three up, but three- and four-port versions are the most common.

One simple example of an application for a circulator is as a transmit-receive switch. The transmitter output is connected to port 1, the antenna to

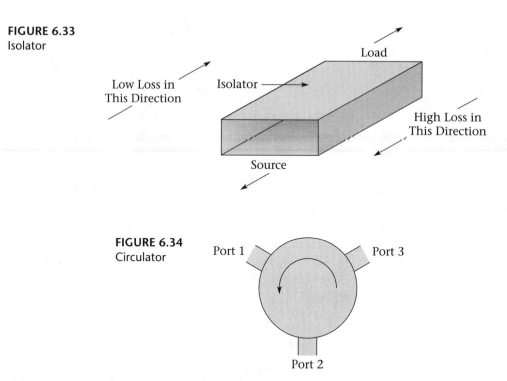

FIGURE 6.33
Isolator

FIGURE 6.34
Circulator

port 2, and the receiver input to port 3. The transmitter output signal is applied to the antenna, and a received signal from the antenna reaches the receiver. The transmitter signal does not reach the receiver, however; if it did, it would probably damage the receiver input circuitry.

Summary

The main points to remember from this chapter are:

- Radio waves are a form of electromagnetic radiation similar to light except for a lower frequency and longer wavelength.

- Any pair of conductors can act as a transmission line. The methods presented in this chapter need to be used whenever the line is longer than approximately one-sixteenth wavelength at the highest frequency in use.

- Any transmission line has a characteristic impedance determined by its geometry and its dielectric. If a transmission line is terminated by an impedance that is different from its characteristic impedance, part of a signal travelling down the line will be reflected. Usually this is undesirable.

- Reflections on lines are characterized by the reflection coefficient and the standing-wave ratio. The latter is easier to measure. A matched line has a reflection coefficient of 0 and an SWR of 1.

- Lines that are terminated by either a short or an open circuit can be used as reactances or as either series- or parallel-resonant circuits, depending on their length.

- Waveguides are a very practical means of transmitting electrical energy at microwave frequencies, as they have much lower losses than coaxial cable. They are not very useful at lower frequencies because they must be too large in cross section.

- Waveguides are generally useful over only a 2:1 frequency range. They have a lower cutoff frequency that depends on their dimensions, and they exhibit dispersion due to multimode propagation at high frequencies.

- Two velocities must be calculated for waveguides. The group velocity, which is lower than the speed of light, is the speed at which signals travel down the guide. The phase velocity, which is greater than the speed of light, is used for calculating the wavelength in the guide.

- Power can be coupled into and out of waveguides using probes, loops, or holes in the guide.

- Ferrites have special properties that make them just as useful at microwave frequencies as at lower frequencies. Applications of ferrites include circulators, isolators, attenuators and resonators.

● Equation List

$$v_p = f\lambda \tag{6.1}$$

$$Z_0 = \sqrt{\frac{L}{C}} \tag{6.3}$$

$$Z_0 = \frac{138}{\sqrt{\epsilon_r}} \log \frac{D}{d} \tag{6.4}$$

$$Z_0 = 276 \log \frac{D}{r} \tag{6.5}$$

$$v_f = \frac{v_p}{c} = \frac{1}{\sqrt{\epsilon_r}} \tag{6.6}$$

$$\phi = \frac{360L}{\lambda} \tag{6.7}$$

$$\Gamma = \frac{V_r}{V_i} \tag{6.8}$$

$$\Gamma = \frac{Z_L - Z_0}{Z_L + Z_0} \tag{6.9}$$

$$SWR = \frac{V_{max}}{V_{min}} \tag{6.10}$$

$$SWR = \frac{1 + |\Gamma|}{1 - |\Gamma|} \tag{6.13}$$

$$|\Gamma| = \frac{SWR - 1}{SWR + 1} \tag{6.14}$$

$$SWR = \frac{Z_L}{Z_0} \tag{6.15}$$

$$SWR = \frac{Z_0}{Z_L} \tag{6.16}$$

$$P_r = \Gamma^2 P_i \tag{6.17}$$

$$P_L = P_i (1 - \Gamma^2) \tag{6.18}$$

$$P_L = \frac{4SWR}{(1 + SWR)^2} P_i \tag{6.19}$$

$$Z = Z_0 \frac{Z_L + jZ_0 \tan \theta}{Z_0 + jZ_L \tan \theta} \tag{6.21}$$

$$Z = jZ_0 \tan \theta \tag{6.22}$$

$$a = \frac{\lambda_c}{2} \tag{6.23}$$

$$f_c = \frac{c}{2a} \tag{6.24}$$

$$v_g = c\sqrt{1 - \left(\frac{f_c}{f}\right)^2} \qquad (6.25)$$

$$v_g v_p = c^2 \qquad (6.26)$$

$$v_p = \frac{c}{\sqrt{1 - \left(\frac{f_c}{f}\right)^2}} \qquad (6.27)$$

$$Z_0 = \frac{377}{\sqrt{1 - \left(\frac{f_c}{f}\right)^2}} \qquad (6.28)$$

$$\lambda_g = \frac{v_p}{f} \qquad (6.29)$$

$$\lambda_g = \frac{\lambda}{\sqrt{1 - \left(\frac{f_c}{f}\right)^2}} \qquad (6.30)$$

● Key Terms

characteristic impedance   ratio between voltage and current on a transmission line, on a waveguide, or in a medium

directional coupler   device that allows signals to travel between ports in one direction only

dispersion   pulse spreading caused by variation of propagation velocity with frequency

group velocity   speed at which signals move down a transmission line or waveguide

microstrip   transmission line consisting of a circuit-board trace on one side of a dielectric substrate (the circuit board) and a ground plane on the other side

phase velocity   the apparent velocity of waves along the wall of a waveguide

stripline   transmission line consisting of a circuit board having a ground plane on each side of the board, with a conducting trace in the center

voltage standing-wave ratio (VSWR)   ratio between maximum and minimum peak or rms voltage on a transmission line

waveguide   metallic tube down which waves propagate

wavelength   distance a wave travels in one period

● Questions

1. Explain why sound waves can have the same wavelength as radio waves at a much lower frequency.

2. Why is an ordinary extension cord not considered a transmission line, while a television antenna cable of the same length would be?

3. Explain the difference between balanced and unbalanced lines, and give an example of each.

4. What is the difference between a microstrip and a stripline?

5. Draw the equivalent circuit for a short section of transmission line, and explain the physical meaning of each circuit element.

6. What is meant by the characteristic impedance of a transmission line?

7. What factors determine the characteristic impedance of a line?

8. Define the velocity factor for a transmission line, and explain why it can never be greater than one.

9. Explain what is meant by the SWR on a line, and state its value when a line is perfectly matched.

10. Why is a high SWR generally undesirable?

11. Why are shorted stubs preferred to open ones for impedance matching?

12. Draw a sketch showing how the impedance varies with distance along a lossless shorted line.

13. What would be the effect of placing a shorted, half-wave stub across a matched line?

14. What are the major contributors to transmission-line loss?

15. How does transmission-line loss vary with frequency, and why?

16. Explain what is meant by the dominant mode for a waveguide, and why the dominant mode is usually the one used.

17. Name the dominant mode for a rectangular waveguide, and explain what is meant by the numbers in the designation.

18. Explain the difference between phase velocity and group velocity in a waveguide. State which one of these is greater than the speed of light, and explain how this is possible.

19. State what is meant by dispersion, and show how it can arise in two different ways in a waveguide.

20. Draw diagrams showing the correct positions in which to install a probe and a loop to launch the dominant mode in a rectangular waveguide.

21. Explain the function of a directional coupler, and draw a sketch of a directional coupler for a waveguide.

22. Explain the operation of a hybrid tee.

23. Sketch a four-port circulator, and show what happens to a signal entering at each port.

● Problems

1. Find the wavelength for radio waves in free space at each of the following frequencies:
   (a) 160 MHz (VHF marine radio range)
   (b) 800 MHz (cell phone range)
   (c) 2 GHz (PCS range)

2. How far does a radio wave travel through space in one microsecond?

3. Visible light has a range of wavelengths from approximately 400 nanometers (violet) to 700 nm (red). Express this as a frequency range.

4. Calculate the characteristic impedance of an open wire transmission line consisting of two wires with diameter 1 mm and separation 1 cm.

5. Calculate the characteristic impedance of a coaxial line with a polyethylene dielectric, if the diameter of the inner conductor is 3 mm and the inside diameter of the outer conductor is 10 mm.

6. Repeat problem 1 for waves on a coaxial cable with a solid polyethylene dielectric.

7. Repeat problem 2 for a radio wave propagating along a coaxial cable with polyethylene foam dielectric.

8. How long a line is required to produce a 45° phase shift at 400 MHz if the dielectric is
   (a) air?
   (b) solid polyethylene?

9. A generator is connected to a short-circuited line 1.25 wavelengths long.
   (a) Sketch the waveforms for the incident, reflected, and resultant voltages at the instant the generator is at its maximum positive voltage.
   (b) Sketch the pattern of voltage standing waves on the line.
   (c) Sketch the variation of impedance along the line. Be sure to note whether the impedance is capacitive, inductive, or resistive at each point.

10. An open-circuited line is 0.75 wavelengths long.
    (a) Sketch the incident, reflected, and resultant voltage waveforms at the instant the generator is at its peak negative voltage.
    (b) Draw a sketch showing how RMS voltage varies with position along the line.

11. A 75-Ω source is connected to a 50-Ω load (a spectrum analyzer) with a length of 75-Ω line. The source produces 10 mW. All impedances are resistive.
    (a) Calculate the SWR.
    (b) Calculate the voltage reflection coefficient.
    (c) How much power will be reflected from the load?

12. A transmitter delivers 50 W into a 600-Ω lossless line terminated with an antenna that has an impedance of 275 Ω, resistive.
    (a) What is the coefficient of reflection?
    (b) How much of the power actually reaches the antenna?

13. A shorted stub acts as a parallel-resonant circuit at 100 MHz. What is its function at 200 MHz?

14. A properly matched transmission line has a loss of 1.5 dB per 100 m. If a signal power of 10 W is supplied to one end of the line, how much power reaches the load, 27 m away?

15. A receiver requires a signal of at least 0.5 µV for satisfactory reception. How strong (in µV) must the signal be at the antenna if the receiver is connected to the antenna by 25 m of matched line having an attenuation of 6 dB per 100 m ?

16. A transmitter with an output power of 50 W is connected to a matched load by means of 32 m of matched coaxial cable. It is found that only 35 W of power is dissipated in the load. Calculate the loss in the cable in dB/100 m.

17. RG-52/U waveguide is rectangular with an inner cross-section of 22.86 by 10.16 mm. Calculate the cutoff frequency for the dominant mode

and the range of frequencies for which single-mode propagation is possible.

18. For a 10-GHz signal in RG-52/U waveguide (described in problem 17), calculate:

    (a) the phase velocity

    (b) the group velocity

    (c) the guide wavelength

    (d) the characteristic impedance

19. A signal with a center frequency of 10 GHz has a bandwidth of 200 MHz. It propagates down 100 m of RG-52/U waveguide (described in problem 17). Calculate the difference in arrival times between the highest-frequency and lowest-frequency components of the signal.

20. A signal with a power of 10 mW enters port 1 of a directional coupler. The output power is 7 mW at port 2 and 100 μW at port 3. Calculate the coupling to each output port and the insertion loss for the device.

21. (a) Draw a sketch showing how a hybrid tee could be used to connect a transmitter and a receiver to the same antenna in such a way that the transmitter power would not reach the receiver.

    (b) Draw a sketch showing how a circulator could be arranged to do the same thing.

    (c) What advantage does the circulator version of this device have over the hybrid-tee version?

22. A transmission line of unknown impedance is terminated with two different resistances and the SWR is measured each time. With a 75-Ω termination the SWR measures 1.5. With a 300-Ω termination it measures 2.67. What is the impedance of the line?

Radio Propagation

7

Objectives

After studying this chapter, you should be able to:

- Describe the propagation of radio waves in free space and over land.

- Calculate power density and electric and magnetic field intensity for waves propagating in free space.

- Calculate free-space attenuation and path loss.

- Perform the necessary calculations to determine the maximum communication range for line-of-sight propagation.

- Calculate path loss in a mobile environment, and explain how such an environment differs from free space.

- Explain rapid fading and calculate the fade period for a moving vehicle.

- Describe the use of repeaters to increase communication range.

- Explain the cellular concept and calculate the signal-to-interference ratio for cellular systems.

- Describe how cell-splitting can increase the capacity of a system.

- Describe means to reduce the effects of fading in mobile systems.

- Distinguish between geostationary satellites and those in lower orbits, and explain the advantages and disadvantages of each.

7.1 Introduction

In the previous chapter we studied the propagation of electromagnetic waves on transmission lines and in waveguides. That is, we looked at the "wired" part of wireless communication. In this chapter we consider the "wireless" part. Waves propagate through space as *transverse electromagnetic (TEM)* waves. This means that the electric field, the magnetic field, and the direction of travel of the wave are all mutually perpendicular. The sketch in Figure 7.1 is an attempt to represent this three-dimensional process in two dimensions. The *polarization* of a wave is the direction of the electric field vector. Polarization may be horizontal or vertical and can also be circular or elliptical if the electric field vector rotates as it moves through space.

Radio waves are generated by electrons moving in a conductor, or set of conductors, called an antenna. Antennas are the subject of the next chapter.

Once launched, electromagnetic waves can travel through free space and through many materials. Any good dielectric will pass radio waves; the material does not have to be transparent to light. The waves do not travel well through lossy conductors, such as seawater, because the electric fields cause currents to flow that dissipate the energy of the wave very quickly. Radio waves reflect from good conductors, such as copper or aluminum.

The speed of propagation of radio waves in free space is the same as that of light, approximately 300×10^6 m/s. In other media, the velocity is lower.

FIGURE 7.1
Transverse electromagnetic waves

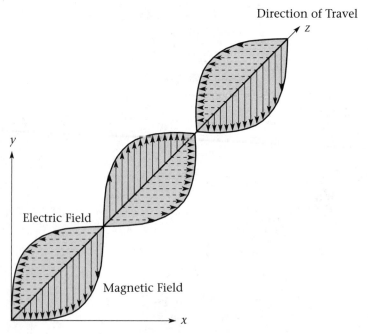

The propagation velocity is given by

$$v_p = \frac{c}{\sqrt{\epsilon_r}} \qquad\qquad (7.1)$$

where

v_p = propagation velocity in the medium

c = 300×10^6 m/s, the propagation velocity in free space

ϵ_r = the relative permittivity (dielectric constant) of the medium

This is the same equation given in Chapter 6 for the velocity of signals along a coaxial cable. One way of thinking of such a line is to imagine an electromagnetic wave propagating through the dielectric, guided by the conductors. It should be no surprise that the wave equation in Chapter 6 also applies to waves in free space and dielectrics. Radio waves are refracted as they pass from one medium to another with a different propagation velocity, just as light is.

7.2   Free-Space Propagation

The simplest source of electromagnetic waves would be a point in space. Waves would radiate equally from this source in all directions. A *wavefront,* that is, a surface on which all the waves have the same phase, would be the surface of a sphere. Such a source is called an **isotropic radiator** and is shown in Figure 7.2. Of course, an actual point source is not possible, but the approximation is good at distances that are large compared with the dimensions of the source. This is nearly always true of radio propagation.

If only a small area on the sphere shown in Figure 7.2 is examined and if the distance from the center of the sphere is large, the area in question

FIGURE 7.2
Isotropic
radiator

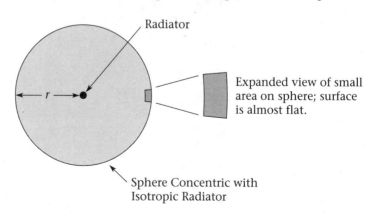

Radiator

Expanded view of small area on sphere; surface is almost flat.

Sphere Concentric with Isotropic Radiator

resembles a plane. In the same way, we experience the earth as flat, though we know it is roughly spherical. Consequently, waves propagating in free space are usually assumed to be plane waves, which are often simpler to deal with than spherical waves.

There is no loss of energy as radio waves propagate in free space, but there is attenuation due to the spreading of the waves. If a sphere were drawn at any distance from the source and concentric with it, all the energy from the source would pass through the surface of the sphere. Since no energy would be absorbed by free space, this would be true for any distance, no matter how large. The energy would be spread over a larger surface as the distance from the source increased.

Since an isotropic radiator radiates equally in all directions, the *power density,* in watts per square meter, is simply the total power divided by the surface area of the sphere. Put mathematically,

$$P_D = \frac{P_t}{4\pi r^2}$$ (7.2)

where

P_D = power density in W/m^2

P_t = transmitted power in W

r = distance from the antenna in meters

Not surprisingly, this is the same "square-law" attenuation that applies to light and sound, and, in fact, to any form of radiation. It is important to realize that this attenuation is not due to any loss of energy in the medium, but only to the spreading out of the energy as it moves farther from the source. Any actual losses will be in addition to this.

EXAMPLE 7.1

A power of 100 W is supplied to an isotropic radiator. What is the power density at a point 10 km away?

SOLUTION

From Equation (7.2),

$$P_D = \frac{P_t}{4\pi r^2}$$

$$= \frac{100\ \text{W}}{4\pi(10 \times 10^3\ \text{m})^2}$$

$$= 79.6\ \text{nW/m}^2$$

Real antennas do not radiate equally in all directions, of course. Equation (7.2) can easily be modified to reflect this, if we define *antenna gain* as follows:

$$G_t = \frac{P_{DA}}{P_{DI}} \tag{7.3}$$

where

G_t = transmitting antenna gain

P_{DA} = power density in a given direction from the real antenna

P_{DI} = power density at the same distance from an isotropic radiator with the same P_t

Antennas are passive devices and do not have actual power gain. They achieve a greater power density in certain directions at the expense of reduced radiation in other directions. In this capacity, an antenna resembles the reflector in a flashlight more than it does an amplifier. The ways in which antennas can be constructed will be examined in the next chapter.

Now we can modify Equation (7.2) to include antenna gain:

$$P_D = \frac{P_t G_t}{4\pi r^2} \tag{7.4}$$

Usually, antenna gain is specified in dBi, where the "i" indicates gain with respect to an isotropic radiator. The gain must be converted to a power ratio to be used with Equation (7.4).

We can define the *effective isotropic radiated power* (*EIRP*) of a transmitting system in a given direction as the transmitter power that would be needed, with an isotropic radiator, to produce the same power density in the given direction. Therefore, it is apparent that

$$EIRP = P_t G_t \tag{7.5}$$

and Equation (7.4) can be modified for use with *EIRP*.

$$P_D = \frac{EIRP}{4\pi r^2} \tag{7.6}$$

The use of Equation (7.4) or Equation (7.6) in a given problem depends on whether the transmitter power and antenna gain are specified separately or combined into one EIRP rating.

EXAMPLE 7.2 ▼

The transmitter of Example 7.1 is used with an antenna having a gain of 5 dBi. Calculate the EIRP and the power density at a distance of 10 km.

SOLUTION

First convert the gain to a power ratio.

$$G_t = \log^{-1}\left(\frac{5}{10}\right)$$

$$= 3.16$$

This means that the EIRP in the given direction is about three times the actual transmitter power. More precisely,

$$EIRP = G_t P_t = 3.16 \times 100 \text{ W} = 316 \text{ W}$$

The power density is

$$P_D = \frac{EIRP}{4\pi r^2}$$

$$= \frac{316}{4\pi(10 \times 10^3)^2}$$

$$= 251.5 \text{ nW/m}^2$$

The strength of a signal is more often given in terms of its electric field intensity than power density, perhaps because the former is easier to measure. There is a simple relationship between electric field intensity and power density. Power density is analogous to power in a lumped-constant system, and electric field intensity is the equivalent of voltage. The familiar equation from basic electricity:

$$P = \frac{V^2}{R}$$

becomes

$$P_D = \frac{\mathscr{E}^2}{\mathscr{Z}}$$

where

$\mathscr{E}$ = electric field intensity in volts per meter

$\mathscr{Z}$ = characteristic impedance of the medium in ohms

The characteristic impedance of free space is 377 Ω so, in free space,

$$P_D = \frac{\mathscr{E}^2}{377} \qquad (7.7)$$

It is quite easy to find a direct relationship between EIRP, distance, and electric field strength. First we rearrange Equation (7.7) so that $\mathcal{E}$ is the unknown:

$$\mathcal{E} = \sqrt{377 P_D} \qquad (7.8)$$

Now we can substitute the expression for P_D found in Equation (7.6) into this equation to get:

$$\mathcal{E} = \sqrt{\frac{377 \, EIRP}{4\pi r^2}} \qquad (7.9)$$

$$= \frac{\sqrt{30 \, EIRP}}{r}$$

EXAMPLE 7.3

Find the electric field intensity for the signal of Example 7.2, at the same distance (10 km) from the source.

SOLUTION

There are two ways to do this. Since P_D is already known, we could use Equation (7.8):

$$\mathcal{E} = \sqrt{377 P_D}$$

$$= \sqrt{377 \times 251.5 \times 10^{-9}}$$

$$= 9.74 \text{ mV/m}$$

Alternatively we could begin again with Equation (7.9):

$$\mathcal{E} = \frac{\sqrt{30 \, EIRP}}{r}$$

$$= \frac{\sqrt{30 \times 316}}{10 \times 10^3}$$

$$= 9.74 \text{ mV/m}$$

Receiving Antenna Gain and Effective Area

A receiving antenna absorbs some of the energy from radio waves that pass it. Since the power in the wave is proportional to the area through which it passes, a large antenna will intercept more energy than a smaller one (other things being equal) because it intercepts a larger area. Antennas are also more efficient at absorbing power from some directions than from others. For instance, a satellite dish would not be very efficient if it were pointed at the ground instead of the satellite. In other words, receiving antennas have gain, just as transmitting antennas do. In fact, the gain is the same whether the antenna is used for receiving or transmitting.

The power extracted from the wave by a receiving antenna depends both on its physical size and on its gain. The effective area of an antenna can be defined as

$$A_{eff} = \frac{P_r}{P_D} \tag{7.10}$$

where

A_{eff} = effective area of the antenna in m^2

P_r = power delivered to the receiver in W

P_D = power density of the wave in W/m^2

Equation (7.10) simply tells us that the effective area of an antenna is the area from which all the power in the wave is extracted and delivered to the receiver. Combining Equation (7.10) with Equation (7.4) gives

$$P_r = A_{eff}P_D \tag{7.11}$$

$$= \frac{A_{eff}P_tG_t}{4\pi r^2}$$

It can be shown that the effective area of a receiving antenna is

$$A_{eff} = \frac{\lambda^2 G_r}{4\pi} \tag{7.12}$$

where

G_r = antenna gain, as a power ratio

λ = wavelength of the signal

Path Loss

Combining Equations (7.11) and (7.12) gives an expression for the receiver power in terms of antenna gain, which is much more commonly found in specifications than is effective area.

$$P_r = \frac{A_{eff} P_t G_t}{4\pi r^2} \tag{7.13}$$

$$= \frac{\lambda^2 G_r P_t G_t}{(4\pi)(4\pi r^2)}$$

$$= \frac{\lambda^2 P_t G_t G_r}{16\pi^2 r^2}$$

While accurate, this equation is not very convenient. Gain and attenuation are usually expressed in decibels rather than directly as power ratios; the distance between transmitter and receiver is more likely to be given in kilometers than meters; and the frequency of the signal, in megahertz, is more commonly used than its wavelength. It is quite easy to perform the necessary conversions to arrive at a more useful equation. The work involved is left as an exercise; the solution follows.

$$P_r = P_t + G_t + G_r - (32.44 + 20 \log d + 20 \log f) \tag{7.14}$$

where

P_r = received power in dBm

P_t = transmitted power in dBm

G_t = transmitting antenna gain in dBi

G_r = receiving antenna gain in dBi

d = distance between transmitter and receiver, in km

f = frequency in MHz

Note that P_t and P_r are the power levels at the transmitting and receiving antennas, respectively. Attenuation due to transmission-line losses or mismatch is not included; these losses (in decibels) can be found separately and subtracted from the result for P_r given above to give the actual received power.

Sometimes it is convenient to have an expression for the free-space attenuation, often called path loss, that is independent of antenna gain. This is easily obtained by extracting the loss part of Equation (7.14), thereby converting Equation (7.14) into two equations:

$$P_r = P_t + G_t + G_r - L_{fs} \tag{7.15}$$

$$L_{fs} = 32.44 + 20 \log d + 20 \log f \tag{7.16}$$

where

L_{fs} = free-space loss in decibels

Using Equation (7.15) allows us to construct a diagram of a radio link, showing all the power levels, gains, and losses, and then merely add and subtract to find the received power. It also allows for the use of a different expression for path loss, necessary if the medium is not free space. The following examples show the use of path loss in calculating received signal strength.

First let us consider a very straightforward application. Then look at Example 7.5, which is a little more complex, involving transmission-line loss and mismatch.

EXAMPLE 7.4

A transmitter has a power output of 150 W at a carrier frequency of 325 MHz. It is connected to an antenna with a gain of 12 dBi. The receiving antenna is 10 km away and has a gain of 5 dBi. Calculate the power delivered to the receiver, assuming free-space propagation. Assume also that there are no losses or mismatches in the system.

SOLUTION

In all problems of this sort, it is a good idea to begin by sketching the system. This example can be done easily enough without such a sketch, but many real-world situations are more complex. See Figure 7.3 for the setup.

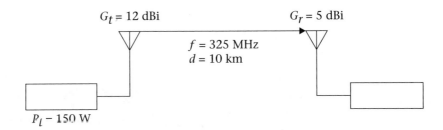

$G_t = 12$ dBi $G_r = 5$ dBi

$f = 325$ MHz
$d = 10$ km

$P_l - 150$ W

FIGURE 7.3

The next step is to convert the transmitter power into dBm:

$$P_t(\text{dBm}) = 10 \log\left(\frac{P_t}{1\,\text{mW}}\right)$$

$$= 10 \log\left(\frac{150\,\text{W}}{0.001\,\text{W}}\right)$$

$$= 51.8\,\text{dBm}$$

Marking the transmitter power and antenna gains on the sketch shows us that the only missing link is the path loss. We can find this from Equation (7.16).

$$L_{fs} = 32.44 + 20 \log d + 20 \log f$$
$$= 32.44 + 20 \log 10 + 20 \log 325$$
$$= 102.7 \text{ dB}$$

Now we can easily find the received power from Equation (7.15).

$$P_r = P_t + G_t + G_r - L_{fs}$$
$$= 51.8 + 12 + 5 - 102.7$$
$$= -33.9 \text{ dBm}$$

EXAMPLE 7.5

A transmitter has a power output of 10 W at a frequency of 250 MHz. It is connected by 20 m of a transmission line having a loss of 3 dB/100 m to an antenna with a gain of 6 dBi. The receiving antenna is 25 km away and has a gain of 4 dBi. There is negligible loss in the receiver feedline, but the receiver is mismatched: the antenna and line are designed for a 50 Ω impedance, but the receiver input is 75 Ω. Calculate the power delivered to the receiver, assuming free-space propagation.

SOLUTION

We begin with the sketch of Figure 7.4. A glance at the sketch shows what needs to be done. First, as in the previous example, we convert the transmitter power to dBm.

$$P_t(\text{dBm}) = 10 \log \left(\frac{P_t}{1 \text{ mW}} \right)$$

$$= 10 \log \left(\frac{10 \text{ W}}{0.001 \text{ W}} \right)$$

$$= 40 \text{ dBm}$$

FIGURE 7.4

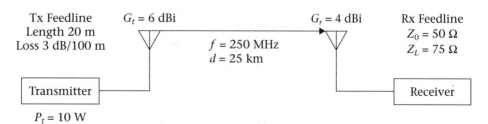

Tx Feedline
Length 20 m
Loss 3 dB/100 m

$G_t = 6$ dBi

$f = 250$ MHz
$d = 25$ km

$G_r = 4$ dBi

Rx Feedline
$Z_0 = 50$ Ω
$Z_L = 75$ Ω

Transmitter

Receiver

$P_t = 10$ W

Next, it is necessary to find the loss at each stage of the system. If all the losses are in decibels, it will only be necessary to add them to get the total loss.

We use Equation (7.16) to find the path loss.

$$
\begin{aligned}
L_{fs} &= 32.44 + 20 \log d + 20 \log f \\
&= 32.44 + 20 \log 25 + 20 \log 250 \\
&= 108.3 \text{ dB}
\end{aligned}
$$

For the transmitter feedline, the loss is

$$
L_{tx} = 20 \text{ m} \times \frac{3 \text{ dB}}{100 \text{ m}} = 0.6 \text{ dB}
$$

The receiver feedline is lossless, but some of the power reflects from the receiver back into the antenna due to the mismatch. This power will be reradiated by the antenna, and will never reach the receiver. Therefore, for our purposes it is a loss. Remember from Chapter 6 that the proportion of power reflected is the square of the reflection coefficient and the reflection coefficient is given by

$$
\begin{aligned}
\Gamma &= \frac{Z_L - Z_0}{Z_L + Z_0} \\
&= \frac{75 - 50}{75 + 50} \\
&= 0.2 \\
\Gamma^2 &= 0.2^2 \\
&= 0.04
\end{aligned}
$$

The proportion of the incident power that reaches the load is

$$
1 - \Gamma^2 = 0.96
$$

In decibels, the loss due to mismatch is

$$
\begin{aligned}
L_{tx} &= -10 \log 0.96 \\
&= 0.177 \text{ dB}
\end{aligned}
$$

Now we can easily find the received signal strength. We start with the transmitter power in dBm, add all the gains, and subtract all the losses (in dB) to get the result. Thus we find that

$$
\begin{aligned}
P_r &= P_t - L_{tx} + G_t - L_{fs} + G_r - L_{rx} \\
&= 40 - 0.6 + 6 - 108.3 + 4 - 0.177 \\
&= -59.1 \text{ dBm}
\end{aligned}
$$

7.3 Terrestrial Propagation

The simplified free-space model described above has some applicability to propagation over the surface of the earth. Air has very little loss at frequencies below about 20 GHz, and its dielectric constant is close to one. The most obvious differences between free-space and terrestrial propagation are that range is often limited by the horizon, signals may reflect from the earth itself, and various obstacles may exist between transmitter and receiver.

At low-to-medium frequencies (up to about 3 MHz), radio waves can follow the curvature of the earth, a phenomenon known as *ground-wave propagation,* and in the high-frequency range (about 3 to 30 MHz) the waves may be returned to earth from an ionized region in the atmosphere called the ionosphere. Both ground-wave and *ionospheric* propagation can result in reception of signals at much greater distances than the horizon, but neither is generally applicable at the frequencies (from VHF up) generally employed in wireless communication systems. At these frequencies, propagation is generally line-of-sight, as explained in the next section.

Line-of-Sight Propagation

The practical communication distance for line-of-sight propagation is limited by the curvature of the earth. In spite of the title of this section, the maximum distance is actually greater than the eye can see because refraction in the atmosphere tends to bend radio waves slightly toward the earth. The dielectric constant of air usually decreases with increasing height, because of the reduction in pressure, temperature, and humidity with increasing distance from the earth. The effect varies with weather conditions, but it usually results in radio communication being possible over a distance approximately one-third greater than the visual line of sight.

Just as one can see farther from a high place, the height above average terrain of both the transmitting and receiving antennas is very important in calculating the maximum distance for radio communication. Figure 7.5 shows the effect of increased antenna height on maximum range. Antenna heights are greatly exaggerated in the figure, of course.

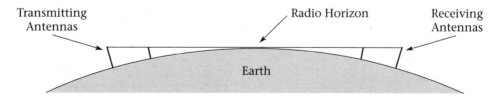

FIGURE 7.5 Line-of-sight propagation

An approximate value for the maximum distance between transmitter and receiver, over reasonably level terrain, is given by the following equation:

$$d = \sqrt{17h_t} + \sqrt{17h_r} \tag{7.17}$$

where

d = maximum distance in kilometers

h_t = height of the transmitting antenna in meters

h_r = height of the receiving antenna in meters

EXAMPLE 7.6

A taxi company uses a central dispatcher, with an antenna at the top of a 15 m tower, to communicate with taxicabs. The taxi antennas are on the roofs of the cars, approximately 1.5 m above the ground. Calculate the maximum communication distance:

(a) between the dispatcher and a taxi

(b) between two taxis

SOLUTION

(a) $d = \sqrt{17h_t} + \sqrt{17h_r}$

$= \sqrt{17 \times 15} + \sqrt{17 \times 1.5}$

$= 21$ km

(b) $d = \sqrt{17h_t} + \sqrt{17h_r}$

$= \sqrt{17 \times 1.5} + \sqrt{17 \times 1.5}$

$= 10.1$ km

The maximum range calculated from Equation (7.17) will only be achieved if the received signal strength, as calculated in the previous section, is sufficient. Maximum range is achieved by using a combination of reasonably large transmitter power and high-gain antennas located as high as possible.

It is not always desirable to achieve maximum distance. There are situations, as in the cellular systems to be described later, where it is common to limit the effective range to a distance smaller than predicted by Equation (7.17).

Multipath Propagation

Although line-of-sight propagation uses a direct path from transmitter to receiver, the receiver can also pick up reflected signals. Probably the simplest case is reflection from the ground, as shown in Figure 7.6. If the ground is rough, the reflected signal is scattered and its intensity is low in any given direction. If, on the other hand, the reflecting surface is relatively smooth—a body of water, for instance—the reflected signal at the receiver can have a strength comparable to that of the incident wave, and the two signals will interfere. Whether the interference is constructive or destructive depends on the phase relationship between the signals: if they are in phase, the resulting signal strength is increased, but if they are 180° out of phase, there is partial cancellation. When the surface is highly reflective, the reduction in signal strength can be 20 dB or more. This effect is called **fading**. The exact phase relationship depends on the difference, expressed in wavelengths, between the lengths of the transmission paths for the direct and reflected signals. In addition, there is usually a phase shift of 180° at the point of reflection.

FIGURE 7.6
Ground reflections

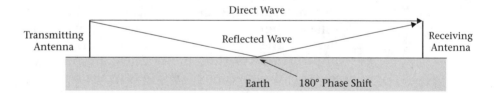

If the transmitter and receiver locations are fixed, the effect of reflections can often be reduced by carefully surveying the proposed route and adjusting the transmitter and receiver antenna heights so that any reflection takes place in wooded areas or rough terrain, where the reflection will be diffuse and weak. Where most of the path is over a reflective surface such as desert or water, fading can be reduced by using either **frequency diversity** or **spatial diversity**. In the former method, more than one frequency is available for use; the difference, in wavelengths, between the direct and incident path lengths will be different for the two frequencies. In spatial diversity there are two receiving antennas, usually mounted one above the other on the same tower. The difference between direct and reflected path length is different for the two antennas.

Diffraction from obstacles in the path can also be a problem for line-of-sight radio links, when the direct wave and the diffracted wave have opposite phase and tend to cancel. Figure 7.7 shows this effect of diffraction.

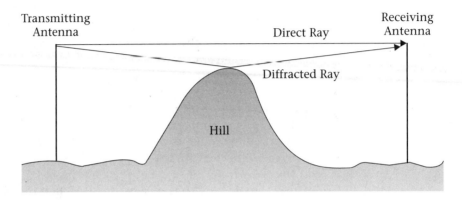

FIGURE 7.7   Diffraction

The solution to the problem of interference due to diffraction is to arrange for the direct and diffracted signals to be in phase. Again, this requires a careful survey of the proposed route and adjustment of the transmitting and receiving antenna heights to achieve this result.

Problems can also occur when the signal reflects from large objects like cliffs or buildings, as shown in Figure 7.8. There may be not only phase cancellation but also significant time differences between the direct and reflected waves. These can cause a type of distortion called, not surprisingly, **multipath distortion**, in FM reception. Directional receiving antennas aimed in the direction of the direct signal can reduce the problem of reflections for fixed receivers.

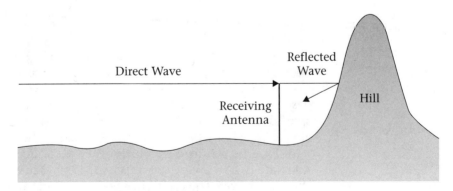

FIGURE 7.8   Multipath reception

The Mobile Environment

In the previous section, we looked at the basics of multipath propagation. In an environment where both transmitter and receiver are fixed, it is often possible, as noted above, to position the antennas in such a way as to reduce, if not eliminate, the effects of multipath interference. In the usual wireless situation, however, either the transmitter, the receiver, or both, are in constant motion, and the multipath environment is therefore in a constant state of flux. The luxury of carefully sited, directional antennas does not exist. In addition, the mobile and portable environment is often very cluttered with the potential for multiple reflections from vehicles and buildings. There may also be places where the direct signal is blocked by a tall building, for example. At such times a reflected signal may actually allow communication to take place where it would otherwise be impossible. Figure 7.9 shows some of the possibilities.

FIGURE 7.9
Multipath reception in a mobile environment

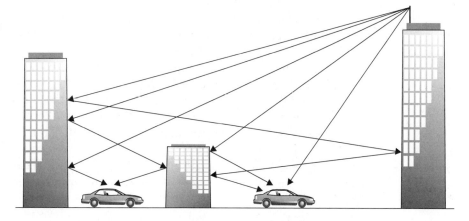

Mobile environments are often so cluttered that the square-law attenuation of free space, which is the basis for Equation (7.16), the path loss equation, no longer applies because of multipath propagation and shadowing by obstacles. In fact, the situation is so complex and so dependent on the actual environment that no one equation can cover all situations. Some approximations are used, however.

To understand how the mobile environment changes path loss, let us first have another look at Equation (7.16):

$$L_{fs} = 32.44 + 20 \log d + 20 \log f$$

Because this equation is set up to give loss in decibels, the square-law attenuation due to distance is represented by its logarithmic equivalent: $20 \log d$. In mobile environments the attenuation increases much more quickly with distance; values range from $30 \log d$ to a more typical $40 \log d$. That is, attenuation is roughly proportional to the fourth power of distance because of reflections and obstacles.

Equation (7.16) had no term for antenna height, since this is irrelevant in free space where there is no ground to measure height from. We looked at antenna height separately in Equation (7.17), as a limiting factor governing the maximum distance for line-of-sight propagation. Even for distances much shorter than the maximum, antenna height is an important determinant of signal strength in mobile systems. The greater the antenna height, the less important ground reflections become, and the more likely the transmitting and receiving antennas are to have an unobstructed path between them. Empirical observations show that doubling the height of a base-station antenna gives about a 6 dB decrease in path loss, while doubling the height of a (much lower) portable or mobile antenna reduces loss by about 3 dB.

Path loss increases with frequency according to Equation (7.16), but only because the effective area of an antenna with a given gain decreases with frequency. For an ideal isotropic radiator and a receiving antenna of constant effective area, there is no dependence of free-space attenuation on frequency. In fact, as we pointed out earlier, the basic square-law attenuation is valid for waves as different as sound and light. The situation is different in a mobile environment because the amount of reflection and diffraction of a wave from an object is a function of the object's dimensions, in wavelengths. At the low-frequency end of the VHF range, only very large objects such as the ground and buildings cause reflections. As the frequency increases, the wavelength becomes smaller and relatively small objects begin to reflect the waves. The increase in multipath interference due to this condition may be compensated for in portable operation because signals at UHF and up have wavelengths small enough to allow them to propagate through windows, improving the results in such situations as the use of a portable wireless phone in a car or building.

There is always a considerable loss in signals as they penetrate buildings. The loss varies with frequency, building construction, size and location of windows, and so on, but a rough estimate would be 20 dB at 800 MHz (cell phone range) in a typical steel-reinforced concrete office building, reduced to about 6 dB if the user is near a window facing the base station.

From the foregoing, it should be obvious that the modeling of terrestrial propagation is very difficult, especially in a mobile/portable environment. There are many models; most of them combine theory with observations of actual propagation in different areas such as open flat terrain, hilly terrain, suburbs, and cities. Usually a computer is used to calculate signal strengths in the coverage region. Even so, the models are all quite approximate; they must be used with care and verified by actual field-strength measurements. Errors of 10 dB or so are common.

As an example, we shall look at a simplified version of a model, proposed by Hata, for propagation in a dense urban mobile environment in the frequency range between 150 and 1000 MHz. A mobile antenna height of 2 m is assumed in this simplified version.

$$L_p = 68.75 + 26.16 \log f - 13.82 \log h \qquad (7.18)$$
$$+ (44.9 - 6.55 \log h) \log d$$

where

L_p = path loss in dB

f = frequency in MHz

h = base station antenna height in m

d = distance in km

Several of the factors we have already mentioned can be seen in this equation. First, note that loss increases with frequency at a greater rate than in the free-space equation. Next, notice that increased antenna height reduces loss. Finally, the loss increases much more quickly with distance than in the free-space model. Antenna height affects this loss factor as well.

EXAMPLE 7.7

Find the propagation loss for a signal at 800 MHz, with a transmitting antenna height of 30 m, over a distance of 10 km, using:

(a) the free-space model (Equation 7.16)

(b) the mobile-propagation model (Equation 7.18)

SOLUTION

(a) From Equation (7.16),
$$L_{fs} = 32.44 + 20 \log d + 20 \log f$$
$$= 32.44 + 20 \log 10 + 20 \log 800$$
$$= 110.5 \text{ dB}$$

(b) From Equation (7.18),
$$L_p = 68.75 + 26.16 \log f - 13.82 \log h + (44.9 - 6.55 \log h) \log d$$
$$= 68.75 + 26.16 \log 800 - 13.82 \log 30$$
$$+ (44.9 - 6.55 \log 30) \log 10$$
$$= 159.5 \text{ dB}$$

There is almost 50 dB more attenuation in the mobile environment than in free space.

Another mobile-specific problem is fast fading. As the mobile user travels through the environment, the signal strength tends to increase and decrease as the mobile moves between areas of constructive and destructive interference. As an example, suppose a mobile receiver moves directly away from the transmitting antenna and toward a reflecting surface, as shown in Figure 7.10. If the two signals are in phase at a given point, they will add. As the mobile moves forward a distance of $\lambda/4$, the direct path is increased and the reflected path is reduced by the same amount, resulting in a total phase shift of 180° and cancellation of the signal. The cancellation will likely not be complete, but it can be quite severe (up to 50 dB in extreme cases). When the vehicle moves another distance of $\lambda/4$, the signals are once again in phase. Thus the fades occur each time the car moves a distance of $\lambda/2$. Given the frequency of the signal and the speed of the vehicle, it is easy to estimate the time between fades. The time between fades is

$$T = \frac{\lambda/2}{v} \qquad (7.19)$$

$$= \frac{\lambda}{2v}$$

$$= \frac{c}{2fv}$$

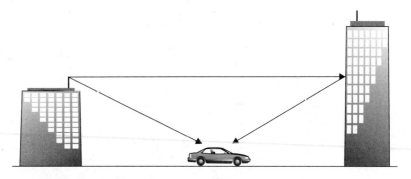

FIGURE 7.10   Fast fading in a mobile environment

EXAMPLE 7.8

An automobile travels at 60 km/hr. Find the time between fades if the car uses:

(a) a cell phone at 800 MHz

(b) a PCS phone at 1900 MHz

SOLUTION

First convert the car's speed to m/s.

$$\frac{60 \text{ km}}{\text{hr}} = \frac{60 \times 10^3 \text{ m}}{3.6 \times 10^3 \text{ s}}$$

$$= 16.7 \text{ m/s}$$

Now we can find the fading period using Equation (7.19).

(a) $T = \dfrac{c}{2fv}$

$$= \frac{300 \times 10^6}{2 \times 800 \times 10^6 \times 16.7}$$

$$= 11.2 \text{ ms}$$

(b) $T = \dfrac{c}{2fv}$

$$= \frac{300 \times 10^6}{2 \times 1900 \times 10^6 \times 16.7}$$

$$= 4.7 \text{ ms}$$

Notice that the rapidity of the fading increases with both the frequency of the transmissions and the speed of the vehicle.

Repeaters and Cellular Systems

The previous two sections have shown that antenna height is important for line-of-sight communication. If it is desired to achieve maximum communication range, increasing antenna height increases the distance to the radio horizon. Even if the distances involved are small, increased antenna height reduces multipath interference and avoids radio shadows.

Unfortunately, in a typical wireless communication system, the users are mobile or portable, and there is not much that can be done about their antenna height. Communication directly between mobile or portable users therefore has limited range and is subject to a great deal of multipath and shadowing, even at close range. A good example of such a system is citizens' band (CB) radio. Even with transmitter power of about 4 W, communication tends to be poor in urban areas.

Modern wireless systems use base stations with elevated antennas. The use of base stations is necessary to provide a connection with the telephone network, and the elevated antennas improve propagation. In addition, it is possible when required, to use higher transmitter power at the base than at the mobile or portable unit. Portable units especially should use as little power as possible to reduce the size and weight of the unit, and especially its batteries.

The simplest form of base station has a transmitter and receiver on the same frequency. The mobiles use the same frequency as well. Typically all units have moderately high transmitter power (on the order of 30 W). The base-station antenna is located as high as practical in order to obtain wide coverage. Figure 7.11 shows the setup, which is commonly used for services such as taxicab dispatching. Most communication is between mobiles and the base, though mobiles can communicate directly with each other if they are close enough together. Communication is half-duplex; that is, each station can both talk and listen, but not at the same time.

FIGURE 7.11
Dispatcher
system

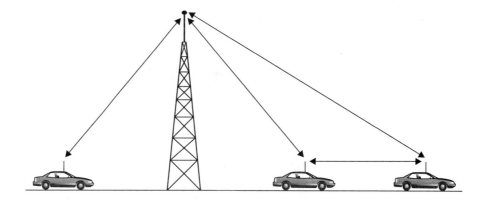

If mobile units need to talk to each other through the base station to achieve greater range, or if they need full-duplex access to the telephone system through the base, at least two frequencies must be used. The base station in this system is called a *repeater,* and must transmit and receive simultaneously on at least two frequencies. Normally the same antenna is used for both transmitting and receiving, and a high-Q filter called a **duplexer**, using resonant cavities, is used to separate the transmit and receive frequencies. Mobiles can communicate with each other through the repeater, and if the repeater is connected to the PSTN, phone calls can be made through the repeater.

There is an extensive network of amateur radio repeaters operating on this principle. The frequencies shown in Figure 7.12 are typical of one of these systems. Early mobile-phone systems used this method, and it is still in common use for fixed microwave links.

FIGURE 7.12
Typical repeater
system

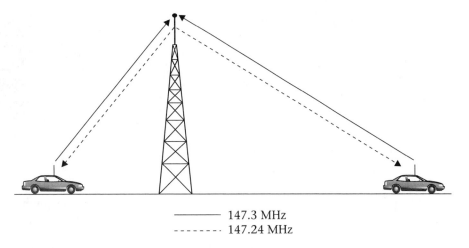

——————— 147.3 MHz
- - - - - - - 147.24 MHz

The major problem with the dispatcher and repeater systems just described is, since they are constructed for maximum coverage using antennas
at large elevations and high-power transmitters, they use spectrum ineffi-
ciently. One telephone conversation ties up two channels over the complete
coverage area of the system, out to the radio horizon. The simplicity of being
able to cover an entire city with one repeater is paid for by this great waste of
spectrum.

Modern cellular systems do not use the radio horizon as the limit of coverage. Antennas may still be mounted quite high in order to reduce
multipath and shadowing, but the range is deliberately limited by using as
low a transmitter power as possible. This enables the same frequencies to be
reused at distances much closer than the radio horizon. Cellular systems are
complex but much more efficient in their use of spectrum.

Figure 7.13 shows a typical cellular system. (The term cellular, here, also
applies to PCS systems, wireless LANs, and so on, because they all use the
same general principle.)

Each repeater is responsible for coverage in a small cell. As shown, the
cells are hexagons with the repeater in the center, but of course in a real situation the antenna patterns will not achieve this precision and the cells will
have irregular shapes with some overlap.

Since each transmitter operates at low power, it is possible to reuse
frequencies over a relatively short distance. Typically a repeating pattern
of either twelve or seven cells is used, and the available bandwidth is divided
among these cells. The frequencies can then be reused in the next pattern.

To find out how many cells are needed in a pattern, it is necessary
to make some assumptions. First, we assume that all signals are well above
the noise level, so that *cochannel interference*—interference from transmitters

FIGURE 7.13
Cellular system

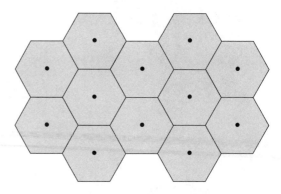

• Cell Site Radio Equipment and Antenna

on the same frequency in other cells—is the range-limiting factor. Next, we assume that only the nearest cells with the same frequency will cause serious problems. This is usually true. Further, assume that all transmitters have equal power. This is actually a worst-case assumption, because most cellular systems reduce transmitter power when possible to reduce interference.

We need a criterion for an acceptable level of interference. Zero interference is not an option, since the next cell with the same frequency is generally above the horizon, and its signal strength may be well above the noise level. The usual assumption is that a signal-to-interference (S/I) ratio of at least 18 dB is sufficient. This number is based on subjective tests of communication quality with FM voice transmission, and may vary with other systems.

We should also know how rapidly signals drop off with distance. The greater the attenuation, within reason, the better in a cellular system because the interfering signal, coming from a greater distance, will be affected more than the desired signal. Earlier in this chapter we saw that square-law attenuation is the rule in free space, but in mobile propagation, especially in urban areas, the attenuation is more likely to be proportional to the fourth power of the distance. Let us make that assumption for now, recognizing that the attenuation may sometimes be less than this, leading to more interference.

With this information and a knowledge of geometry, we can analyze any repeating pattern of cells to find the probable S/I ratio. In particular, let us look at a pattern of seven cells, as shown in Figure 7.14, since it is the one most often used in practice.

In a seven-cell pattern, there will be six interfering signals with equal distances from our chosen cell. The interference powers from these six cells will add. All other interfering signals will be much farther away and can be

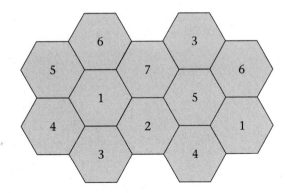

FIGURE 7.14
7-Cell repeating pattern

ignored. Let us place ourselves in the center of a cell, at the cell site, and assume that a mobile at the edge of our cell, a distance r from the center, is being interfered with by transmitters in the six nearest cells with the same number. Since the interfering signals can be anywhere in these cells, we will assume that they are all at the center of their respective cells, at a distance d from us. From geometry, the ratio between these distances, which is usually called q, is

$$q = \frac{d}{r} = 4.6$$

Now we can find the signal-to-interference ratio, S/I. Assuming equal transmitter powers, and fourth-law attenuation

$$\frac{S}{I} = \frac{d^4}{6r^4}$$

$$= \frac{q^4}{6}$$

$$= \frac{4.6^4}{6}$$

$$= 74.6$$

$$= 18.7 \text{ dB}$$

The result appears satisfactory at first, but it is actually marginal. Fading of the desired signal or less attenuation for any of the interfering signals could easily degrade the signal to the point where it is unusable. The situation can be improved by reusing the frequencies less often, for instance, using a twelve-cell repeating pattern. For such a pattern, $q = 6.0$, there are eleven significant interfering signals, and the theoretical performance is

$$\frac{S}{I} = \frac{d^4}{11r^4}$$

$$= \frac{q^4}{11}$$

$$= \frac{6.0^4}{11}$$

$$= 117.8$$

$$= 20.7 \text{ dB}$$

This represents about a 2 dB improvement, but at the expense of much less frequency reuse. Another possibility, widely employed in practice, is to use **sectorization**. Three directional antennas are located at each base station in such a way that each antenna covers an angle of 120 degrees. Each sector uses a different set of channels. The effect of this is to reduce the number of interfering signals that must be considered from six to two. The effect on the S/I ratio is quite dramatic:

$$\frac{S}{I} = \frac{d^4}{2r^4}$$

$$= \frac{q^4}{2}$$

$$= \frac{4.6^4}{2}$$

$$= 223.8$$

$$= 23.5 \text{ dB}$$

Another way of seeing this is to note that S/I improves by

$$10 \log\left(\frac{6}{2}\right)$$

$$= 10 \log 3$$

$$= 4.8 \text{ dB}$$

The division of cells into sectors has a downside. Since each cell now has three sectors with different channels, each cell acts like three cells with only one-third the number of channels. This gives the equivalent in frequency reuse of a 21-cell repeating pattern; so it is actually less efficient in terms of frequency reuse than the 12-cell pattern just described. Since cell sites are expensive, however, the 7-cell pattern is more economical to build.

As the number of users increases, cell sizes can be made smaller by installing more cell sites, and the frequencies can be reused over closer intervals. This technique is called cell-splitting, and gives cellular systems great flexibility to adapt to changes in demand over both space and time.

Here is how cell-splitting works. The number of cells required for a given area is given by:

$$N = \frac{A}{a} \tag{7.20}$$

where

N = number of cells
A = total area to be covered
a = area of one cell

This assumes that all cells are of equal size and that there is no overlap between cells, but it serves as a reasonably good estimate.

If hexagonal cells are used, there need be no overlap, at least in theory. (In practice, the cells will not be perfectly hexagonal.) The area of a regular hexagon (one with all six sides equal) is given by the equation

$$A = 3.464r^2 \tag{7.21}$$

where

A = the area of the hexagon
r = the radius of a circle inscribed in the hexagon, as in Figure 7.15

FIGURE 7.15
Area of hexagon

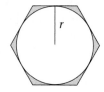

The required number of cells is:

$$N = \frac{A}{a} \tag{7.22}$$

$$= \frac{A}{3.464r^2}$$

The number of cells required is inversely proportional to the square of the cell radius. Reducing the cell radius by a factor of two increases the number of cells and therefore the system capacity by a factor of four. Unfortunately, reducing the cell radius by a factor of two also increases the number of cell sites required by a factor of four. As cell sizes are reduced, the available spectrum is used more efficiently, but cell-site equipment is not, since the maximum number of users per cell remains constant.

EXAMPLE 7.9

A metropolitan area of 1000 square km is to be covered by cells with a radius of 2 km. How many cell sites would be required, assuming hexagonal cells?

SOLUTION

From Equation (7.22),

$$N = \frac{A}{3.464r^2}$$

$$= \frac{1000}{3.464 \times 2^2}$$

$$= 73$$

Users of earlier mobile-radio systems could use the same frequencies throughout a conversation, but since the cells in a cellular radio system are relatively small, many calls from or to moving vehicles must be transferred or *handed off* from one cell site to another as the vehicles proceed. This requires a change in frequency, since frequencies are not reused in adjacent cells. The system has to instruct the mobile units to change frequency. The details will be covered later, but it is worth noting here that cellular schemes were not practical for portable/mobile use until microprocessors were cheap and small enough to be incorporated in the phones themselves.

Control of Fading in Mobile Systems

Though mobile and portable communication is more problematic than for fixed locations, there are many techniques that can be used to reduce the problem of fading. Probably the most obvious is to increase the transmitter power. If the signal is subject to 20-dB fades, then increasing power by 20 dB should solve the problem. The trouble with this is that a 20-dB increase requires multiplying transmitter power by 100. This may be practical for base stations, but size and battery-life considerations make it a highly undesirable solution for mobile and especially portable equipment. A typical portable cell phone produces about 700 mW of RF power. Increasing this to 70 W would be completely impractical: the equipment would have to be mounted in a vehicle. Increasing power can also cause interference problems.

Frequency diversity of the conventional type, where two channels are used in place of one in each direction, is similarly impractical. Mobile systems seldom have enough bandwidth to afford this luxury. There is an exception, however: spread spectrum systems achieve frequency diversity without an increase in bandwidth. They do this by, in effect, sharing multiple RF channels among multiple voice channels so that each voice channel has its data bits distributed over many frequencies. Fading of a narrow band of frequencies causes some loss of data, but this can often be corrected by error-correcting codes.

CDMA spread-spectrum systems have another advantage in the presence of multipath interference. Using a special receiver called a **rake receiver**, they can receive several data streams at once. These actually contain the same data, displaced in time because of the different propagation times that arise from the reflections. The receiver combines the power from the various streams. This adds another possible type of diversity to the system.

Space diversity can be used with portable and mobile systems. Usually this consists of positioning two receiving antennas at the base station (two per sector in a sectorized system), though it is possible to use space diversity on a vehicle by installing an antenna at each end. Since multipath interference is very dependent on the exact phase difference between signals traveling along different paths, moving the antenna, which changes the path length for both direct and reflected signals, can often cause the interference to become constructive rather than destructive.

As a vehicle or pedestrian moves, all the path lengths are constantly changing, causing the signal strength to vary rhythmically as the interference cycles between constructive and destructive. The frequency of these fades depends on the signal wavelength and the vehicle's speed. Spreading a digital signal in time by transmitting it more than once or by scrambling bits so that consecutive bits in the baseband signal are not transmitted in the same order, can help to reduce the destructive effects of multipath fading in a mobile environment.

7.4 Satellite Propagation

Satellites have been used for many years as repeaters for broadcast and point-to-point transmission, but their use in wireless communication is relatively new. Several wireless satellite systems are now in operation, however.

In this chapter we look at the basics of propagation to and from satellites. The details of practical systems will be covered in Chapter 12.

Many communication satellites use the **geosynchronous orbit**: that is, they occupy a circular orbit above the equator, at a distance of 35,784 km above the earth's surface. At this height, the satellite's orbital period is equal to the time taken by the earth to rotate once, approximately 24 hours. If the direction of the satellite's motion is the same as that of the earth's rotation, the satellite appears to remain almost stationary above one spot on the earth's surface. The satellite can then be said to be **geostationary**, as illustrated in Figure 7.16.

Geostationary satellites are convenient in several respects. Antennas at fixed earth stations can be aimed once and left in place. Satellites are usable at all times; they never disappear below the horizon.

Geostationary satellites have many disadvantages for mobile use, however. Their distance from earth is so great that high-gain antennas, which are impossible to use in handheld phones, are highly desirable. The transmitter power required puts a strain on batteries. The path length also causes a time delay of about 0.25 seconds for a round trip, which is inconvenient. Some mobile communication systems do employ geostationary satellites, but most of the new proposals use constellations of satellites at much lower altitudes where signals are stronger and propagation time is shorter.

Satellites that are not geostationary are called **orbital satellites**, though of course all satellites are in orbit. They are further categorized by the height of their orbit above the earth. **Low earth orbit (LEO)** satellites have orbits below about 1500 km while **medium earth orbit (MEO)** systems are from

FIGURE 7.16
Geostationary satellite orbit

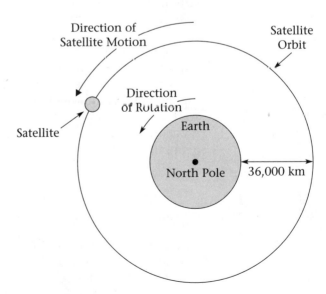

10,000 to 15,000 km in altitude. LEO satellites, because they are much closer to earth than the others, can have relatively strong signals and short propagation times. However, they do not stay above the horizon very long, so many satellites are needed for continuous coverage. MEO satellites are a compromise between LEOs and geostationary satellites.

Satellite propagation generally obeys the free-space equation, provided the mobile unit has a clear view of the satellite. At VHF and UHF frequencies the ionosphere tends to change the axis of polarization of radio waves in a random way so circular polarization is commonly used at these frequencies.

An example will demonstrate the differences among LEO, MEO, and geostationary satellites in terms of path loss.

EXAMPLE 7.10

Calculate the free-space path loss at a frequency of 1 GHz, for each of the following paths, to a point on the earth directly under the satellite:

(a) from a LEO satellite at an altitude of 500 km

(b) from a MEO satellite at an altitude of 12,000 km

(c) from a geostationary satellite at 36,000 km

SOLUTION

(a) From Equation (7.16),

$$L_{fs} = 32.44 + 20 \log d + 20 \log f$$
$$= 32.44 + 20 \log 500 + 20 \log 1000$$
$$= 146.4 \text{ dB}$$

(b) Using the same equation, but $d = 12,000$ km:

$$L_{fs} = 32.44 + 20 \log d + 20 \log f$$
$$= 32.44 + 20 \log 12,000 + 20 \log 1000$$
$$= 174 \text{ dB}$$

(c) Now use $d = 36,000$ km.

$$L_{fs} = 32.44 + 20 \log d + 20 \log f$$
$$= 32.44 + 20 \log 36,000 + 20 \log 1000$$
$$= 184 \text{ dB}$$

Note the great difference in loss between the LEO satellite and the other two.

 **Summary** The main points to remember from this chapter are:

- Radio waves are transverse electromagnetic waves very similar, except for frequency, to light waves.
- The impedance of a medium is the ratio of the electrical and magnetic field strengths in that medium. The impedance of free space is 377 Ω.
- The polarization of a radio wave is the direction of its electric field vector.
- As a radio wave propagates in space, its power density reduces with distance due to spreading of the waves. Directional antennas can partially offset this effect.
- Reflection, refraction, and diffraction are possible for radio waves and follow rules similar to those for light.
- Line-of-sight propagation is the system most commonly used at VHF and above. The radio range is actually somewhat greater than the visible range and is very dependent on the height of the transmitting and receiving antennas.
- In mobile systems, reflections and radio shadows generally cause signal strength to decrease much more rapidly with distance from the transmitter than in free space.
- In cellular systems, propagation distances are deliberately kept shorter than line-of-sight range in order to allow frequency reuse at shorter distances.
- Propagation via satellite is useful for long-distance communication. Geostationary systems are convenient but the very large path loss and propagation time limit their usefulness for wireless communication. Low-earth-orbit satellites reduce these problems, but several are needed for continuous service.

Equation List

$$v_p = \frac{c}{\sqrt{\epsilon_r}}$$ (7.1)

$$P_D = \frac{EIRP}{4\pi r^2}$$ (7.6)

$$\mathcal{E} = \frac{\sqrt{30EIRP}}{r}$$ (7.9)

$$P_r = P_t + G_t + G_r - L_{fs} \tag{7.15}$$

$$L_{fs} = 32.44 + 20 \log d + 20 \log f \tag{7.16}$$

$$d = \sqrt{17h_t} + \sqrt{17h_r} \tag{7.17}$$

$$L_p = 68.75 + 26.16 \log f - 13.82 \log h + (44.9 - 6.55 \log h) \log d \tag{7.18}$$

$$T = \frac{c}{2fv} \tag{7.19}$$

$$N = \frac{A}{3.464r^2} \tag{7.22}$$

● Key Terms

duplexer   combination of filters to separate transmit and receive signals when both use the same antenna simultaneously

fading   reduction in radio signal strength, usually caused by reflection or absorption of the signal

frequency diversity   use of two or more RF channels to transmit the same information, usually done to avoid fading

geostationary orbit   satellite orbit in which the satellite appears to remain stationary at a point above the equator

geosynchronous orbit   satellite orbit in which the satellite's period of revolution is equal to the period of rotation of the earth

isotropic radiator   an antenna that radiates all power applied to it, equally in all directions. It is a theoretical construct and not a practical possibility.

low-earth-orbit (LEO) satellite   an artificial satellite orbiting the earth at an altitude less than about 1500 kilometers

medium-earth-orbit (MEO) satellite   a satellite in orbit at a distance above the earth's surface of between approximately 8,000 and 20,000 km

multipath distortion   distortion of the information signal resulting from the difference in arrival time in signals arriving via multiple paths of different lengths

orbital satellite   any satellite that is not in a geostationary orbit

rake receiver   a radio receiver that is capable of combining several received signals with different time delays into one composite signal

sectorization   division of a cell into several sections all radiating outward from a cell site

spatial diversity   use of two or more physically separated antennas at one end of a communication link

● Questions

1. What are the similarities between radio waves and light waves?

2. What is meant by the characteristic impedance of a medium? What is the characteristic impedance of free space?

3. State the difference between power and power density. Explain why power density decreases with the square of the distance from a source.

4. A radio wave propagates in such a way that its magnetic field is parallel with the horizon. What is its polarization?

5. What is an isotropic radiator? Could such a radiator be built? Explain.

6. State three factors that determine the amount of power extracted from a wave by a receiving antenna.

7. The equation given for calculating path loss in decibels shows the loss increasing with frequency. Why is this?

8. State two undesirable effects that can be caused by reflections in line-of-sight communication and explain how they arise.

9. Why is the attenuation greater for mobile communication than for free space?

10. What is meant by space diversity? How can it be used to improve the reliability of a communication system?

11. Explain how spread-spectrum systems automatically take advantage of frequency diversity.

12. What is a rake receiver? How does it reduce the effects of multipath propagation?

13. What is fast fading and how is it caused?

14. Explain how cellular systems allow for frequency reuse.

15. What is meant by cell-splitting, and why is it done?

16. Why are all geostationary communication satellites at the same distance from the earth?

17. Why is circular polarization commonly used with VHF and UHF satellite communication?

18. Why are high-gain antennas needed with geostationary satellites?

19. Explain the difference between LEO, MEO, and geostationary satellites.

20. What advantages and disadvantages do LEO satellite systems have for mobile communication?

● Problems

1. Find the propagation velocity of radio waves in glass with a relative permittivity of 7.8.

2. Find the wavelength, in free space, for radio waves at each of the following frequencies:
   (a) 50 kHz
   (b) 1 MHz
   (c) 23 MHz
   (d) 300 MHz
   (e) 450 MHz
   (f) 12 GHz

3. An isotropic source radiates 100 W of power in free space. At a distance of 15 km from the source, calculate the power density and the electric field intensity.

4. A certain antenna has a gain of 7 dB with respect to an isotropic radiator.
   (a) What is its effective area if it operates at 200 MHz?
   (b) How much power would it absorb from a signal with a field strength of 50 μV/m?

5. Find the characteristic impedance of glass with a relative permittivity of 7.8.

6. A transmitter has an output power of 50 W. It is connected to its antenna by a feedline that is 25 meters long and properly matched. The loss in the feedline is 5 dB/100 m. The antenna has a gain of 8.5 dBi.
   (a) How much power reaches the antenna?
   (b) What is the EIRP in the direction of maximum antenna gain?
   (c) What is the power density 1 km from the antenna in the direction of maximum gain, assuming free space propagation?
   (d) What is the electric field strength at the same place as in (c)?

7. A satellite transmitter operates at 4 GHz with an antenna gain of 40 dBi. The receiver, 40,000 km away, has an antenna gain of 50 dBi. If the transmitter has a power of 8 W (ignoring feedline losses and mismatch), find:

(a) the EIRP in dBW

(b) the power delivered to the receiver

8. A paging system has a transmitting antenna located 50 m above average terrain. How far away could the signal be received by a pager carried 1.2 m above the ground?

9. A boat is equipped with a VHF marine radio, which it uses to communicate with other nearby boats and shore stations. If the antenna on the boat is 2.3 m above the water, calculate the maximum distance for communication with:

(a) another similar boat

(b) a shore station with an antenna on a tower 22 m above the water level

(c) another boat, but using the shore station as a repeater

10. A PCS signal at 1.9 GHz arrives at an antenna via two paths differing in length by 19 m.

(a) Calculate the difference in arrival time for the two paths.

(b) Calculate the phase difference between the two signals. (Hint: 360° is the same as 0° as far as phase is concerned, so multiples of 360° can be ignored.)

11. Use the mobile-propagation model given in Equation (7.18) to calculate the loss over a path of 5 km, with a base antenna 25 m above the ground, for a

(a) cellular telephone at 800 MHz

(b) PCS at 1900 MHz

12. How rapidly would the signal fade if a cell phone (at 800 MHz) is in a car moving at 100 km/h?

13. Suppose that cells have a 3 km radius and that the vehicle in the previous problem is moving directly across a cell. How long will it remain in that cell before it has to be handed off to the next cell?

14. How many hexagonal cells with a 3 km radius would be needed to cover an area of 600 km^2?

15. Two points 50 km apart on earth can communicate using direct radio link, with a LEO satellite of 500 km or a geostationary satellite with a path length to the satellite of 36,000 km. Calculate the one-way time

delay for each mode assuming that each satellite is directly over the center of the path.

16. For each situation in the preceding question, calculate the path loss at a frequency of 1 GHz. Assume free-space propagation.

17. A map indicates that a certain spot on the earth is in the 40 dBW contour of a satellite beam, that is, the EIRP from the satellite is 40 dBW in that direction. What is the actual signal strength at the earth's surface, in W/m^2, if the satellite is 37,000 km away?

Antennas

Objectives

After studying this chapter, you should be able to:

- Explain the basic principles of operation of antenna systems.
- Define radiation resistance and use it to calculate the efficiency of an antenna.
- Define antenna gain, beamwidth, and front-to-back ratio, and determine them from a plot of an antenna's radiation pattern.
- Convert between antenna gains given in dBi and dBd, and use either to calculate effective isotropic radiated power (EIRP) and effective radiated power (ERP) for an antenna-transmitter combination.
- Calculate the dimensions of simple practical antennas for a given frequency.
- Identify, explain the operation of, and sketch the approximate radiation patterns for common types of antennas and antenna arrays.
- Calculate the gain and beamwidth for selected antenna types.
- Explain the use of diversity and downtilt in cellular base station antennas.

8.1 Introduction

So far, we have considered the ways in which electromagnetic waves can propagate along transmission lines and through space. The antenna is the interface between these two media and is a very important part of the communication path. In this chapter we will study the basic operating principles of antennas and look at some of the parameters that describe their performance. Some representative examples of practical antennas used in wireless communication will be analyzed.

Before we begin, you should have two ideas firmly in mind. First, antennas are passive devices. Therefore, the power radiated by a transmitting antenna cannot be greater than the power entering from the transmitter. In fact, it is always less because of losses. We will speak of antenna gain, but remember that gain in one direction results from a concentration of power and is accompanied by a loss in other directions. Antennas achieve gain the same way a flashlight reflector increases the brightness of the bulb: by concentrating energy.

The second concept to keep in mind is that antennas are reciprocal devices; that is, the same design works equally well as a transmitting or a receiving antenna and in fact has the same gain. In wireless communication, often the same antenna is used for both transmission and reception.

Essentially, the task of a transmitting antenna is to convert the electrical energy travelling along a transmission line into electromagnetic waves in space. This process, while difficult to analyze in mathematical detail, should not be hard to visualize. The energy in the transmission line is contained in the electric field between the conductors and in the magnetic field surrounding them. All that is needed is to *launch* these fields (and the energy they contain) into space.

At the receiving antenna, the electric and magnetic fields in space cause current to flow in the conductors that make up the antenna. Some of the energy is thereby transferred from these fields to the transmission line connected to the receiving antenna, and thence to the receiver.

8.2 Simple Antennas

The simplest antenna, in terms of its radiation pattern, is the isotropic radiator that was introduced in Chapter 7. Recall that it has zero size, is perfectly efficient, and radiates power equally in all directions. Though merely a theoretical construct, the isotropic radiator makes a good reference with which to compare the gain and directionality of other antennas. That is because, even though this antenna cannot be built and tested, its characteristics are simple and easy to derive.

The *half-wave dipole antenna,* on the other hand, is a simple, practical antenna which is in common use. An understanding of the half-wave dipole is important both in its own right and as a basis for the study of more complex antennas. A half-wave dipole is sketched in Figure 8.1.

FIGURE 8.1
Half-wave dipole
antenna

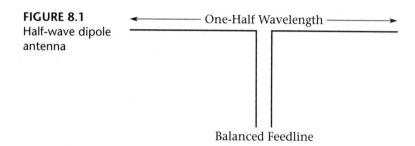

The word **dipole** simply means it has two parts, as shown. A dipole antenna does not have to be one-half wavelength in length like the one shown in the figure, but this length is handy for impedance matching, as we shall see. Actually, in practice its length should be slightly less than one-half the free-space wavelength to allow for capacitive effects. A half-wave dipole is sometimes called a Hertz antenna, though strictly speaking the term Hertzian dipole refers to a dipole of infinitesimal length. This, like the isotropic radiator, is a theoretical construct; it is used in the calculation of antenna radiation patterns.

Typically the length of a half-wave dipole, assuming that the conductor diameter is much less than the length of the antenna, is 95% of one-half the wavelength measured in free space. Recall that the free-space wavelength is given by

$$\lambda = \frac{c}{f}$$

where

λ = free-space wavelength in meters
c = 300×10^6 m/s
f = frequency in hertz

Therefore, the length of a half-wave dipole, in meters, is approximately

$$L = 0.95 \times 0.5 \frac{c}{f}$$

$$= 0.95 \times 0.5 \times \frac{300 \times 10^6}{f}$$

$$= \frac{142.5 \times 10^6}{f}$$

If the frequency is given in megahertz, as is the usual case, this equation becomes

$$L = \frac{142.5}{f} \qquad (8.1)$$

where

L = length of a half-wave dipole in meters

f = frequency in megahertz

For length measurements in feet, the equivalent equation is

$$L = \frac{468}{f} \qquad (8.2)$$

where

L = length of a half-wave dipole in feet

f = operating frequency in megahertz

EXAMPLE 8.1

Calculate the length of a half-wave dipole for an operating frequency of 200 MHz.

SOLUTION

From Equation (8.1),

$$L = \frac{142.5}{f}$$

$$= \frac{142.5}{200}$$

$$= 0.7125 \text{ m}$$

One way to think about the half-wave dipole is to consider an open-circuited length of parallel-wire transmission line, as shown in Figure 8.2(a). The line has a voltage maximum at the open end, a current maximum one-quarter wavelength from the end, and a very high standing-wave ratio. Now, suppose that the two conductors are separated at a point one-quarter wavelength from the end, as in Figure 8.2(b). The drawing shows how the electric field seems to stretch away from the wires. If the process continues, as in Figure 8.2(c), some of the field detaches itself from the antenna and helps to form electromagnetic waves that propagate through space.

FIGURE 8.2
Development of the
half-wave dipole

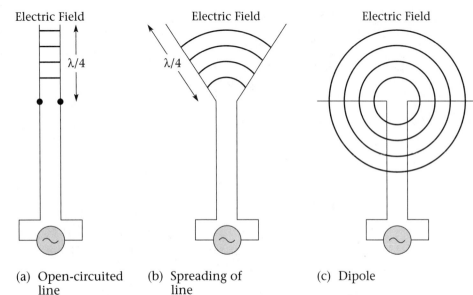

(a) Open-circuited
line

(b) Spreading of
line

(c) Dipole

Radiation Resistance

The radiation of energy from a dipole is quite apparent if we measure the impedance at the *feedpoint* in the center of the antenna. An open-circuited lossless transmission line, as described in Chapter 6, would look like a short circuit at a distance of one-quarter wavelength from the open end. At distances slightly greater than or less than one-quarter wavelength, the line would appear reactive. There would never be a resistive component to the feedpoint impedance, since an open-circuited line has no way of dissipating power.

A lossless half-wave dipole does not dissipate power either, but it does radiate power into space. The effect on the feedpoint impedance is the same as if a loss had taken place. Whether power is dissipated or radiated, it disappears from the antenna and therefore causes the input impedance to have a resistive component. The half-wave dipole looks like a resistance of about 70 Ω at its feedpoint.

The portion of an antenna's input impedance that is due to power radiated into space is known, appropriately, as the **radiation resistance**. It is important to understand that this resistance does not represent loss in the conductors that make up the antenna.

The ideal antenna just described radiates all the power supplied to it into space. A real antenna, of course, has losses in the conductor and therefore has an efficiency less than one. This efficiency can be defined as

$$\eta = \frac{P_r}{P_t} \tag{8.3}$$

where

P_r = radiated power

P_t = total power supplied to the antenna

Recalling that $P = I^2 R$, we have

$$\eta = \frac{I^2 R_r}{I^2 R_t}$$

$$= \frac{R_r}{R_t} \tag{8.4}$$

where

R_r = radiation resistance, as seen from the feedpoint

R_t = total resistance, as seen from the feedpoint

EXAMPLE 8.2

A dipole antenna has a radiation resistance of 67 Ω and a loss resistance of 5 Ω, measured at the feedpoint. Calculate the efficiency.

SOLUTION

From Equation (8.4),

$$\eta = \frac{R_r}{R_t}$$

$$= \frac{67}{67 + 5}$$

$$= 0.93 \text{ or } 93\%$$

Unlike the isotropic radiator, the half-wave dipole does not radiate uniformly in all directions. The field strength is at its maximum along a line at a right angle to the antenna and is zero off the ends of the antenna. In order to describe more precisely the radiation of this and other more complex antennas, we need some way of graphing radiation patterns. These techniques will be developed in the next section.

 8.3 # Antenna Characteristics

Now that two simple antennas have been described, it is already apparent that antennas differ in the amount of radiation they emit in various directions. This is a good time to introduce some terms to describe and quantify the directional characteristics of antennas and to demonstrate methods of graphing some of them. These terms will be applied to the isotropic and half-wave dipole antennas here and will also be applied to other antenna types as each is introduced.

Radiation Pattern The diagrams used in this book follow the three-dimensional coordinate system shown in Figure 8.3. As shown in the figure, the x–y plane is horizontal, and the angle ϕ is measured from the x axis in the direction of the y axis. The z axis is vertical and the angle θ is usually measured from the horizontal plane toward the zenith. This vertical angle, measured upward from the ground, is called the **angle of elevation**. Most work with antennas uses positive angles of elevation, but sometimes (as when the transmitting antenna is on a tall tower and the receiving antenna is close to it and much lower) we are interested in angles below the horizon. Different manufacturers handle below-horizon angles differently as shown in Figure 8.4 and described in the following paragraphs.

FIGURE 8.3
Three-dimensional
coordinate system

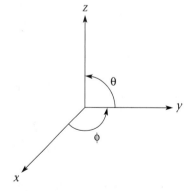

Figure 8.4 shows two ways in which the radiation pattern of a dipole can be represented. The three-dimensional picture in Figure 8.4(a) is useful in showing the general idea and in getting a feel for the characteristics of the antenna. The two views in Figures 8.4(b) and (c), on the other hand, are less intuitive but can be used to provide quantitative information about the

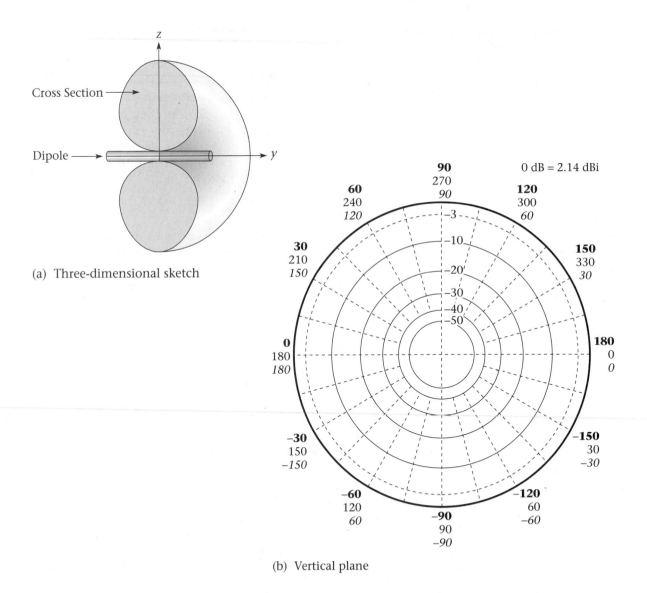

(a) Three-dimensional sketch

(b) Vertical plane

FIGURE 8.4 Radiation pattern of half-wave dipole

FIGURE 8.4
(*continued*)

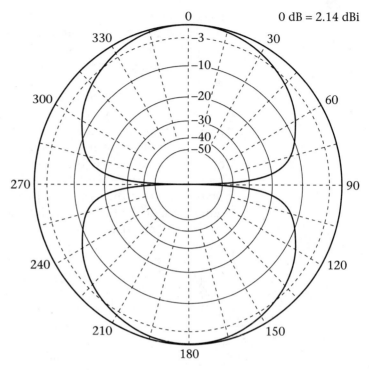

(c) Horizontal plane

performance of the antenna. It is the latter type of antenna pattern that is usually found in manufacturers' literature, and it is this type of depiction that will be used in this book.

Notice that polar graph paper is used in Figures 8.4(b) and (c). The angle is measured in one plane from a reference axis. Because the three-dimensional space around the antenna is being represented in two dimensions, at least two views are required to give the complete picture. It is important to choose the axes carefully to take advantage of whatever symmetry exists. Usually, the horizontal and vertical planes are used; most often the antenna is mounted either vertically or horizontally, so the axis of the antenna itself can be used as one of the reference axes.

The two graphs in Figures 8.4(b) and (c) can be thought of as slices of the three-dimensional pattern of Figure 8.4(a). By convention, the horizontal-plane diagram is oriented so that the direction of maximum radiation (or one of these directions, if the antenna is bidirectional like a dipole) is toward the top of the page, and the vertical-plane drawing, not surprisingly, has the zenith at the top. All manufacturers draw the horizontal scale in the same way, with 0° at the top, but practices differ for the vertical angle scale. The three numbering methods shown on Figure 8.4(b) are all in use by various manufacturers. No confusion should result if you simply remember that the top of the diagram represents the zenith ("up" in the drawing represents "up" in real life). Sometimes the radiation plot for angles below the horizon is omitted.

In Figure 8.4(a), the dipole itself is drawn to help visualize the antenna orientation. This is not done in the polar plots, however. These radiation patterns are valid only in the **far-field region**; an observer must be far enough away from the antenna that any local capacitive or inductive coupling is negligible. In practice this means a distance of at least several wavelengths, and generally an actual receiver is at a much greater distance than that. From this distance the antenna would be more accurately represented as a dot in the center of the graph. The space close to the antenna is called the **near-field region** and does not have the same directional characteristics.

The distance from the center of the graph represents the strength of the radiation in a given direction. The scale is usually in decibels with respect to some reference. Often the reference is an isotropic radiator, as in Figure 8.4. Note that the farthest point on the graph from the center is at 2.14 dBi; that is, the gain of a lossless dipole, in its direction of maximum radiation, is 2.14 decibels with respect to an isotropic radiator. Usually, but not always, the pattern is drawn so that the point of maximum radiation is at the outside of the chart and the reference level for that point is stated, as shown in the figure.

The half-wave dipole itself is sometimes used as a reference. In that case, the gain of an antenna may be expressed in decibels with respect to a half-wave dipole, or dBd for short. Since the gain of a dipole is 2.14 dBi, the gain of any antenna in dBd is 2.14 dB less than the gain of the same antenna expressed in dBi. That is,

$$G(\text{dBd}) = G(\text{dBi}) - 2.14 \text{ dB} \tag{8.5}$$

where

$G(\text{dBd})$ = gain of antenna in decibels with respect to a half-wave dipole

$G(\text{dBi})$ = gain of the same antenna in decibels with respect to an isotropic radiator

Obviously, when comparing antennas, it is important to know which reference antenna was used in gain calculations.

EXAMPLE 8.3 ▽

Two antennas have gains of 5.3 dBi and 4.5 dBd, respectively. Which has greater gain?

SOLUTION

Convert both gains to the same standard. In this case, let us use dBi. Then, for the second antenna,

$$G = 4.5 \text{ dBd}$$
$$= 4.5 + 2.14 \text{ dBi}$$
$$= 6.64 \text{ dBi}$$

Therefore, the second antenna has higher gain.

Gain and Directivity

The sense in which a half-wave dipole antenna can be said to have gain can be seen from Figure 8.5. This sketch shows the pattern of a dipole, from Figure 8.4(c), superimposed on that of an isotropic radiator. It can be seen that while the dipole has a gain of 2.14 dBi in certain directions, in others its gain is negative. If the antennas were to be enclosed by a sphere that would absorb all the radiated power, the total radiated power would be found to be the same for both antennas. Remember that for antennas, power gain in one direction is at the expense of losses in others.

FIGURE 8.5
Isotropic and dipole antennas

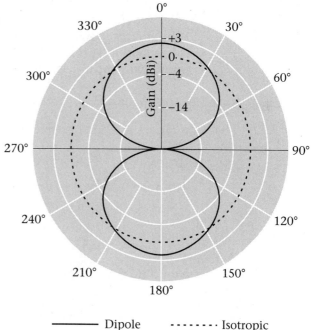

Sometimes the term **directivity** is used. This is not quite the same as gain. Directivity is the gain calculated assuming a lossless antenna. Real antennas have losses, and gain is simply the directivity multiplied by the efficiency of the antenna, that is:

$$G = D\eta \qquad\qquad (8.6)$$

where

D = directivity, as a ratio (not in dB)

G = gain, as a ratio (not in dB)

η = efficiency

When an antenna is used for transmitting, the total power emitted by the antenna is less than that delivered to it by the feedline. Earlier we defined the efficiency as

$$\eta = \frac{P_r}{P_t}$$

where

η = antenna efficiency

P_r = radiated power

P_t = power supplied to the antenna

The figure of 2.14 dBi we have been using for the gain of a lossless dipole is also the directivity for *any* dipole. To find the gain of a real (lossy) dipole, it is necessary first to convert the decibel directivity to a power ratio, then to multiply by the efficiency.

EXAMPLE 8.4

A dipole antenna has an efficiency of 85%. Calculate its gain in decibels.

SOLUTION

The directivity of 2.14 dBi can be converted to a power ratio:

$$D = \log^{-1}\frac{2.14}{10}$$
$$= 1.638$$

Now, find the gain:

$$G = D\eta$$
$$= 1.638 \times 0.85$$
$$= 1.39$$

This gain can easily be converted to dBi:

$$G(\text{dBi}) = 10 \log 1.39$$
$$= 1.43 \text{ dBi}$$

Beamwidth

Just as a flashlight emits a beam of light, a directional antenna can be said to emit a beam of radiation in one or more directions. The width of this beam is defined as the angle between its half-power points. These are also the points at which the power density is 3 dB less than it is at its maximum point. An inspection of Figures 8.4(b) and (c) will show that the half-wave dipole has a **beamwidth** of about 78° in one plane and 360° in the other. Many antennas are much more directional than this, with a narrow beam in both planes.

Front-to-Back Ratio

As you might expect, the direction of maximum radiation in the horizontal plane is considered to be the *front* of the antenna, and the *back* is the direction 180° from the front. For a dipole, the front and back have the same radiation, but this is not always the case. Consider the unidirectional antenna shown in Figure 8.6: there is a good deal more radiation from the front of

FIGURE 8.6
Unidirectional antenna

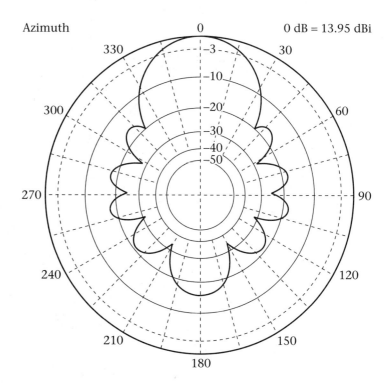

this antenna than from the back. The ratio between the gains to the front and back is the *front-to-back ratio*. It is generally expressed in dB, in which case it can be found by subtracting the gains in dBi or dBd.

Major and Minor Lobes   The dipole antenna of Figure 8.4 viewed in the horizontal plane has two equal lobes of radiation. The more complex antenna of Figure 8.6, however, has one *major lobe* and a number of minor ones. Each of these lobes has a gain and beamwidth which can be found separately using the diagram.

EXAMPLE 8.5

For the antenna pattern sketched in Figure 8.6, find:

(a) the antenna gain in dBi and dBd

(b) the front-to-back ratio in dB

(c) the beamwidth for the major lobe

(d) the angle, gain and beamwidth for the most important minor lobe

SOLUTION

(a) Since the major lobe just reaches the outer ring of the chart, which is specified as 13.95 dBi, that is the antenna gain. Antenna plots are usually, but not always, done this way. The gain of a dipole is 2.14 dBi so the gain of this antenna with respect to a dipole is:

$$G(dBd) = G(dBi) - 2.14 \text{ dB}$$
$$= 13.95 \text{ dBi} - 2.14 \text{ dB}$$
$$= 11.81 \text{ dBd}$$

(b) Since the major lobe reaches 0 dB on the chart, all we have to do is look at the decibel reading for the point 180° from the first point and change the sign. Thus the front-to-back ratio is about 15 dB.

(c) The beamwidth is the angle between the two points at which the major lobe is 3 dB down from its maximum. The 3 dB down point is shown as a dashed circle. From this we can see that the 3 dB beamwidth is about 44°.

(d) The most important minor lobe is to the rear. We discovered in part (b) that its gain is 15 dB less than for the major lobe. Therefore the gain for this lobe is 13.95 − 15 = −1.05 dB. Its beamwidth is about 20°.

Effective Isotropic Radiated Power and Effective Radiated Power

In a practical situation we are usually more interested in the power emitted in a particular direction than in the total radiated power. Looking from a distance, it is impossible to tell the difference between a high-powered transmitter using an isotropic antenna and a transmitter of lower power working into an antenna with gain. In Chapter 7 we defined effective isotropic radiated power (EIRP), which is simply the actual power going into the antenna multiplied by its gain with respect to an isotropic radiator.

$$EIRP = P_t G_t \tag{8.7}$$

Another similar term that is in common use is **effective radiated power (ERP)**, which represents the power input multiplied by the antenna gain measured with respect to a half-wave dipole. Since an ideal half-wave dipole has a gain of 2.14 dBi, the EIRP is 2.14 dB greater than the ERP for the same antenna-transmitter combination. That is,

$$EIRP = ERP + 2.14 \text{ dB} \tag{8.8}$$

where

$EIRP$ = effective isotropic radiated power for a given transmitter and antenna

ERP = effective radiated power for the same transmitter and antenna

The path loss equations in Chapter 7 require EIRP, but they can easily be used with ERP values. Simply add 2.14 dB to any ERP value to convert it to EIRP. Convert the power to dBm or dBW first, if it is not already expressed in such units.

EXAMPLE 8.6

The ERP of a transmitting station is specified as 17 W in a given direction. Express this as an EIRP in dBm so that it can be used with the path loss equations in Chapter 7.

SOLUTION

First convert the ERP to dBm, then add 2.14 dB.

$$ERP(\text{dBm}) = 10 \log \frac{ERP}{1 \text{ mW}}$$

$$= 10 \log (17 \times 10^3)$$
$$= 42.3 \text{ dBm}$$

$$EIRP(\text{dBm}) = ERP(\text{dBm}) + 2.14 \text{ dB}$$
$$= 42.3 + 2.14$$
$$= 44.44 \text{ dBm}$$

Impedance   The radiation resistance of a half-wave dipole situated in free space and fed at the center is approximately 70 Ω. The impedance is completely resistive at resonance, which occurs when the length of the antenna is about 95% of the calculated free-space half-wavelength value. The exact length depends on the diameter of the antenna conductor relative to the wavelength. If the frequency is above resonance, the feedpoint impedance has an inductive component; if the frequency is lower than resonance, the antenna impedance is capacitive. Another way of saying the same thing is that an antenna that is too short appears capacitive, while one that is too long is inductive. Figure 8.7 shows graphically how reactance varies with frequency.

FIGURE 8.7
Variation of dipole
reactance with
frequency

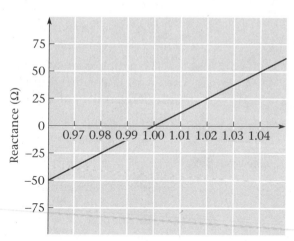

Operating Frequency ÷ Resonant Frequency

A center-fed dipole is a balanced device, and should be used with a balanced feedline. If coaxial cable is used, a balun (balanced-to-unbalanced) transformer should be connected between the cable and the antenna. Balun transformers are described in Appendix A.

A half-wave dipole does not have to be fed at its midpoint. It is possible to feed the antenna at some distance from the center in both directions, as shown in Figure 8.8. This is called a delta match and allows the impedance

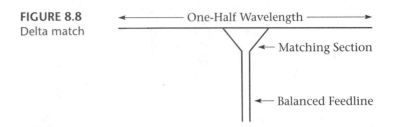

FIGURE 8.8
Delta match

to be adjusted to match a transmission line. The greater the separation between the connections to the antenna, the higher the impedance. Just as for a center-fed antenna, the impedance is resistive only if the antenna length is accurate.

Polarization   Recall from Chapter 7 that the polarization of a radio wave is the orientation of its electric field vector. The polarization of the radiation from a half-wave dipole is easy to determine: it is the same as the axis of the wire. That is, a horizontal antenna produces horizontally polarized waves, and a vertical antenna gives vertical polarization.

It is important that the polarization be the same at both ends of a communication path. Wireless communication systems usually use vertical polarization because this is more convenient for use with portable and mobile antennas.

8.4 Other Simple Antennas

The half-wave dipole is simple, useful, and very common, but it is by no means the only type of antenna in use. In this section some other simple antennas will be introduced. Later, ways of combining antenna elements into arrays with specific characteristics will be examined.

Folded Dipole   Figure 8.9 shows a folded dipole. It is the same length as a standard half-wave dipole, but it is made with two parallel conductors, joined at both ends and separated by a distance that is short compared with the length of the antenna. One of the conductors is broken in the center and connected to a balanced feedline.

The folded dipole differs in two ways from the ordinary half-wave dipole described above. It has a wider *bandwidth,* that is, the range of frequencies within which its impedance remains approximately resistive is larger than

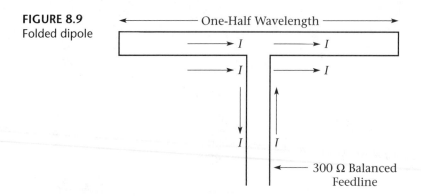

FIGURE 8.9
Folded dipole

for the single-conductor dipole. For this reason, it is often used—alone or with other elements—for television and FM broadcast receiving antennas. It also has approximately four times the feedpoint impedance of an ordinary dipole. This accounts for the extensive use of 300-Ω balanced line, known as twinlead, in TV and FM receiving installations.

It is easy to see why the impedance of a folded dipole is higher than that of a standard dipole. First, suppose that a voltage V and a current I are applied to an ordinary half-wave dipole. Then, at resonance when the feedpoint impedance is resistive, the power supplied is

$P = VI$

Now consider the folded dipole of Figure 8.9. Looking at the center of the dipole, the length of the path from the center of the lower conductor to the center of the upper conductor is one-half wavelength. Therefore, the currents in the two conductors will be equal in magnitude. If the points were one-half wavelength apart on a straight transmission line we would say that they were equal in magnitude but out of phase. Here, however, because the wire has been folded, the two currents, flowing in opposite directions with respect to the wire, actually flow in the same direction in space and contribute equally to the radiation from the antenna.

If a folded dipole and a regular dipole radiate the same amount of power, the current in each must be the same. However, the current at the feedpoint of a folded dipole is only one-half the total current. If the feedpoint current is reduced by one-half, yet the power remains the same, the feedpoint voltage must be doubled. That is,

$P = VI$

$$= 2V\left(\frac{I}{2}\right)$$

Assuming equal power is provided to both antennas, the feedpoint voltage must be twice as great for the folded dipole. The resistance at the feedpoint of the ordinary dipole is, by Ohm's Law,

$$R = \frac{V}{I}$$

For the folded dipole, it will be

$$R' = \frac{2V}{I/2}$$

$$= \frac{4V}{I}$$

$$= 4R$$

Since the current has been divided by two and the voltage multiplied by two, the folded dipole has four times the feedpoint impedance as the regular version.

It is also possible to build folded dipoles with different-sized conductors, and with more than two conductors. In this way, a wide variety of feedpoint impedances can be produced.

Monopole Antenna Many wireless applications require antennas on vehicles. The directional effects of a horizontal dipole would be undesirable. A vertical dipole is possible, but awkward to feed in the center and rather long at some frequencies. Similar results can be obtained by using a vertical quarter-wave **monopole** antenna. It is mounted on a ground plane, which can be the actual ground or an artificial ground such as the body of a vehicle. The monopole is fed at the lower end with coaxial cable. The ground conductor of the feedline is connected to the ground plane. See Figure 8.10(a).

The radiation pattern of a quarter-wave monopole in the vertical plane has the same shape as that of a vertical half-wave dipole in free space. Only half the pattern is present, however, since there is no underground radiation. In the horizontal plane, of course, a vertical monopole is omnidirectional. Since, assuming no losses, all of the power is radiated into one-half the pattern of a dipole, this antenna has a power gain of two (or 3 dB) over a dipole in free space. See Figures 8.10(b) and (c) for the radiation patterns for a monopole.

The input impedance at the base of a quarter-wave monopole is one-half that of a dipole. This can be explained as follows: with the same current, the antenna produces one-half the radiation pattern of a dipole, and therefore one-half the radiated power. The radiated power is given by

$$P_r = I^2 R_r$$

where

P_r = radiated power
I = antenna current at the feedpoint
R_r = radiation resistance measured at the feedpoint

If the radiated power decreases by a factor of two for a given current, then so must the feedpoint radiation resistance.

In some mobile and portable applications a quarter wavelength is too long to be convenient. In that case, the electrical length of the antenna can be increased by adding inductance to the antenna. This can be done at the

FIGURE 8.10
Monopole antenna

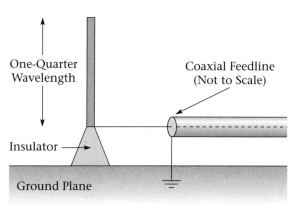

(a) Antenna

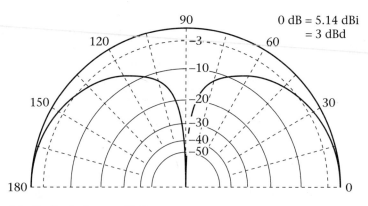

(b) Vertical plane radiation

FIGURE 8.10
(*continued*)

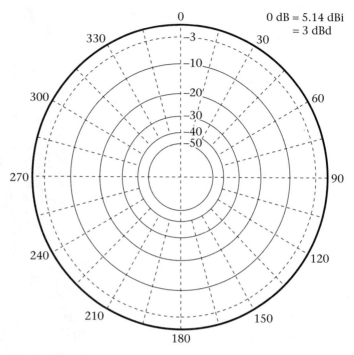

0 dB = 5.14 dBi
= 3 dBd

(c) Horizontal plane radiation

base or at the center, or the whole antenna can be coiled. The "rubber duckie" antennas on many handheld transceivers use this technique. Inductors used to increase the effective length of antennas are called *loading coils*. Figure 8.11 shows the three types of inductive loading just described.

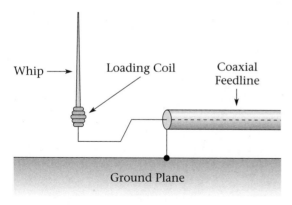

(a) Base loading

(b) Center loading

FIGURE 8.11   Inductive loading

FIGURE 8.11
(*continued*)

(c) Helical "rubber duckie" antenna

The Five-Eighths Wavelength Antenna

This antenna is often used vertically as either a mobile or base antenna in VHF and UHF systems. Like the quarter-wave monopole, it has omni-directional response in the horizontal plane. However, the radiation is concentrated at a lower angle, resulting in gain in the horizontal direction, which is often useful. In addition, it has a higher feedpoint imped-ance and therefore does not require as good a ground, because the current at the feedpoint is less. The impedance is typically lowered to match that of a 50-Ω feedline by the use of an impedance-matching section. The circu-lar section at the base of the antenna in the photograph in Figure 8.12(a) is an impedance-matching device. Figure 8.12(a) shows a 5/8 wave-length antenna, and Figure 8.12(b) shows its radiation pattern in the vertical plane.

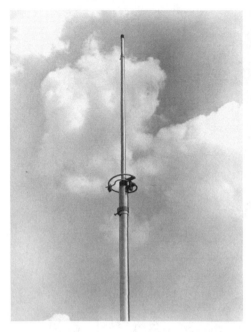

(a) Photograph of 5/8 wavelength
    antenna

(Courtesy of Cushcraft Corporation)

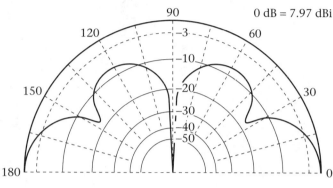

(b) Vertical radiation pattern

FIGURE 8.12   Five-Eighths wavelength antenna

The Discone Antenna

The rather unusual-looking antenna shown in Figure 8.13 is known appropriately as the *discone*. It is characterized by very wide bandwidth, covering approximately a ten-to-one frequency range, and an omnidirectional pattern in the horizontal plane. The signal is vertically polarized and the gain is comparable to that of a dipole. The feedpoint impedance is approximately 50 Ω; the feedpoint is located at the intersection of the disk and the cone. The disk-cone combination acts as a transformer to match the feedline impedance to the impedance of free space, which is 377 Ω. Typically the length measured along the surface of the cone is about one-quarter wavelength at the lowest operating frequency.

The wide bandwidth of the discone makes it a very popular antenna for general reception in the VHF and UHF ranges. It is a favorite for use with *scanners*. These receivers can tune automatically to a large number of channels in succession and are often used for monitoring emergency services. The discone can be used for transmitting but seldom is. Most transmitting stations operate at one frequency or over a narrow band of frequencies.

FIGURE 8.13
Discone antenna
(Courtesy of Tandy
Corporation)

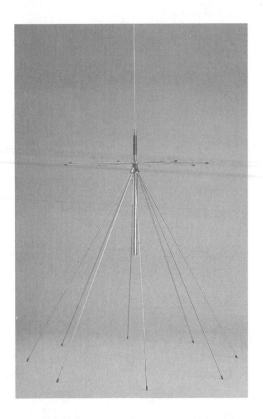

Simpler antennas with equivalent performance or equally elaborate antennas with better performance, are available when wide bandwidth is not required.

Helical Antennas

A helical antenna is a spiral, usually several wavelengths long. Such an antenna is shown in Figure 8.14. Typically the circumference of each turn is about one wavelength and the turns are about a one-quarter wavelength apart.

FIGURE 8.14
Helical antenna

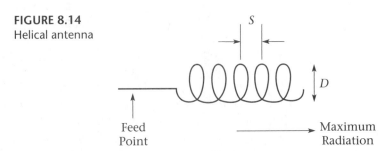

Helical antennas produce circularly polarized waves whose sense is the same as that of the helix. A helical antenna can be used to receive circularly polarized waves with the same sense and can also receive plane-polarized waves with the polarization in any direction. Helical antennas are often used with VHF and UHF satellite transmissions. Since they respond to any polarization angle, they avoid the problem of Faraday rotation, which makes the polarization of waves received from a satellite impossible to predict.

The gain of a helical antenna is proportional to the number of turns. An approximate expression for the gain, as a power ratio with respect to an isotropic radiator is:

$$G = \frac{15NS(\pi D)^2}{\lambda^3} \tag{8.8}$$

where

G = gain (as a ratio, not in dB), with respect to an isotropic radiator
N = number of turns in the helix, $N > 3$
S = turn spacing in meters, $S \cong \lambda/4$
D = diameter of the helix in meters, $D \cong \lambda/\pi$
λ = wavelength in meters

The radiation pattern for this antenna has one major lobe and several minor lobes. For the major lobe, the 3 dB beamwidth (in degrees) is approximately

$$\theta = \frac{52\lambda}{\pi D} \sqrt{\frac{\lambda}{NS}} \tag{8.9}$$

EXAMPLE 8.7

A helical antenna with 8 turns is to be constructed for a frequency of 1.2 GHz.

(a) Calculate the optimum diameter and spacing for the antenna and find the total length of the antenna.

(b) Calculate the antenna gain in dBi.

(c) Calculate the beamwidth.

SOLUTION

(a) $\quad \lambda = \dfrac{c}{f}$

$$= \dfrac{300 \times 10^6}{1200 \times 10^6}$$

$$= 0.25 \text{ m}$$

$$D = \dfrac{\lambda}{\pi}$$

$$= \dfrac{0.25}{\pi}$$

$$= 0.08 \text{ m}$$

$$= 80 \text{ mm}$$

$$S = \dfrac{\lambda}{4}$$

$$= \dfrac{0.25}{4}$$

$$= .0625 \text{ m}$$

$$= 62.5 \text{ mm}$$

The length is just the number of turns multiplied by the turn spacing:

$$L = NS$$
$$= 8 \times 62.5 \text{ mm}$$
$$= 500 \text{ mm}$$

(b) $\quad G = \dfrac{15NS(\pi D)^2}{\lambda^3}$

$$= \dfrac{15 \times 8 \times 0.0625(\pi \times 0.08)^2}{0.25^3}$$

$$= 30.3$$

$$= 14.8 \text{ dBi}$$

(c) $$\theta = \frac{52\lambda}{\pi D}\sqrt{\frac{\lambda}{NS}}$$

$$= \frac{52 \times 0.25}{\pi \times 0.08}\sqrt{\frac{0.25}{8 \times 0.0625}}$$

$$= 36.6°$$

Slot Antenna

Figure 8.15 shows a slot antenna, which is actually just a hole in a waveguide. The length of the slot is generally one-half wavelength. Its radiation pattern and gain are similar to those of a dipole with a plane reflector behind it. It therefore has much less gain than, for instance, a horn antenna. It is seldom used alone but is usually combined with many other slots to make a phased array (see the next section).

FIGURE 8.15
Slot antenna

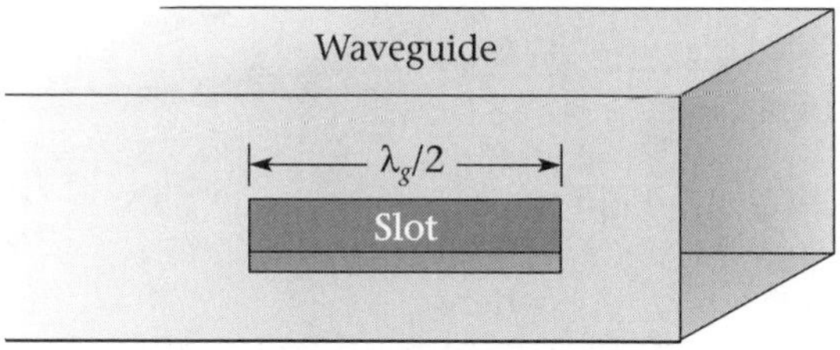

Horn Antenna

Horn antennas, like those shown in Figure 8.16, can be viewed as impedance transformers that match waveguide impedances to that of free space. The examples in the figure represent the most common types. The E- and H-plane sectoral horns are named for the plane in which the horn flares; the pyramidal horn flares in both planes. The conical horn is most appropriate with circular waveguide.

The gain and directivity of horn antennas depend on the type of horn and its dimensions. Let us examine the pyramidal horn, since that is the most common type. Its gain is proportional to both of the aperture dimensions, shown on the diagram as d_E (for the E-plane) and d_H (for the H-plane). The flare angle is limited by impedance-matching considerations, so high gain requires a long, unwieldy horn. The equation for gain is

$$G = \frac{7.5 d_E d_H}{\lambda^2} \tag{8.10}$$

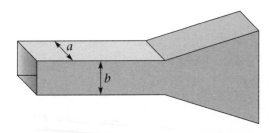

(a) E-plane

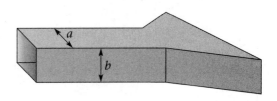

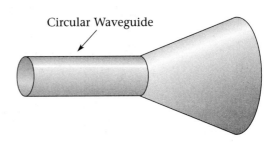

(b) H-plane

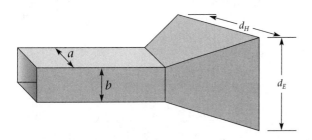

(c) Pyramidal

(d) Conical

FIGURE 8.16 Horn antennas

where

> G = gain as a power ratio, with respect to an isotropic radiator
> d_E = E-plane aperture
> d_H = H-plane aperture
> λ = wavelength

The beamwidth is different in the two directions. In the H-plane it is

$$\theta_H = \frac{70\lambda}{d_H} \tag{8.11}$$

where

> θ_H = H-plane beamwidth in degrees
> λ = wavelength
> d_H = H-plane aperture

and in the E-plane it is

$$\theta_E = \frac{56\lambda}{d_E} \tag{8.12}$$

where

θ_E = E-plane beamwidth in degrees

λ = wavelength

d_E = E-plane aperture

For practical horns, the gain is often in the vicinity of 20 dBi with a beamwidth of about 25°. The bandwidth is about the same as that of the associated waveguide. That is, it works over a frequency range of approximately 2:1.

EXAMPLE 8.8 ▼

A pyramidal horn has an aperture (opening) of 58 mm in the E plane and 78 mm in the H plane. It operates at 10 GHz. Calculate:

(a) its gain in dBi

(b) the beamwidth in the H plane

(c) the beamwidth in the E plane

SOLUTION

(a) At 10 GHz the wavelength is:

$$\lambda = \frac{c}{f}$$

$$= \frac{300 \times 10^6}{10 \times 10^9}$$

$$= 0.03 \text{ m}$$

From Equation (8.8),

$$G = \frac{7.5 d_E d_H}{\lambda^2}$$

$$= \frac{7.5 \times 0.058 \times 0.078}{0.03^2}$$

$$= 37.7$$

$$= 15.8 \text{ dBi}$$

(b) From Equation (8.9),

$$\theta_H = \frac{70\lambda}{d_H}$$

$$= \frac{70 \times 0.03}{0.078}$$

$$= 26.9°$$

(c) From Equation (8.10),

$$\theta_E = \frac{56\lambda}{d_E}$$

$$= \frac{56 \times 0.03}{0.058}$$

$$= 29°$$

As well as being used with a parabolic reflector, the horn antenna can be and often is used alone as a simple, rugged antenna with moderate gain.

Patch Antenna   A patch antenna consists of a thin metallic patch placed a small fraction of a wavelength above a conducting ground plane. The patch and ground plane are separated by a dielectric. Generally, a piece of double-sided circuit board (with all the copper coating left on one side for the ground plane and the patch photo-etched on the other) is used. For best results a low-loss circuit board material should be used. Patch antennas are low in cost, compact at UHF and microwave frequencies, and have gain on the order of 6 dBi.

The patch conductor can have any shape, but simple geometries are most common; this simplifies the design and analysis of the antenna. Most of the radiation is in the direction perpendicular to the plane of the antenna and on the patch side of the antenna. The ground plane prevents the patch from radiating very much in the opposite direction.

The half-wave rectangular patch is the most commonly used microstrip antenna. It is characterized by its length L, width W and thickness d, as shown in Figure 8.17. The length is approximately equal to one-half wavelength in the dielectric; the width is not critical but the antenna is often made square. Radiation is from fringing fields between the patch and the ground plane. This works best with a relatively thick substrate (a few millimeters) having a fairly low relative permittivity, normally between 1 and 4.

FIGURE 8.17
Patch antenna

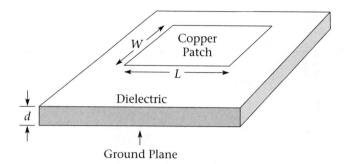

The patch antenna can be fed by a coaxial cable. The inner conductor of the coaxial line (sometimes referred to as a *probe*) is connected to the radiating patch, while the outer conductor is connected to the ground-plane. An impedance match for the cable used is obtained by moving the connection point: the impedance is high (several hundred ohms) at the edge, declining to zero at the center. Another feed method is use of a microstrip line, photo-etched on the same substrate as the antenna and connected to its edge.

EXAMPLE 8.9

Calculate the approximate dimensions for a square patch antenna, for a frequency of 2 GHz, on a substrate with a relative permittivity of 2.

SOLUTION

Recall that wavelength is given by

$$\lambda = \frac{v_p}{f} \tag{8.13}$$

where

λ = wavelength in m
v_p = propagation velocity in m/s
f = frequency in Hz

In a dielectric, the propagation velocity is

$$v_p = \frac{c}{\sqrt{\varepsilon_r}} \tag{8.14}$$

where

c = 300×10^6 m/s
ε_r = relative permittivity

Substituting Equation (8.14) into Equation (8.13) we find

$$\lambda = \frac{c}{f\sqrt{\varepsilon_r}}$$

$$= \frac{300 \times 10^6}{2 \times 10^9 \sqrt{2}}$$

$$= 0.106 \text{ m}$$

The antenna width and length are each approximately half of this, or 50 mm.

8.5 Antenna Arrays

The simple elements described above can be combined to build a more elaborate antenna. The radiation from the individual **elements** will combine, resulting in reinforcement in some directions and cancellation in others to give greater gain and better directional characteristics. For instance, it is often desirable to have high gain in only one direction, something that is not possible with the simple antennas previously described. In mobile and portable applications, often an omnidirectional pattern in the vertical plane is wanted, but radiation upward and downward from the antenna is of no use. In both of these situations, a properly designed array could redirect the unwanted radiation in more useful directions.

Arrays can be classified as *broadside* or *end-fire,* according to their direction of maximum radiation. If the maximum radiation is along the main axis of the antenna (which may or may not coincide with the axis of its individual elements), the antenna is an end-fire array. If the maximum radiation is at right angles to this axis, the array has a broadside configuration. Figure 8.18 shows the axes of each type.

Antenna arrays can also be classified according to the way in which the elements are connected. A phased array has all its elements connected to the feedline. There may be phase-shifting, power-splitting, and impedance-matching arrangements for individual elements, but all receive power from the feedline (assuming a transmitting antenna). Since the transmitter can be said to *drive* each element by supplying power, these are also called *driven arrays.* On the other hand, in some arrays only one element is connected to the feedline. The others work by absorbing and reradiating power radiated from the driven element. These are called *parasitic elements,* and the antennas are known as *parasitic arrays.*

FIGURE 8.18
Broadside and
end-fire arrays

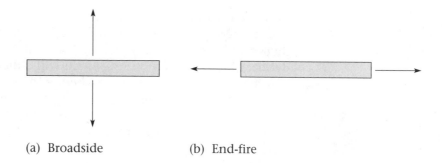

(a) Broadside   (b) End-fire

The following sections describe several common types of arrays that are useful in wireless communication, and give some general characteristics of each. Detailed analysis of antenna arrays is usually done with the aid of a computer.

Phased Arrays   Phased arrays can be made by connecting together any of the simple antenna types already discussed. Depending on the geometry of the array and the phase and current relationships between the elements, the array can be either broadside or end-fire.

Collinear Array   Figure 8.19 shows one type of broadside array using half-wave dipoles. This is called a *collinear array* because the axes of the elements are all along the same line.

Suppose that the collinear antenna in the figure is used for transmitting, and imagine a receiving antenna placed along the main axis of the antenna.

FIGURE 8.19
Collinear array

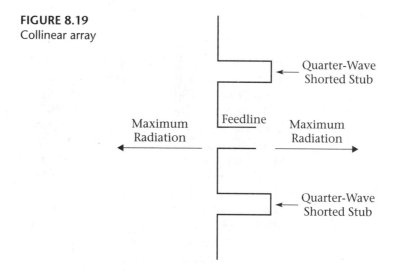

None of the individual elements radiate any energy in this direction, so of course there will be no signal from the array in this direction. Now move the hypothetical receiving antenna to a point straight out to one side of the antenna. All of the individual elements will radiate a signal in this direction, but it is necessary to determine whether these signals add constructively or destructively. If the elements are in phase, the signals will add. A quick check shows that this is indeed the case. The half-wave dipoles are linked by quarter-wave transmission-line sections that provide a phase reversal between adjacent ends. Therefore, all the dipoles will be in phase.

Collinear antennas are often mounted with the main axis vertical. They are then omnidirectional in the horizontal plane but have a narrow angle of radiation in the vertical plane. Thus, they make good base-station antennas for mobile radio systems. Many cellular radio and PCS base stations use collinear antennas.

Broadside Array Although only one of many broadside arrays (the collinear array above also qualifies, for example), the configuration of dipoles shown in Figure 8.20 is often given that name.

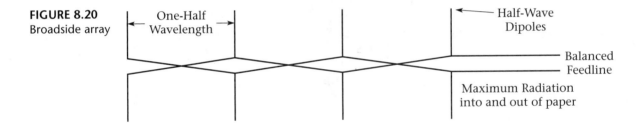

FIGURE 8.20
Broadside array

One-Half Wavelength

Half-Wave Dipoles

Balanced Feedline

Maximum Radiation into and out of paper

This time the elements are not collinear, but they are still in phase. That may not be obvious because of the crossed lines connecting them, but notice that the separation between the elements is one-half wavelength. This would cause a 180° phase shift as the signal travels along the feedline from one element to the next. This phase shift is cancelled by the crossing of the transmission line section that joins the dipoles.

Although this antenna, like the previous one, is a broadside array, its pattern is not identical. There is no radiation off the end of any of the elements, so of course there is no radiation in the equivalent direction from the array. If this antenna were erected with its main axis vertical, it would *not* be omnidirectional in the horizontal plane. In addition to having a narrow vertical pattern, it would have a bidirectional pattern in the horizontal plane. As for radiation off the end of the antenna, each dipole provides a signal in this direction, but the signals from adjacent elements cancel due to the one-half wavelength difference in path lengths between adjacent elements. Figure 8.21 shows a comparison of the radiation patterns in the horizontal plane for the two types of antennas.

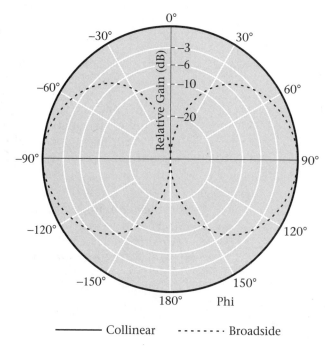

FIGURE 8.21
Collinear-broadside comparison

——— Collinear   ┄┄┄┄┄ Broadside

Still another difference between these two broadside arrays is in their polarization. The polarization for a dipole is in the plane of the antenna: that is, a horizontal dipole has horizontal polarization, and a vertical dipole has vertical polarization. If each of the two dipole combinations just mentioned has its main axis vertical, then the collinear array has vertical polarization and the broadside array has horizontal polarization. Either can be useful, but remember that the polarization must be the same at both ends of a radio link. If one end is a portable or mobile unit, then vertical polarization is the usual choice.

End-fire Array   Figure 8.22 shows an end-fire array using dipoles. It is identical to the broadside array in Figure 8.20 except that the feedline between elements is not crossed this time. This causes alternate elements to be 180° out of phase. Therefore, the radiation from one element cancels that from the next in the broadside direction. Off the end of the antenna, however, the radiation from all the elements adds, since the 180° phase shift between adjacent elements is cancelled by the one-half wavelength physical separation

FIGURE 8.22
End-fire array

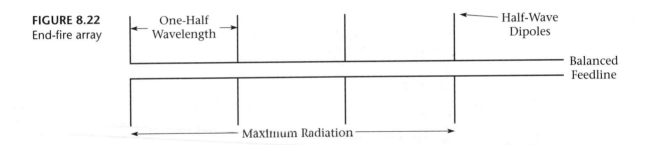

between them. Once again the polarization of the array is determined by the orientation of its constituent elements; this antenna is normally mounted with its main axis horizontal for radiation in one direction. It can have either horizontal or vertical polarization depending on whether it is set up with the dipoles horizontal or vertical.

Turnstile Antenna arrays are not always designed to give directionality and gain. The turnstile array illustrated in Figure 8.23(a) is a simple combination of two dipoles designed to give omnidirectional performance in the horizontal plane, with horizontal polarization. The dipoles are fed 90° out of phase.

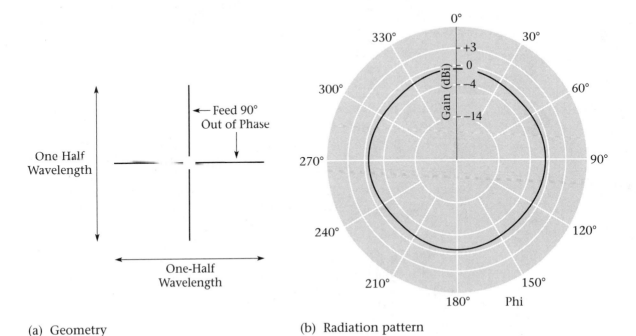

(a) Geometry

(b) Radiation pattern

FIGURE 8.23 Turnstile antenna

The gain of a turnstile antenna is actually about 3 dB less than that of a single dipole in its direction of maximum radiation, because each of the elements of the turnstile receives only one-half the transmitter power. Figure 8.23(b) shows the radiation pattern for a typical turnstile antenna.

Turnstile antennas are often used for FM broadcast reception, where they give reasonable performance in all directions without the need for a rotor.

Log-Periodic Dipole Array   The log-periodic array derives its name from the fact that the feedpoint impedance is a periodic function of the operating frequency. Although log-periodic antennas take many forms, perhaps the simplest is the dipole array, illustrated in Figure 8.24. The log-periodic dipole array is the most popular antenna for television reception.

FIGURE 8.24
Log-periodic
dipole array

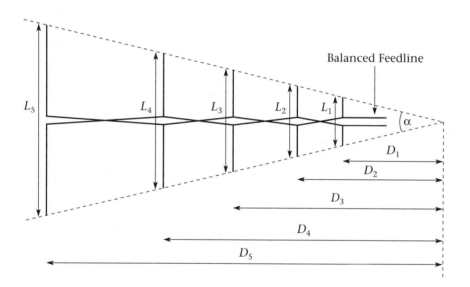

The elements are dipoles, with the longest at least one-half wavelength in length at the lowest operating frequency, and the shortest less than one-half wavelength at the highest. The ratio between the highest and lowest frequencies can be 10:1 or more. A balanced feedline is connected to the narrow end, and power is fed to the other dipoles via a network of crossed connections as shown. The operation is quite complex, with the dipoles that are closest to resonance at the operating frequency doing most of the radiation. The gain is typically about 8 dBi.

The design of a log-periodic antenna is based on several equations. A parameter τ is chosen, with a value that must be less than 1 and is typically between 0.7 and 0.9. A value toward the larger end of the range gives an antenna with better performance but more elements. The variable τ is the ratio between the lengths and spacing of adjacent elements. That is,

$$\tau = \frac{L_1}{L_2} = \frac{L_2}{L_3} = \frac{L_3}{L_4} = \cdots \tag{8.15}$$

where

$$L_1, L_2 \cdots = \text{lengths of the elements, in order from shortest to longest}$$

and

$$\tau = \frac{D_1}{D_2} = \frac{D_2}{D_3} = \frac{D_3}{D_4} = \cdots \tag{8.16}$$

where

$$D_1, D_2 \cdots = \text{spacings between the elements and the apex of the angle enclosing them, in order from shortest to longest.}$$

The angle designated α in the figure is typically about 30°. From simple trigonometry it can be shown that

$$\frac{L_1}{2D_1} = \tan\frac{\alpha}{2} \tag{8.17}$$

EXAMPLE 8.10 ▼

Design a log-periodic antenna to cover the frequency range from 100 to 300 MHz. Use $\tau = 0.7$ and $\alpha = 30°$.

SOLUTION

In order to get good performance across the frequency range of interest, it is advisable to design the antenna for a slightly wider bandwidth. For the longest element, we can use a half-wave dipole cut for 90 MHz, and for the shortest, one designed for 320 MHz.

From Equation (8.1),

$$L = \frac{142.5}{f}$$

For a frequency of 90 MHz

$$L = \frac{142.5}{90}$$

$$= 1.58 \text{ m}$$

Similarly, for 320 MHz,

$$L = 0.445 \text{ m}$$

Because of the way the antenna is designed, it is unlikely that elements of exactly these two lengths will be present, but we can start with the shorter one and simply make sure that the longest element has at least the length calculated above.

Starting with the first element and using Equation (8.17),

$$\frac{L_1}{2D_1} = \tan \frac{\alpha}{2}$$

We have specified $L_1 = 0.445$ m and $\alpha = 30°$, so

$$D_1 = \frac{L_1}{2 \tan \dfrac{\alpha}{2}}$$

$$= \frac{0.445}{2 \tan 15°}$$

$$= 0.83 \text{ m}$$

From Equation (8.15),

$$\tau = \frac{L_1}{L_2} = \frac{L_2}{L_3} = \frac{L_3}{L_4} = \cdots$$

τ and L_1 are known, so L_2 can be calculated.

$$L_2 = \frac{L_1}{\tau}$$

$$= \frac{0.445}{0.7}$$

$$= 0.636 \text{ m}$$

We continue this process until we obtain an element length that is greater than 1.58 m.

$$L_3 = 0.909 \text{ m}, \ L_4 = 1.30 \text{ m}, \ L_5 = 1.85 \text{ m}$$

L_5 is longer than necessary, so the antenna will need five elements.

The spacing between the elements can be found from Equation (8.16):

$$\tau = \frac{D_1}{D_2} = \frac{D_2}{D_3} = \frac{D_3}{D_4} = \cdots$$

Since τ and D_1 are known, it is easy to find D_2 and then the rest of the spacings, in the same way as the lengths were found above. We get $D_2 = 1.19$ m, $D_3 = 1.69$ m, $D_4 = 2.42$ m, and $D_5 = 3.46$ m.

Phased arrays are not restricted to combinations of simple dipoles. Any of the other elementary antennas described in the previous section, such as monopoles or patch antennas, can be combined into an array. For that matter, an array can be used as one element of a more complex array.

Parasitic Arrays   The Yagi-Uda array shown in Figure 8.25 is the most popular type of parasitic array. It has one driven element, one *reflector* behind the driven element, and one or more *directors* in front of the driven element. The driven element is a half-wave dipole or folded dipole. The reflector is slightly longer than one-half wavelength, and the directors are slightly shorter. The spacing between elements varies, but is typically about 0.2 wavelength. The Yagi-Uda antenna is often called just the Yagi array.

The Yagi antenna is a unidirectional end-fire array with a single main lobe in the direction shown in Figure 8.25. The antenna pattern for a typical Yagi is shown in Figure 8.26. This one has eight elements: one driven, one reflector, and six directors. Yagis are often constructed with five or six directors for a gain of about 10 to 12 dBi, but higher gains, up to about 16 dBi, can be achieved by using more directors.

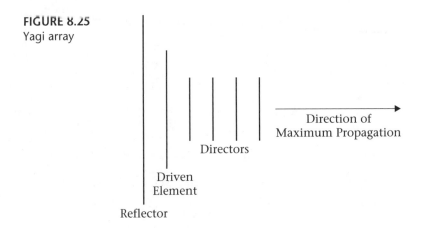

FIGURE 8.25
Yagi array

Direction of
Maximum Propagation

Directors

Driven
Element

Reflector

FIGURE 8.26
Radiation pattern for
8-element Yagi

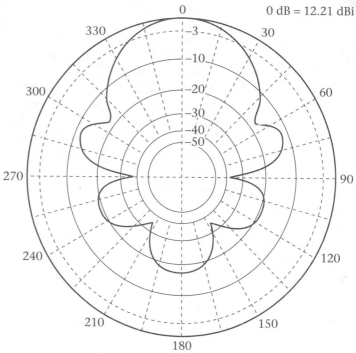

0 dB = 12.21 dBi

The Yagi is a relatively narrow-band antenna. When optimized for gain, its usable bandwidth is only about two percent of the operating frequency. Wider bandwidth can be obtained by varying the length of the directors, making them shorter as the distance from the driven element increases.

8.6 Reflectors

The antennas and arrays described in the preceding section can often be used with reflecting surfaces to improve their performance. A reflector may consist of one or more planes, or it may be parabolic in shape. In order to reduce wind and snow loads, reflectors are often constructed of mesh or closely-spaced rods. As long as the spacing is small compared with a wavelength, the effect on the antenna pattern, compared with a solid reflector, is negligible.

Plane and Corner Reflectors

A plane reflector acts in a similar way to an ordinary mirror. Like a mirror, its effects can be predicted by supposing that there is an "image" of the antenna on the opposite side of the reflecting surface at the same distance from it as the source. Reflection changes the phase angle of a signal by 180°. Whether the image antenna's signal aids or opposes the signal from the real antenna depends on the spacing between the antenna and the reflector and on the location of the receiver. In Figure 8.27 the antenna is one-quarter wavelength from the reflector, and the signals aid in the direction shown. The reflected signal experiences a 180° phase shift on reflection and another 180° shift because it must travel an additional one-half wavelength to reach the receiver. The magnitude of the electric field in the direction shown is thus increased by a factor of two. The power density in this direction is increased by a factor of four, or 6 dB, because power is proportional to the square of voltage.

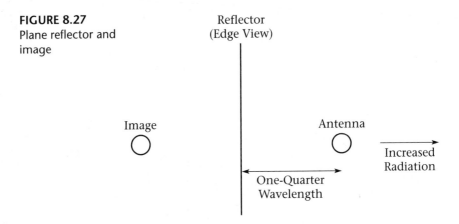

FIGURE 8.27
Plane reflector and image

It is possible to use a plane reflector with almost any antenna. For example, a reflector can be placed behind a collinear antenna, as shown in Figure 8.28. The antenna becomes directional in both the horizontal and vertical planes. Base antennas for cellular radio systems are often of this type.

The corner reflector creates two images, as shown in Figure 8.29, for a somewhat sharper pattern. Corner reflectors are often combined with Yagi arrays in UHF television antennas.

Parabolic Reflector

Parabolic reflectors have the useful property that any ray originating at a point called the focus and striking the reflecting surface will be reflected parallel to the axis of the parabola. That is, a *collimated* beam of radiation will be produced. The *parabolic dish* antenna, familiar from backyard satellite-receiver installations, consists of a small antenna at the focus of a large parabolic reflector, which focuses the signal in the same way as the reflector of a searchlight focuses a light beam. Figure 8.30 shows a typical example. Of course the antenna is reciprocal: radiation entering the dish along its axis will be focused by the reflector.

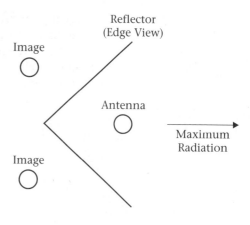

FIGURE 8.29 Corner reflector and images

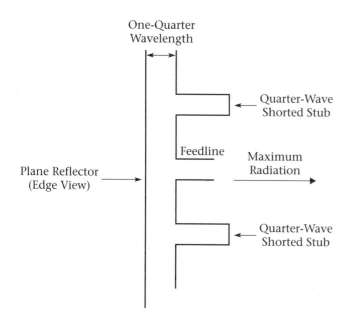

FIGURE 8.28 Collinear array with plane reflector

FIGURE 8.30 Parabolic antenna
(Courtesy of Andrew Corporation)

Ideally the antenna at the feedpoint should illuminate the entire surface of the dish with the same intensity of radiation and should not "spill" any radiation off the edges of the dish or in other directions. If that were the case, the gain and beamwidth of the antenna could easily be calculated. The equation for beamwidth is

$$\theta = \frac{70\lambda}{D}$$
(8.18)

where

θ = beamwidth in degrees at the 3 dB points
λ = free-space wavelength in m
D = diameter of the dish in m

The radiation pattern for a typical parabolic antenna is shown in Figure 8.31. The width of the beam measured between the first nulls is approximately twice the 3 dB beamwidth.

For gain, the equation is

$$G = \frac{\eta\pi^2 D^2}{\lambda^2}$$
(8.19)

FIGURE 8.31
Pattern for typical parabolic antenna

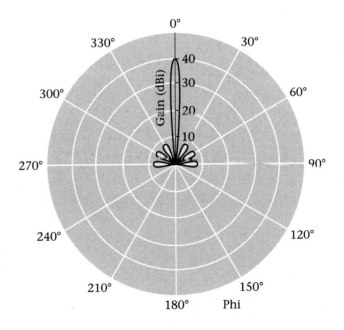

where

G = gain with respect to an isotropic antenna, as a power ratio (not in dB)

D = diameter of the dish in m

λ = the free-space wavelength in m

η = the antenna efficiency

The effect of uneven illumination of the antenna, of losses, or of any radiation from the feed antenna spilling off at the edges, is to reduce the efficiency, and therefore, the gain. To include these effects in gain calculations, it is necessary to include a constant η, the efficiency of the antenna. η for a typical antenna is between 0.5 and 0.7.

EXAMPLE 8.11

A parabolic antenna has a diameter of 3 m, an efficiency of 60%, and operates at a frequency of 4 GHz. Calculate its gain and beamwidth.

SOLUTION

The free-space wavelength is

$$\lambda = \frac{c}{f}$$

$$= \frac{300 \times 10^6}{4 \times 10^9}$$

$$= 0.075 \text{ m}$$

Substituting this into Equation (8.18), the beamwidth is

$$\theta = \frac{70\lambda}{D}$$

$$= \frac{70 \times 0.075}{3}$$

$$= 1.75°$$

The gain is given by Equation (8.19):

$$G = \frac{\eta \pi^2 D^2}{\lambda^2}$$

$$= \frac{0.6 \pi^2 D^2}{0.075^2}$$

$$= 9475$$

$$= 39.8 \text{ dBi}$$

Any type of antenna can be used with a parabolic reflector. In the microwave portion of the spectrum, where parabolic reflectors are most useful because they can have a practical size, a horn antenna provides a simple and efficient method to feed power to the antenna.

Besides the simple horn feed shown in Figure 8.30, there are several ways to get power to the parabolic reflector. For example, the Gregorian feed shown in Figure 8.32(a) uses a feed horn in the center of the dish itself, which radiates to a reflector at the focus of the antenna. This reflects the signal to the main parabolic reflector. By removing the feedhorn from the focus, this system allows any waveguide or electronics associated with the feedpoint to be placed in a more convenient location. The strange-looking antenna shown in Figure 8.32(b) is a combination of horn and parabolic antennas called a hog-horn; it is often used for terrestrial microwave links.

FIGURE 8.32
Parabolic antenna variations

(Courtesy of Andrew Corporation)

(a) Gregorian feed

(b) Hog-horn antenna

8.7 Cellular and PCS Antennas

This section considers antennas for 800-MHz cellular radio and 1900-MHz PCS systems. The requirements are similar, though of course the PCS antennas are physically smaller. Dual-band antennas also exist, both for base and portable units.

Cell-Site Antennas

The antennas used for cellular radio systems have to fulfill the requirements described in Chapter 7. There is a need for omnidirectional antennas and for antennas with beamwidths of 120° and less for sectorized cells. Narrower beamwidths are also useful for filling in dead spots. Typical cellular antennas use variations of the collinear antenna described earlier for omni-directional patterns and either collinear antennas backed by reflectors or log-periodic antennas for directional patterns. Figure 8.33 shows patterns and specifications for a typical omnidirectional antenna, and Figure 8.34 shows specifications for an antenna with a beamwidth of 120°. As expected, the directional antenna has higher gain.

When cell sizes are small, in high-traffic areas, directional base-station antennas are often tilted downwards in order to reduce the distance the signal travels and the interference to neighboring cells. This *downtilt* can be done mechanically by mounting the antenna so that it aims downwards at a slight angle, but it can also be done electrically. Figure 8.35 shows the vertical pattern for a cellular antenna with a built-in downtilt of 9°.

Cellular and PCS base-station receiving antennas are usually mounted in such a way as to obtain space diversity. For an omnidirectional pattern, typically three antennas are mounted at the corners of a tower with a triangular cross section, as shown in Figure 8.36(a). When the cell is divided into three 120° sectors, it is usual to mount two antennas for each sector on the sides of the tower, as shown in Figure 8.36(b). It is possible for a single antenna to be used for both receiving and transmitting using a duplexer, but often the transmitting antenna is located separately. Only one transmitting antenna is needed for a cell with an omnidirectional pattern; otherwise, one is needed per sector. Of course, other mounting locations, such as the walls of buildings, require variations of these arrangements.

A recent development that can reduce the number of antennas required for diversity is the use of dual-polarization antennas. We noted earlier that vertical polarization is usual for portable and mobile systems because this polarization is much easier to implement in the mobile units. However, in a cluttered mobile environment, signal polarization may be randomized by reflections. In that case, diversity can be achieved by using two polarizations, typically at 45° angles to the vertical. Dual-polarization antennas can considerably reduce the number of visible structures needed at a cell site.

			ANDREW	SPECIFICATIONS		

Antenna Type: CT1D0F-0080-009
Description: omni, 9 dBi, no downtilt, 7-16 DIN Female
Frequency Band, MHz: 824-896

Gain (dBi)	9.0
Gain (dBd)	7.0
Azimuth Beamwidth (deg)	360
Polarization Type	Single/Vertical
Intermodulation	<−150 dBc
Downtilt (deg)	0
Impedance (ohms)	50
Return Loss, dB (VSWR)	>14.0 (<1.5)
Maximum Input Power (watts)	500
Lightning Protection	DC Ground
Input Connector Type	7-16 DIN Female
Dimensions, L × W × D, mm (inches)	2540 (100) × 76 (3.0)
Weight, kg (lb)	11.3 (25)
Radome Material	FIBERGLASS
Radome Color	Gray
Survival Wind Speed, km/h (mph)	200 (125)
Temperature Range, C, (F)	−40/+70 (−40/+158)
Humidity	Up to 100%
Traditional Weatherproofing Kit	221213
Omni Mount Part Number	600033
Connector type	1
Pattern Type	0
Polarization	0
Elevation Beamwidth (deg)	10.0
Flat Plate Area, sq cm (sq in)	1135 (176)
Omni Wind Load, N (lbf)	280 (63)

(a) Specifications

FIGURE 8.33 Omnidirectional cellular base antenna
(Courtesy of Andrew Corporation)

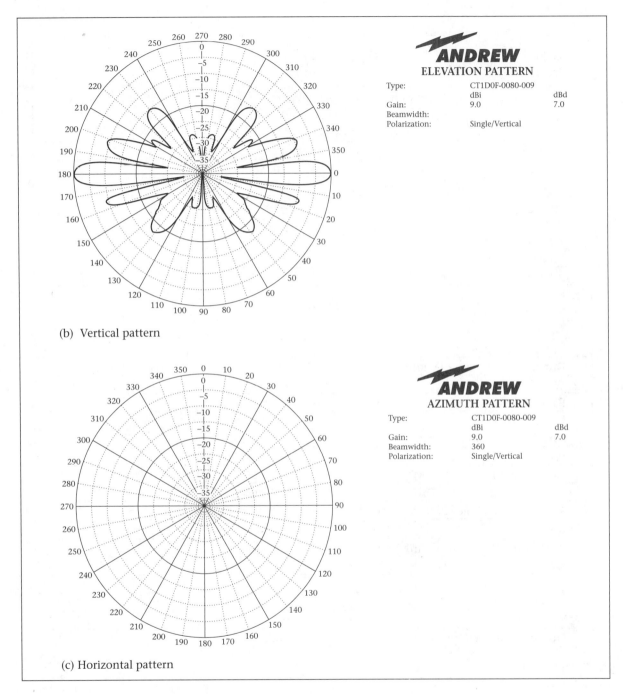

(b) Vertical pattern

ANDREW
ELEVATION PATTERN

Type:	CT1D0F-0080-009	
	dBi	dBd
Gain:	9.0	7.0
Beamwidth:		
Polarization:	Single/Vertical	

(c) Horizontal pattern

ANDREW
AZIMUTH PATTERN

Type:	CT1D0F-0080-009	
	dBi	dBd
Gain:	9.0	7.0
Beamwidth:	360	
Polarization:	Single/Vertical	

FIGURE 8.33 (*continued*)

SPECIFICATIONS

Antenna Type: CTS08-12010-0D
Description: 120 deg., 12 dBi, no downtilt, 7-16 DIN Female
Frequency Band, MHz: 824-896

Gain (dBi)	12.0
Gain (dBd)	10.0
Azimuth Beamwidth (deg)	120
Polarization Type	Single/Vertical
Intermodulation	<−150 dBc
Downtilt (deg)	0
Impedance (ohms)	50
Return Loss, dB (VSWR)	>15.5 (<1.4)
Front to Back Ratio (dB)	20
Maximum Input Power (watts)	500
Lightning Protection	DC Ground
Input Connector Type	7-16 DIN Female
Dimensions, L × W × D, mm (inches)	1140 (44.9) × 267 (10.5) × 127 (5)
Weight, kg (lb)	8 (18)
Radome Material	PVC
Radome Color	Gray
Survival Wind Speed, km/h (mph)	200 (125)
Wind Load, Frontal, N (lbf)	431 (97)
Wind Load, Lateral, N (lbf)	109 (24)
Wind Load, Back, N (lbf)	565 (127)
Temperature Range, C, (F)	−40/+70 (−40/+158)
Humidity	Up to 100%
Traditional Weatherproofing Kit	221213
Tilt Mount Part Number	600899A
Connector type	1
Pattern Type	1
Polarization	0
Elevation Beamwidth (deg)	18.0

(a) Specifications

FIGURE 8.34 Directional cellular base antenna

(Courtesy of Andrew Corporation)

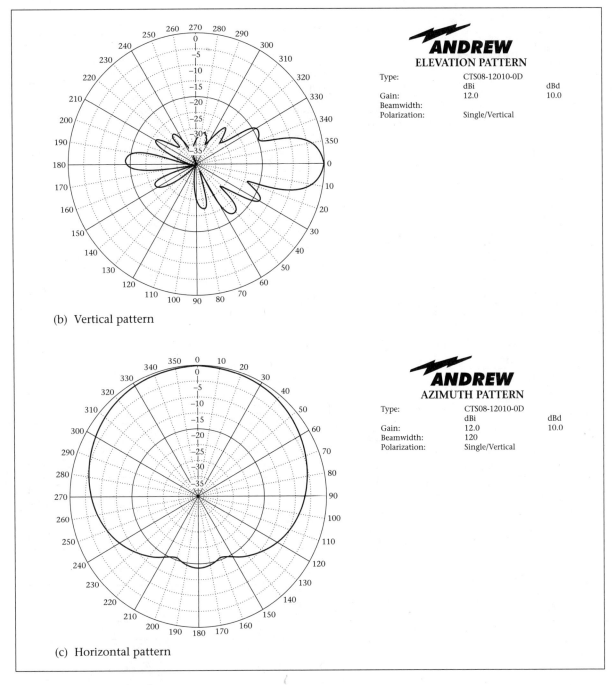

(b) Vertical pattern

ANDREW
ELEVATION PATTERN

Type:	CTS08-12010-0D	
	dBi	dBd
Gain:	12.0	10.0
Beamwidth:		
Polarization:	Single/Vertical	

(c) Horizontal pattern

ANDREW
AZIMUTH PATTERN

Type:	CTS08-12010-0D	
	dBi	dBd
Gain:	12.0	10.0
Beamwidth:	120	
Polarization:	Single/Vertical	

FIGURE 8.34 (*continued*)

SPECIFICATIONS

Antenna Type: CTS08-06013-9D
Description: 60 deg., 15 dBi. 9 degree downtilt, 7-16 DIN Female
Frequency Band, MHz: 824-896

Gain (dBi)	15.0
Gain (dBd)	13.0
Azimuth Beamwidth (deg)	60
Polarization Type	Single/Vertical
Intermodulation	<–150 dBc
Downtilt (deg)	9
Impedance (ohms)	50
Return Loss, dB (VSWR)	>15.5 (<1.4)
Front to Back Ratio (dB)	25
Maximum Input Power (watts)	500
Lightning Protection	DC Ground
Input Connector Type	7-16 DIN Female
Dimensions, L × W × D, mm (inches)	1390 (54.7) × 267 (5) × 127 (5)
Weight, kg (lb)	9 (21)
Radome Material	UV Protected PVC
Radome Color	Gray
Survival Wind Speed, km/h (mph)	200 (125)
Wind Load, Frontal, N (lbf)	525 (118)
Wind Load, Lateral, N (lbf)	133 (30)
Wind Load, Back, N (lbf)	688 (155)
Temperature Range, C, (F)	–40/+70 (–40/+158)
Humidity	Up to 100%
Traditional Weatherproofing Kit	221213
Tilt Mount Part Number	600899
Connector type	1
Pattern Type	1
Polarization	0
Elevation Beamwidth (deg)	16.0

(a) Specifications

FIGURE 8.35 Cellular base antenna with downtilt

(Courtesy of Andrew Corporation)

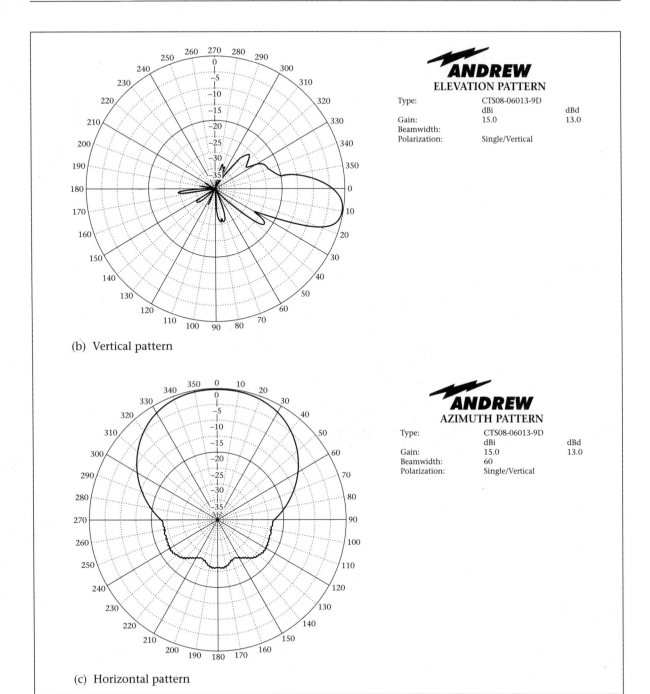

(b) Vertical pattern

ANDREW
ELEVATION PATTERN

Type:	CTS08-06013-9D	
	dBi	dBd
Gain:	15.0	13.0
Beamwidth:		
Polarization:	Single/Vertical	

(c) Horizontal pattern

ANDREW
AZIMUTH PATTERN

Type:	CTS08-06013-9D	
	dBi	dBd
Gain:	15.0	13.0
Beamwidth:	60	
Polarization:	Single/Vertical	

FIGURE 8.35 (*continued*)

FIGURE 8.36
Cell-site antenna
mounting

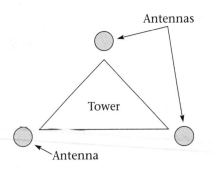

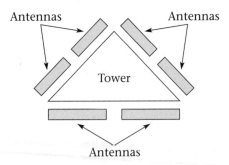

(a) Using omnidirectional antennas   (b) Using directional antennas

Mobile and Portable Antennas

The portable and mobile antennas used with cellular and PCS systems have to be omnidirectional and small, especially in the case of portable phones. The latter requirement is, of course, easier to achieve at 1900 MHz than at 800 MHz. Many PCS phones must double as 800-MHz cell phones, however, so they need an antenna that works well at 800 MHz.

The simplest suitable antenna is a quarter-wave monopole, and these are the usual antennas supplied with portable phones. For mobile phones, where compact size is not quite as important, a very common configuration consists of a quarter-wave antenna with a half-wave antenna mounted collinearly above it. The two are connected by a coil which matches impedances. Figure 8.37 shows such an antenna. This combination has a gain of about 3 dB compared with the quarter-wave monopole.

FIGURE 8.37
Collinear mobile
antenna

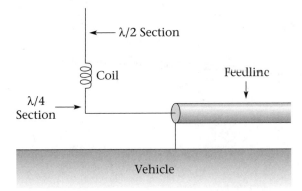

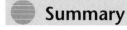 **Summary** The main points to remember from this chapter are:

- Antennas are reciprocal passive devices that couple electrical energy between transmission lines and free space.

- The isotropic radiator is convenient for calculations because it emits all the energy supplied to it equally in all directions. Antenna gain is usually calculated with reference to an isotropic radiator.

- Gain is obtained in an antenna when more power is emitted in certain directions than in others.

- The beamwidth of a directional antenna is the angle between the half-power points of the main lobe.

- The half-wave dipole is a simple practical antenna with a bidirectional pattern, a small gain over an isotropic radiator, and a feedpoint impedance of about 70 Ω. The dipole is sometimes used instead of an isotropic antenna as a reference for antenna gain.

- The radiation resistance of an antenna is the equivalent resistance that appears at its feedpoint due to the radiation of energy into space.

- The polarization of most simple antennas is the same as the axis of the antenna.

- Simple antennas can be used as elements of arrays, in order to obtain specified values of gain and directivity.

- Plane and parabolic reflectors can be used to increase the gain and directivity of antennas.

- Cellular systems use multiple antennas for space diversity and cell sectorization. Sometimes polarity diversity is used as well.

Equation List

$$L = \frac{142.5}{f} \tag{8.1}$$

$$\eta = \frac{R_r}{R_t} \tag{8.4}$$

$$G = \frac{15NS(\pi D)^2}{\lambda^3} \tag{8.8}$$

$$\theta = \frac{52\lambda}{\pi D}\sqrt{\frac{\lambda}{NS}} \tag{8.9}$$

$$G = \frac{7.5 d_E d_H}{\lambda^2} \qquad (8.10)$$

$$\theta_H = \frac{70\lambda}{d_H} \qquad (8.11)$$

$$\theta_E = \frac{56\lambda}{d_E} \qquad (8.12)$$

$$\tau = \frac{L_1}{L_2} = \frac{L_2}{L_3} = \frac{L_3}{L_4} = \cdots \qquad (8.15)$$

$$\tau = \frac{D_1}{D_2} = \frac{D_2}{D_3} = \frac{D_3}{D_4} = \cdots \qquad (8.16)$$

$$\frac{L_1}{2D_1} = \tan \frac{\alpha}{2} \qquad (8.17)$$

$$\theta = \frac{70\lambda}{D} \qquad (8.18)$$

$$G = \frac{\eta \pi^2 D^2}{\lambda^2} \qquad (8.19)$$

● Key Terms

angle of elevation   angle measured upward from the horizon. Used to describe antenna patterns and directions

array   combination of several antenna elements

beamwidth   angle between points in an antenna pattern at which radiation is 3 dB down from its maximum

dipole   any antenna with two sections

directivity   gain of an antenna with losses ignored

effective radiated power (ERP)   transmitter power that would, if used with a lossless dipole oriented for maximum gain, produce the same power density in a given direction as a given transmitting installation

element   an antenna used as part of an array

far-field region   distance from an antenna great enough to avoid local magnetic or electrical coupling, and great enough for the antenna to resemble a point source

monopole   an antenna with only one conductor, generally using ground or a ground plane to represent a second conductor

near-field region   the region of space close to an antenna, where the radiation pattern is disturbed by induced, as well as radiated, electric and magnetic fields

radiation resistance   representation of energy lost from an antenna by radiation as if it were dissipated in a resistance

● Questions

1. Explain how an antenna can have a property called gain, even though it is a passive device.

2. Sketch the radiation patterns of an isotropic antenna and of a half-wave dipole in free space, and explain why a dipole has gain over an isotropic radiator.

3. What is the significance of radiation resistance, and what would be the effect on an antenna's efficiency of reducing the radiation resistance, all other things being equal?

4. What is the difference between gain and directivity?

5. Suggest two reasons for using a directional antenna.

6. Distinguish between the near and far field of an antenna. Why is it necessary to use the far field for all antenna measurements?

7. How can an antenna's electrical length be increased without increasing its physical length?

8. What advantages does a five-eighths wavelength antenna have over a quarter-wave antenna for mobile use? Does it have any disadvantages?

9. Name two types of antennas that are noted for having particularly wide bandwidth. Which of these is directional?

10. Name one type of antenna that produces circularly polarized waves. Under what circumstances is circular polarization desirable?

11. What polarization is generally used with portable and mobile radio systems, and why?

12. Distinguish between end-fire and broadside arrays. Name and sketch one example of each.

13. Distinguish between parasitic and phased arrays. Name and sketch one example of each. What is another term for a phased array?

14. What is the purpose of using a plane reflector with an antenna array?

15. What is meant by a collinear antenna? Sketch such an antenna and its radiation pattern, and classify it as broadside or end-fire, phased or parasitic.

16. How is a patch antenna constructed?

17. Why are parabolic dish antennas impractical at frequencies below the microwave region?

18. What is meant by space diversity? How is it implemented in a typical cell site?

19. How is polarization diversity possible in cellular radio, and why is it useful?

20. Why are base-station antennas sometimes tilted downward in cellular systems?

● **Problems**

1. Calculate the length of a practical half-wave dipole for a frequency of 150 MHz.

2. Calculate the efficiency of a dipole with a radiation resistance of 68 Ω and a total feedpoint resistance of 75 Ω.

3. Given that a half-wave dipole has a gain of 2.14 dBi, calculate the electric field strength at a distance of 10 km in free space in the direction of maximum radiation from a half-wave dipole that is fed, by means of lossless, matched line, by a 15 W transmitter.

4. Refer to the plot in Figure 8.38 and find the gain and beamwidth for the antenna shown.

FIGURE 8.38

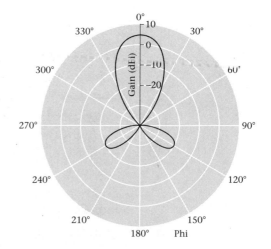

5. Calculate the EIRP in dBW for a 25 W transmitter operating into a dipole with 90% efficiency.

6. Calculate the length of a quarter-wave monopole antenna for a frequency of 900 MHz.

7. Calculate the optimum length of an automobile FM broadcast antenna, for operation at 100 MHz.

8. Draw a dimensioned sketch of a discone antenna that will cover the VHF range from 30 to 300 MHz.

9. A helical antenna consists of 10 turns with a spacing of 10 cm and a diameter of 12.7 cm.

    (a) Calculate the frequency at which this antenna should operate.

    (b) Calculate the gain in dBi at the frequency found in part (a).

    (c) Calculate the beamwidth at the frequency found in part (a).

10. Assuming the aperture of a pyramidal horn is square, how large does it have to be to have a gain of 18 dBi at 12 GHz?

11. Calculate the beamwidth in each plane of the antenna referred to in problem 10.

12. Calculate the size of a square patch antenna for the ISM band at 2.4 GHz, if the circuit board used has a relative permittivity of 3.

13. Characterize each of the arrays sketched in Figure 8.39 as phased or parasitic, and as broadside or end-fire. Show with an arrow the direction(s) of maximum output for each antenna.

FIGURE 8.39

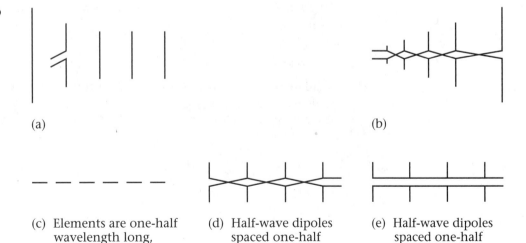

(a)

(b)

(c) Elements are one-half wavelength long, fed in phase

(d) Half-wave dipoles spaced one-half wavelength apart

(e) Half-wave dipoles spaced one-half wavelength apart

14. Design a log-periodic antenna to cover the low-VHF television band, from 54 to 88 MHz. Use $\alpha = 30°$, $\tau = 0.7$, $L_1 = 1.5$ m.

15. Two horizontal Yagis are stacked with one of them one-half wavelength above the other and fed in phase. What will be the effect of the stacking on the radiation

    (a) in the horizontal plane?

    (b) in the vertical plane?

16. Calculate the minimum diameter for a parabolic antenna with a beam-width of 2° at a frequency of:

    (a) 4 GHz

    (b) 12 GHz

17. Calculate the gain of each of the antennas in the previous question, assuming an efficiency of 65%.

18. Calculate the effective area of a 3 m dish, with an efficiency of 0.7, at 3 GHz and at 12 GHz. Explain your results.

19. A Yagi antenna has a gain of 10 dBi and a front-to-back ratio of 15 dB. It is located 15 km from a transmitter with an ERP of 100 kW at a frequency of 100 MHz. The antenna is connected to a receiver via a matched feedline with a loss of 2 dB. Calculate the signal power supplied to the receiver if the antenna is

    (a) pointed directly toward the transmitting antenna.

    (b) pointed directly away from the transmitting antenna.

20. A lossless half-wave dipole is located in a region with a field strength of 150 μV/m. Calculate the power that this antenna can deliver to a receiver if the frequency is:

    (a) 100 MHz

    (b) 500 MHz

21. Calculate the gain, with respect to an isotropic antenna, of a half-wave dipole with an efficiency of 90% that is placed one-quarter wavelength from a plane reflector that reflects 100% of the signal striking it.

Transmitter and Receiver Circuitry

9

Objectives

After studying this chapter, you should be able to:

- Discuss the requirements and specifications for transmitters and determine whether a given transmitter is suitable for a particular application.

- Draw block diagrams for several types of transmitters and explain their operation.

- Analyze the operation of transmitter circuits.

- Explain the basic operation of frequency synthesizers and mixers.

- Describe the characteristics of Class A, B, and C amplifiers and decide which type is the most suitable for a given application.

- Describe the ways in which FM transmitters and receivers differ from those for AM and explain why they differ.

- Analyze the operation of analog and digital modulators.

- Describe the basic superheterodyne system, and explain why it is the preferred design for most receivers.

- Explain the requirements for each stage in a receiver.

- Analyze the operation of demodulators for analog and digital signals.

- Analyze specifications for receivers and use them to determine suitability for a given application.

- Explain how transceivers differ from separate transmitters and receivers.

9.1 Introduction

The transmitter and receiver are very important parts of any radio communication link. Sometimes, as at most base stations, they are separate components; sometimes they are combined into a transceiver. In the latter case, it may be necessary for the unit to transmit and receive simultaneously, as in a cellular phone; or it may be sufficient to do one at a time with manual or voice-actuated switchover, as with most marine and aircraft radios. We begin by looking separately at the requirements for transmitters and receivers, then we examine transceivers.

We will find that all transmitters and all receivers have certain similarities. A thorough familiarity with the basic topologies for each will help the reader understand the many variations that appear. New types of transmitters and receivers appear almost daily, but the underlying structures change much more slowly.

9.2 Transmitters

The function of a transmitter is to generate a modulated signal with sufficient power, at the right frequency, and to couple that signal into an antenna feedline. The modulation must be done in such a way that the demodulation process at the receiver can yield a faithful copy of the original modulating signal.

All of the foregoing applies to any transmitter. Differences among transmitters result from variations in the required power level, carrier frequency, and modulation type, as well as from special requirements such as portability and the ability to be controlled remotely.

Basic Topologies   Figure 9.1 shows the block diagrams of some typical transmitters. Real transmitters have almost infinite variety, but most are variations on these structures. We will look at some of the variations and the reasons for them, as we proceed through this chapter and in other chapters that examine particular systems in more detail.

An overview of the figure shows that in all cases a modulated RF signal is generated and transmitted. In Figure 9.1(a), which represents a typical transmitter for full-carrier AM, the carrier is generated by a **frequency synthesizer** and amplified to its full output power before modulation takes place. The optional **frequency multiplier** would be used if the required carrier frequency were higher than could be conveniently generated by the synthesizer.

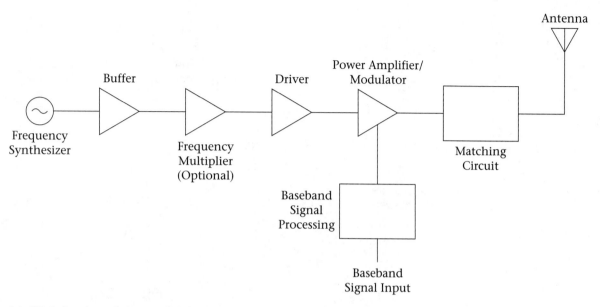

(a) High-level modulation (AM)

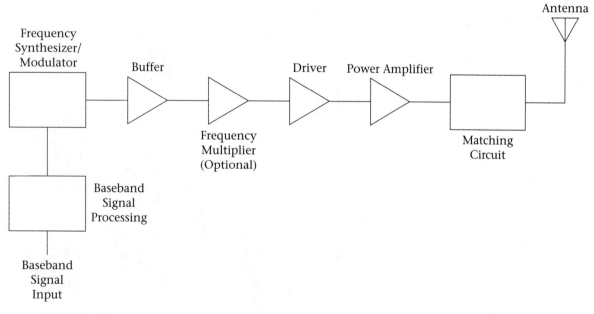

(b) Low-level modulation of synthesizer (FM, FSK)

FIGURE 9.1   Typical transmitter topologies

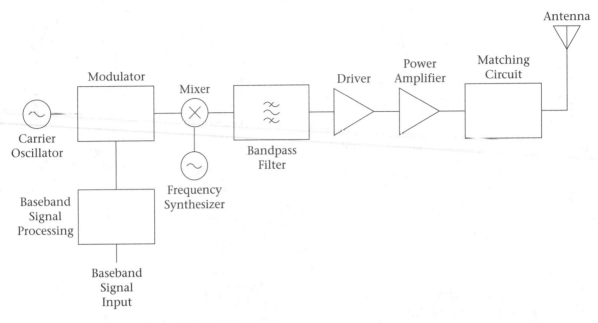

(c) Heterodyne system (PSK, QAM, SSBSC AM)

FIGURE 9.1 (*continued*)

By delaying modulation as long as possible, this topology allows all the RF amplifier stages in the transmitter to operate in a nonlinear mode for greater efficiency. High-power AM modulators are relatively easy to build, so this is the preferred topology for AM.

When the modulation involves changing the transmitted frequency as in FM and FSK, it is usual to modulate the carrier oscillator. In the example shown in Figure 9.1(b), that oscillator is a frequency synthesizer. The optional frequency multiplier would multiply the frequency deviation as well as the carrier frequency, which makes it useful in case the modulated oscillator can not achieve the required deviation.

Figure 9.1(c) shows a slightly more complex design, generally used where it is more convenient to modulate the signal at a fixed frequency. The modulated signal is then moved to the required output frequency by an oscillator-mixer combination, with the oscillator usually taking the form of a frequency synthesizer. A bandpass filter eliminates undesired mixing components and sends the signal to an amplifier chain.

All three transmitter designs end with a multistage amplifier. Two stages are shown: the **driver** and power amplifier. More may be required, especially for high output power levels. The matching circuit matches the power amplifier to the antenna feedline impedance, which is usually 50 Ω, and also removes harmonics and other spurious signals from the transmitter output. The power amplifier must be linear for any signal that has variable amplitude unless, as with Figure 9.1(a), the modulation is done at the transmitter output.

Next, let us examine the requirements for each stage of our generic transmitters in more detail.

Frequency Synthesizer Crystal oscillators are very stable but work at only one frequency unless crystals are changed. **Variable-frequency oscillators (VFOs)**, where the operating frequency is controlled by LC resonant circuits, are available but generally quite unstable in frequency. A frequency synthesizer uses a phase-locked loop to set the frequency of an oscillator to a multiple of a crystal-controlled reference frequency, combining most of the advantages of both crystal-controlled and LC oscillators.

Figure 9.2 shows how a simple frequency synthesizer works. A phase-locked loop controls the frequency of a voltage-controlled oscillator so that it is always a multiple of a crystal-controlled reference frequency.

$$f_o = Nf_{ref} \qquad\qquad (9.1)$$

where

f_o = output frequency
f_{ref} = reference frequency
N = divider modulus (must be an integer)

By changing the modulus N of the programmable divider, the output frequency can be varied in steps equal to the reference frequency f_{ref}. Since f_{ref} must often be quite small (a few kilohertz), usually a fixed divider is used to reduce the frequency of a crystal oscillator to the required value.

FIGURE 9.2
Basic frequency synthesizer

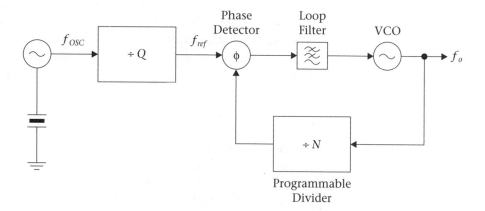

EXAMPLE 9.1 ▼

Using a 25.6-MHz crystal for the reference oscillator, configure a frequency synthesizer to generate frequencies from 100 to 200 MHz at 100-kHz intervals.

SOLUTION

The step size is 100 kHz so let us use that value for f_{ref}. Then

$$Q = \frac{25.6 \times 10^6}{100 \times 10^3}$$

$$= 256$$

To find the range of values of N, look at the extremes of the frequency range. At the low end,

$$N = \frac{100 \times 10^6}{100 \times 10^3}$$

$$= 1000$$

Similarly, at the high end, $N = 2000$.

▲

Carrier Oscillator

The main requirements for the carrier oscillator are frequency stability and spectral purity. Almost always this means crystal control; that is, the frequency of the oscillator is determined by the mechanical resonant frequency of a small slab of quartz. Crystal oscillators are available in frequencies from about 100 kHz to tens of megahertz. Their only real drawbacks are that the frequency is fixed (unless the crystal is changed) and they are limited in power output. Neither is a significant problem here, as the frequency can be moved by the synthesizer-mixer combination and the required power level is achieved by amplification, which takes place after modulation and mixing.

Analog Modulation

The type of modulator required depends on the modulation scheme in use. This can be either analog or digital, involving modulation of amplitude, frequency or phase, or a combination of amplitude and phase. The various types of modulation were described in Chapters 2 and 4, and digital baseband processing was described in Chapter 3. Let us look at suitable modulators for both analog and digital systems, beginning with analog. While we do this, we should note for each system whether there are amplitude variations in the modulated signal; that is, whether the modulated signal has an envelope. Signals with envelopes require linear amplification after the modulator, while those without amplitude variation can be amplified more efficiently using Class C or switching amplifiers.

Analog systems use full-carrier AM, its variation SSBSC AM, or FM. Full-carrier AM is a "mature," perhaps obsolescent, technology, but it is still in use for CB radio, VHF aircraft radio, and of course standard AM-band and short-wave broadcasting. SSBSC AM is used in military, commercial, and amateur HF communication and in some point-to-point microwave systems. A variation of it is also used for the video portion of terrestrial television broadcasting; DSBSC AM is used in FM and television stereo and television color signals. FM is very widely used in cellular radio, cordless phones, VHF marine radio, commercial and public-service mobile radio, and broadcasting.

Amplitude Modulation

The usual way to achieve full-carrier AM is to use the baseband signal to vary the power-supply voltage to a nonlinear (usually Class C) amplifier, as shown in Figure 9.3. Since the output voltage of such an amplifier is proportional to its supply voltage rather than to the input signal voltage, this achieves the desired result. If the amplifier is the final stage before the antenna, then all RF stages in the transmitter can be nonlinear for greater efficiency.

FIGURE 9.3
AM modulator

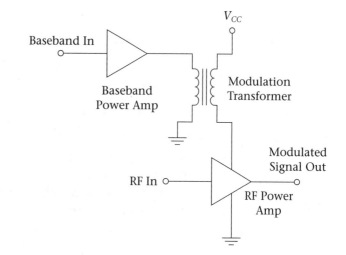

Single-Sideband Suppressed-Carrier AM Generating an SSB signal is much more complicated than for AM, and is nearly always done at low power and at only one frequency. The frequency is then shifted by mixing, as in Figure 9.1(c). The usual way to generate SSB is to first develop a double-sideband suppressed-carrier signal (DSBSC) using a balanced modulator (otherwise known as a multiplier).

It can be shown that if the carrier and baseband signals are multiplied together the result will contain upper and lower sidebands but no carrier. Suppose that a carrier and a modulation signal, each sine waves with 1 V peak amplitude, are applied to a multiplier as shown in Figure 9.4.

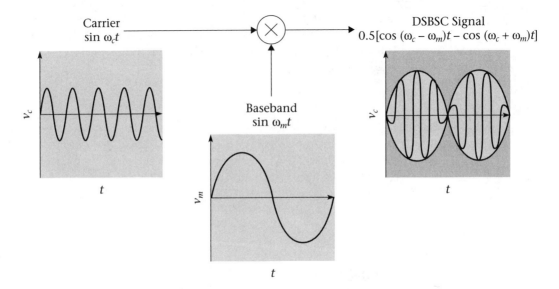

FIGURE 9.4 DSBSC generation

The output of the multiplier will be

$$v_o = \sin \omega_m t \sin \omega_c t$$
$$= 0.5[\cos(\omega_c - \omega_m)t - \cos(\omega_c + \omega_m)t] \qquad (9.2)$$

which is a DSBSC signal.

The DSBSC signal can be converted into SSB by using a bandpass filter to pass the desired sideband while rejecting the other sideband. Figure 9.5 illustrates the idea.

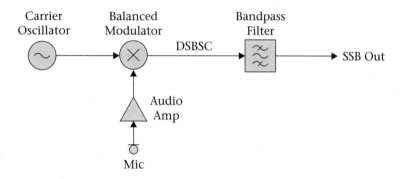

(a) Block diagram

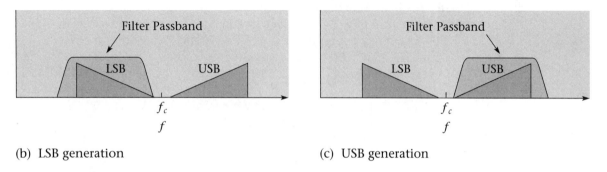

(b) LSB generation   (c) USB generation

FIGURE 9.5 SSBSC generation

EXAMPLE 9.2

An SSBSC generator of the type shown in Figure 9.5 has the following specifications:

Filter center frequency:	5.000 MHz
Filter bandwidth:	3 kHz
Carrier oscillator frequency:	4.9985 MHz

(a) Which sideband will be passed by the filter?

(b) What frequency should the carrier oscillator have if it is required to generate the other sideband?

SOLUTION

(a) Since the carrier frequency is at the low end of the filter passband, the upper sideband will be passed.

(b) To generate the lower sideband, the carrier frequency should be moved to the high end of the filter passband, at 5.0015 MHz.

▲ ───

Frequency Modulation   Frequency modulation generally involves modulating an oscillator, since that is the only way to directly change the transmitted frequency. (There is an indirect way, via phase modulation, but that is seldom used nowadays.)

Any crystal or *LC* oscillator can be turned into a **voltage-controlled oscillator (VCO)** by using a varactor diode as part of the frequency-determining circuit. A varactor is a reverse-biased diode whose junction capacitance varies with the applied bias voltage. The baseband signal modulates the bias voltage to generate FM.

Unfortunately, VCOs using crystal oscillators have a very small maximum frequency deviation and are able to operate at only one frequency (a significant drawback). On the other hand, *LC* oscillators tend to be unstable. However, by making the modulated VCO part of a frequency synthesizer, it is possible to build a stable signal source whose frequency can be varied over a wide range. Figure 9.6 shows how it can be done.

FIGURE 9.6
PLL FM
modulator

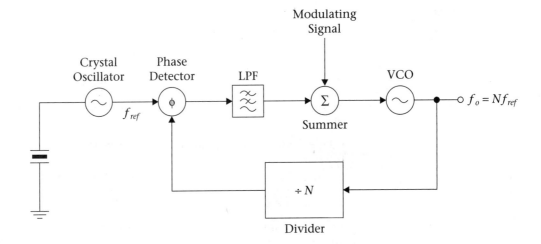

As in the frequency synthesizer described earlier, the carrier frequency is equal to Nf_{ref}. The baseband signal is added to the VCO control voltage from the phase detector. The change in VCO frequency due to modulation is detected by the phase detector, but the low-pass filter prevents the correction signal generated by the phase detector from reaching the VCO and canceling out the modulation. We thus have a combined frequency synthesizer and FM modulator. Most FM transmitters used for wireless communication use this scheme or a variation of it.

Digital Modulation

Digital modulation in wireless communication usually uses either FSK or QPSK. QAM can be used in fixed links but tends not to be robust enough for mobile and portable communication paths.

FSK modulation is identical to analog FM except that there are usually only two transmitted frequencies rather than an infinite number. A special case of FSK, mentioned earlier, is AFSK (audio frequency-shift keying). In that system, an analog FM modulator is supplied with two audio tones, one for mark and one for space. AFSK is a good choice for sending data over existing analog voice FM systems.

PSK Modulators   The easiest way to understand QPSK modulation is to begin with *binary phase-shift keying* (*BPSK*). This system uses two phase angles 180° apart. It can easily be generated with a multiplier, also known as a balanced modulator. All that is necessary to create a working modulator is to apply the carrier to one input and a voltage to the other input, which is positive for one binary state and negative for the other. For example, let the data input be +1V for logic 1, and −1V for logic 0, as shown in Figure 9.7(a). For simplicity, assume the carrier amplitude is also 1V (peak).

The balanced modulator output is the product of the two inputs. If the carrier signal is

$$e_c = \sin \omega_c t$$

then the output for a binary 1 will be the same,

$$v(t) = \sin \omega_c t$$

and the output for a binary 0 is

$$v(t) = -\sin \omega_c t$$

which is the same signal with a 180° phase shift.

Now let's look at QPSK. First, of course, we note that this is a dibit system; that is, it transmits two bits simultaneously so the serial data has to be converted to parallel using a shift register. Assume this has been done. Now we can apply each of the two bits to a separate balanced modulator, as shown in Figure 9.7(b). We could do it either way, but let us supply the first bit to the top balanced modulator. We call this the in-phase (I) bit because the carrier has not been shifted. The second bit we call the quadrature

FIGURE 9.7
PSK modulators

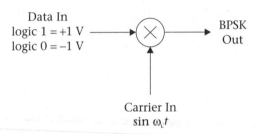

(a) BPSK Modulator

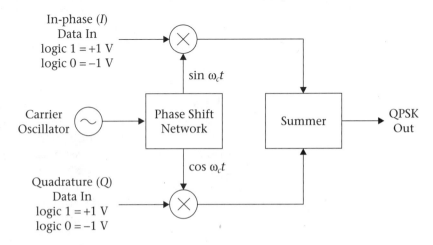

(b) QPSK Modulator

(Q) bit and apply it to the lower modulator, along with the carrier, shifted 90° in phase, as shown. The BPSK output signals from the two balanced modulators are summed to give the output. We can show mathematically that the output is indeed QPSK by looking at each of the four possible inputs, as follows.

The first of our dibits is applied to the upper balanced modulator. For the first bit equal to 1, the output is $\sin \omega_c t$, and for the first bit equal to 0, the output is $-\sin \omega_c t$, as before. Similarly when the second bit is 1, the lower balanced modulator outputs $\cos \omega_c t$, and for -1, $-\cos \omega_c t$. Now all we have to do is add the two outputs for each bit and see what happens. A little trigonometry lets us simplify the resulting expressions. A truth table (Table 9.1) might help.

If we plot these four possibilities in a vector diagram, we see that we have indeed achieved QPSK. See Figure 9.8.

TABLE 9.1   Truth Table for QPSK Modulator

Input	Top BM Out	Bottom BM Out	Sum	Simplified Sum
00	$-\sin \omega_c t$	$-\cos \omega_c t$	$-\sin \omega_c t - \cos \omega_c t$	$\sin (\omega_c t - 135°)$
01	$-\sin \omega_c t$	$\cos \omega_c t$	$-\sin \omega_c t + \cos \omega_c t$	$\sin (\omega_c t + 135°)$
10	$\sin \omega_c t$	$-\cos \omega_c t$	$\sin \omega_c t - \cos \omega_c t$	$\sin (\omega_c t - 45°)$
11	$\sin \omega_c t$	$\cos \omega_c t$	$\sin \omega_c t + \cos \omega_c t$	$\sin (\omega_c t + 45°)$

FIGURE 9.8
QPSK vector diagram

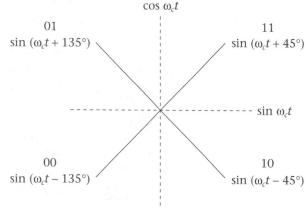

Mixing   The frequency of a signal, whether modulated or not, can be changed by combining it with another signal in a circuit called a **mixer**. The multiplier described in the previous section is an example of a mixer, as is the amplitude modulator described earlier. In general, any nonlinear circuit will combine two input frequencies to produce the sum and the difference of the two input frequencies. In the case of the multiplier (also called a balanced mixer), those are the only frequencies produced. In the more general case, the two input frequencies will appear along with their harmonics and other spurious frequencies. Usually, either the sum or the difference frequency signal is the desired one, and all other output frequencies must be removed by filtering.

A little math will show how this works. In general a nonlinear device produces a signal at its output that can be represented by a power series, for example

$$v_o = Av_i + Bv_i^2 + Cv_i^3 + \cdots \qquad (9.3)$$

where

$$v_o = \text{instantaneous output voltage}$$
$$v_i = \text{instantaneous input voltage}$$
$$A, B, C, \cdots = \text{constants}$$

With a single input frequency, the output contains all the harmonics as well as the fundamental. As the order of the harmonics increases, the magnitude decreases. This is the principle of the frequency multiplier and also the cause of harmonic distortion.

When the input contains two different frequencies we get cross products that represent sum and difference frequencies. If the two frequencies are f_1 and f_2 we get $mf_1 \pm nf_2$, where m and n are integers. Usually the most important are $f_1 + f_2$ and $f_1 - f_2$, where f_1 is assumed to be higher than f_2. When the cross products are not wanted they are called intermodulation distortion.

Two important special cases of mixers are the balanced mixer already described, and the square-law mixer. The operation of the latter is described by a truncated version of Equation (9.3):

$$v_o = Av_i + Bv_i^2 \tag{9.4}$$

To see how a square-law mixer works, let us apply two signals to this circuit, summing them at the input. The signals will be sine waves of different frequencies. For convenience we can let each signal have an amplitude of 1 V peak. Then

$$v_i = \sin \omega_1 t + \sin \omega_2 t$$

and

$$
\begin{aligned}
v_o &= Av_i + Bv_i^2 \\
&= A(\sin \omega_1 t + \sin \omega_2 t) + B(\sin \omega_1 t + \sin \omega_2 t)^2 \\
&= A \sin \omega_1 t + A \sin \omega_2 t + B \sin^2 \omega_1 t + B \sin^2 \omega_2 t \\
&\quad + 2B \sin \omega_1 t \sin \omega_2 t
\end{aligned}
\tag{9.5}
$$

The first two terms in Equation (9.5) are simply the input signals multiplied by A, which is a gain factor. The next two terms involve the squares of the input signals, which are signals at twice the input frequency (plus a dc component). This can be seen from the trigonometric identity

$$\sin^2 A = \tfrac{1}{2} - \tfrac{1}{2} \cos 2A$$

The final term is the interesting one. We can use the trigonometric identity

$$\sin A \sin B = \tfrac{1}{2} \{\cos (A - B) - \cos (A + B)\}$$

to expand this term.

$$2B \sin \omega_1 t \sin \omega_2 t = 2 \ (B/2)\{\cos (\omega_1 - \omega_2)t - \cos (\omega_1 + \omega_2)t\}$$
$$= B \ \{\cos (\omega_1 - \omega_2)t - \cos (\omega_1 + \omega_2)t\}$$

As predicted, there are output signals at the sum and difference of the two input frequencies, in addition to the input frequencies themselves and their second harmonics. In a practical application, either the sum or the difference frequency would be used and the others would be removed by filtering.

EXAMPLE 9.3

Sine-wave signals with frequencies of 10 MHz and 11 MHz are applied to a square-law mixer. What frequencies appear at the output?

SOLUTION

Let $f_1 = 11$ MHz and $f_2 = 10$ MHz. Then the output frequencies are as follows.

$$f_1 = 11 \text{ MHz}$$
$$f_2 = 10 \text{ MKz}$$
$$2f_1 = 22 \text{ MHz}$$
$$2f_2 = 20 \text{ MHz}$$
$$f_1 + f_2 = 21 \text{ MHz}$$
$$f_1 - f_2 = \ \ 1 \text{ MHz}$$

Power Amplification

The type of power amplification used depends on whether the signal has an envelope. If it does, a linear amplifier (Class A or AB) is required. These are discussed in any analog electronics text. As we have seen, all varieties of AM including SSBSC and QAM, have envelope variations and require linear amplification. Perhaps surprisingly, QPSK in its classic form does require linearity. A second look at Figure 9.8 will show why. Suppose that the signal changes from the 11 state to the 00 state. This requires that the signal pass through the origin, that is, its amplitude goes through zero.

If the signal has a constant amplitude, as with frequency modulation whether analog or digital (FSK), nonlinear amplifiers like Class C or Class D (switching) amplifiers are more efficient and quite satisfactory. If a transmitter is designed for more than one modulation scheme, it needs a linear power amplifier unless all of the modulation types used are free from linearity requirements.

In order to understand nonlinear amplifiers, it would be useful to look at a very simple Class C amplifier using a bipolar transistor. The transistor is biased beyond cutoff, that is, it is off for most of the input cycle. Yet for maximum efficiency, the transistor must almost saturate at peaks of the input cycle. This type of operation minimizes power dissipation in the transistor, which is zero when the transistor is cut off, low when it is saturated, and much higher when it is in the normal, linear operating range. Since Class C amplifiers are biased beyond cutoff, they would be expected to have zero power dissipation with no input. This is indeed the case, provided the bias is independent of the input signal.

There remains the problem of distortion. Figure 9.9(a), a simplified circuit for a Class C amplifier, shows how the distortion is kept to a reasonable level. The process can be explained using either the time or the frequency domain. In the time domain, the output tuned circuit is excited once per cycle by a pulse of collector current. This keeps oscillations going in the resonant circuit. They are damped oscillations, of course, but the Q of the circuit will be high enough to ensure that the amount of damping that takes place in one cycle is negligible, and the output is a reasonably accurate sine wave. The frequency-domain explanation is even simpler: the resonant circuit constitutes a bandpass filter that passes the fundamental frequency and attenuates harmonics and other spurious signals. In many cases, especially where the amplifier is the final stage of a transmitter, additional filtering after the amplifier further reduces the output of harmonics.

FIGURE 9.9
Class C amplifier

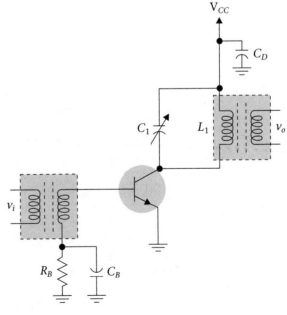

(a) Circuit

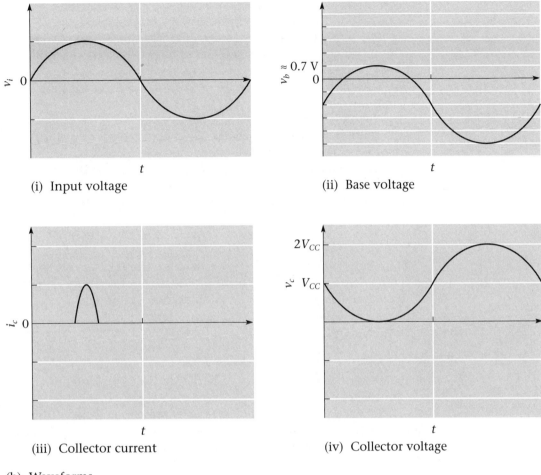

(i) Input voltage

(ii) Base voltage

(iii) Collector current

(iv) Collector voltage

(b) Waveforms

FIGURE 9.9 (*continued*)

From the foregoing description, it might seem that Class C amplifiers would have to have narrow bandwidth. Many of them do, but this is not a requirement as long as the output is connected to a low-pass filter that will attenuate all the harmonics that are generated.

Let us look at the circuit of Figure 9.9(a) a little more closely. We begin with the bias circuit, which is unconventional. Far from being biased in the middle of the linear operating range, as for Class A, a Class C amplifier must be biased beyond cutoff. For a bipolar transistor, that can mean no base bias at all, since a base-to-emitter voltage of about 0.7 V is needed for conduction. With no signal at its input, the transistor will have no base current and no collector current and will dissipate no power.

If a signal with a peak voltage of at least 0.7 V is applied to the input of the amplifier, the transistor will turn on during positive peaks. It might seem at first glance that an input signal larger than this would turn on the transistor for more and more of the cycle, so that with large signals the amplifier would approach Class B operation, but this is not the case. The base-emitter junction rectifies the input signal, charging C_B to a negative dc level that increases with the amplitude of the input signal. This self-biasing circuit will allow the amplifier to operate in Class C with a fairly large range of input signals.

The collector circuit can be considered next. At the peak of each input cycle, the transistor turns on almost completely. This effectively connects the bottom end of the tuned circuit to ground. Current flows through the coil L_1.

When the input voltage decreases a little, the transistor turns off. For the rest of the cycle, the transistor represents an open circuit between the lower end of the tuned circuit and ground. The current will continue to flow in the coil, decreasing gradually until the stored energy in the inductor has been transferred to capacitor C_1, which becomes charged. The process then reverses, and oscillation takes place. Energy is lost in the resistance of the inductor and capacitor, and of course energy is transferred to the load, so the amplitude of the oscillations would gradually be reduced to zero, except for one thing: Once each cycle, the transistor turns on, another current pulse is injected, and, because of this, oscillations continue indefinitely at the same level.

Figure 9.9(b) shows some of the waveforms associated with the Class C amplifier. Part (i) shows the input signal, and part (ii) shows the actual signal applied to the base. Note the bias level that is generated by the signal itself. Part (iii) shows the pulses of collector current, and part (iv), the collector voltage. It should not be surprising that this reaches a peak of almost $2V_{CC}$. The peak voltage across the inductor must be nearly V_{CC}, since, when the transistor is conducting, the top end of the coil is connected to V_{CC} and the bottom end is almost at ground potential ("almost" because, even when saturated, there will be a small voltage across the transistor). At the other peak of the cycle, the inductor voltage will have the same magnitude but opposite polarity and will add to V_{CC} to make the peak collector voltage nearly equal to $2V_{CC}$.

This description implies that the output tuned circuit must be tuned fairly closely to the operating frequency of the amplifier, and that is indeed the case. Since the transistor must swing between cutoff and something close to saturation for Class C operation, it is also implicit in the design that this amplifier will be nonlinear; that is, doubling the amplitude of the input signal will not double the output.

Class C amplification can be used with either field-effect or bipolar transistors. It is also quite commonly used with vacuum tubes in large transmitters. Class D is similar but even more extreme, with the active device acting strictly as a switch.

Transmitter Specifications Transmitters for different functions have some different types of specifications, but there are a number of types of ratings that are common to all or most transmitters. Some of the most common are described in the remainder of this section.

Frequency Accuracy and Stability The exact requirements for accuracy and stability of the transmitter frequency vary with the use to which the transmitter is put and are set by government regulatory bodies: the Federal Communications Commission in the United States and the Communications Branch of Industry Canada in Canada, for example. Depending on the application, frequency accuracy and stability may be specified in hertz or as a percentage of the operating frequency. It is easy to convert between the two methods, as the following example shows.

EXAMPLE 9.4

A crystal oscillator is accurate within ±0.005%. How far off frequency could its output be at 270 MHz?

SOLUTION

The frequency could be out by 0.005% of 270 MHz which is

$$\Delta f = \frac{0.005}{100} \times 270 \times 10^6$$

$$= 13.5 \times 10^3 \text{ Hz}$$

$$= 13.5 \text{ kHz}$$

Frequency Agility *Frequency agility* refers to the ability to change operating frequency rapidly, without extensive retuning. In a broadcast transmitter this is not a requirement, since such stations rarely change frequency. When they do it is a major, time-consuming operation involving extensive changes to the antenna system as well as the retuning of the transmitter.

With other services, for example citizens' band radio, the situation is different. Rapid retuning to any of the 40 available channels is essential to any modern CB transceiver. Cellular and PCS phones use many more frequencies and must change frequencies quickly under microprocessor control. In addition to a frequency synthesizer for setting the actual transmitting frequency, such transmitters are required to use broadband techniques throughout so that frequency changes can be made instantly with no retuning. The cellular and PCS base stations, on the other hand, can use a separate transmitter for each channel.

Spectral Purity   All transmitters produce **spurious signals**. That is, they emit signals at frequencies other than those of the carrier and the sidebands required for the modulation scheme in use. Spurious signals are often harmonics of the operating frequency or of the carrier oscillator if it operates at a different frequency.

All amplifiers produce harmonic distortion. Class C and switching amplifiers, especially, produce a large amount of harmonic energy. All frequencies except the assigned transmitting frequency must be filtered out to avoid interference with other transmissions.

The filtering of harmonics can never be perfect, of course, but in modern, well-designed transmitters it is very effective. Usually harmonics and other spurious signals are specified in decibels with respect to the unmodulated carrier level. The abbreviation for this is dBc.

EXAMPLE 9.5

An FM transmitter has a power output of 10 W and a specified maximum level for its second harmonic of −60 dBc. A reading of −40 dBm for the second harmonic is obtained using a spectrum analyzer. Is this reading within specifications?

SOLUTION

FM is a constant-power modulation scheme so the unmodulated carrier power is 10 W, which corresponds to 40 dBm. The maximum allowable second-harmonic power is 60 dB less than this, or −20 dBm. The measured reading of −40 dBm is less than this, so the transmitter is within specifications.

Power Output   Wireless communication systems often require careful calibration of transmitter power. The old "more is better" idea has been completely discarded in cellular radio systems.

Different modulation schemes require different power measurements. AM and FM transmitters for analog communication and all digital systems are generally rated in terms of carrier power output with no modulation. This specification would be useless for single-sideband suppressed-carrier systems, so their power is measured at the peak of the transmitted envelope and known as *peak envelope power* (*PEP*).

Measurement of transmitter power is easy when the signal is unmodulated. Many instruments will even measure power while the transmitter is connected to an antenna. The Thruline™ wattmeter shown in Figure 9.10 is an example. With modulated signals that have an envelope, however, many power meters (including the one shown in the figure) become inaccurate. True-average power reading meters are available: the classic method is to use a thermistor to measure the rise in temperature in a resistive load and then calculate the power from that measurement.

FIGURE 9.10
Thruline wattmeter
(Courtesy of Bird
Electronics)

When testing communication transmitters, the technologist should be aware of the rated duty cycle of the transmitter. Many transmitters designed for two-way voice communication are not rated for continuous operation at full power, since it is assumed that the operator will normally be talking for less than half the time, and probably only for a few seconds to a minute at a time. This is easy to forget when testing, and the result can be overheated transistors or other components. Any full-duplex system, as used in cellular telephone systems, for instance, must be rated for continuous duty.

Efficiency Transmitter efficiency is important for two reasons. The most obvious one is energy conservation. This is especially important where very large power levels are involved as in broadcasting, or at the other extreme of the power-level range, where hand-held operation from batteries is required.

Another reason for achieving high efficiency becomes apparent when we consider what happens to the power that enters the transmitter from the power supply but does not exit via the antenna. It is converted into heat in the transmitter, and this heat must be dissipated. Large amounts of heat require large components, heat sinks, fans, and in the case of some high-powered transmitters, even water cooling. All of these add to the cost of the equipment. Miniaturized transmitters like those in cell phones have difficulty removing heat quickly enough because of their small surface area, so high efficiency is very important even at quite low power levels.

When discussing efficiency it is important to distinguish between the efficiency of an individual stage and that of the entire transmitter. Knowing the efficiency of an amplifier stage is useful in designing cooling systems and sizing power supplies. In calculating energy costs, on the other hand, what is important is the overall efficiency of the system. This is the ratio of output power to power input from the primary power source, whether it be the ac power line or a battery. Overall efficiency is reduced by factors such as losses in the power supply.

Efficiency is easy to measure. The output power must be measured accurately (see the previous section). The input power can be found from the power supply current and voltage. The efficiency is simply

$$\eta = \frac{P_o}{P_s} \tag{9.6}$$

where

η = efficiency

P_o = output power

P_s = power from power supply

EXAMPLE 9.6

A mobile transmitter produces 24 W as measured at the antenna. It draws 3.4 A of current when connected to a 13.8 V power source. (Mobile equipment is usually rated for approximately this voltage, which corresponds to the battery voltage of a vehicle with a 12 V battery when the engine is running.) Calculate its efficiency.

SOLUTION

First calculate the supply power

$$P_s = 13.8 \text{ V} \times 3.4 \text{ A}$$
$$= 46.92 \text{ W}$$

Now calculate the efficiency.

$$\eta = \frac{P_o}{P_s}$$

$$= \frac{24}{46.92}$$

$$= 0.511 \text{ or } 51.1\%$$

Modulation Fidelity   As mentioned in Chapter 1, an ideal communication system allows the original information signal to be recovered exactly, except for a time delay. Any distortion introduced at the transmitter is likely to remain; in most cases it will not be possible to remove it at the receiver. It might be expected then, that a transmitter would be capable of handling any baseband frequency to preserve the information signal as much as possible. In practice, however, the baseband spectrum must be restricted in order to keep the transmitted bandwidth within legal limits.

In addition, some form of compression, where low-level baseband signals are amplified more than high-level signals, is often used to keep the modulation index high. This distorts the original signal by reducing the dynamic range, which is the ratio between the levels of the loudest and the quietest passages in the audio signal. The result is an improved signal-to-noise ratio at the receiver, at the expense of some distortion. The effects of compression, while unpleasant when used with music, are not usually objectionable in voice communication.

The effects of compression on dynamic range can be removed by applying an equal and opposite expansion at the receiver. Such expansion would involve giving more gain to signals at higher levels. The combination of compression and expansion, called *companding,* is quite common in communication systems.

Other kinds of distortion, such as harmonic and intermodulation distortion, also have to be kept within reasonable limits. As would be expected, low distortion levels are more important in the broadcast service than in the mobile-radio service.

9.3   Receivers

The receiver performs an inverse function to that of the transmitter. It must separate the desired signal from others present at the antenna, amplify it greatly (often 100 dB or more), and demodulate it to recover the original baseband signal. There are many ways to do this, but almost all of them use the **superheterodyne principle**, invented by Edwin Armstrong in 1918.

The Basic Superheterodyne Receiver

Figure 9.11 shows the basic topology of a superheterodyne receiver. As for transmitters, there are infinite variations of this basic layout.

The signal chain begins with one or more stages of RF amplification. Low-cost receivers sometimes omit the RF amplifier, but they do include some sort of input filter, such as a tuned circuit. The input filter and RF amplifier are sometimes referred to as the *front end* of a receiver.

The next stage is a mixer. The signal frequency is mixed with a sine-wave signal generated by an associated stage called the *local oscillator*. A difference frequency is created, which is called the *intermediate frequency* (*IF*). The local oscillator is tunable, so the IF is fixed regardless of the signal frequency. The combination of mixer and local oscillator is known as a *converter*.

The mixer is followed by the *IF amplifier,* which provides most of the receiver's gain and **selectivity**. Selectivity is the ability to separate signals from interference and noise. Generally there are two or more IF stages, with selectivity provided either by resonant circuits or, in recent designs, by a crystal or ceramic filter. The use of a fixed IF greatly simplifies the problem of achieving adequate gain and selectivity.

The remainder of the receiver is straightforward. There is a *demodulator* followed by whatever baseband processing is required to restore the original signal.

The *automatic gain control* (*AGC*) adjusts the gain of the IF—and sometimes the RF—stages in response to the strength of the received signal,

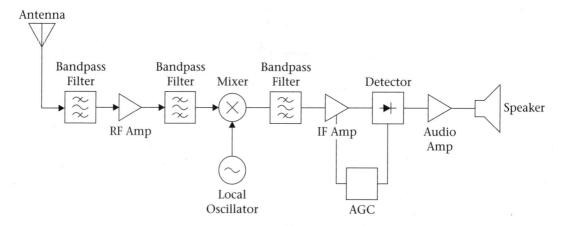

FIGURE 9.11 Superheterodyne receiver

providing more gain for weak signals. This allows the receiver to cope with the very large variation in signal levels found in practice. This range may be more than 100 dB for a communications receiver.

Now that we have looked at the general structure of a typical superheterodyne receiver, it is time to consider the design of the various stages in a little more detail, starting at the input and working our way through to the output.

The Front End

Inexpensive receivers often omit the RF amplifier stage, especially if they are designed for operation at low to medium frequencies where atmospheric noise entering the receiver with the signal is likely to be more significant than noise generated within the receiver itself. Nearly all other receivers use a single RF stage to provide gain before the relatively noisy mixer.

It is not necessary for the front end to have a small enough bandwidth to separate the desired signal from others on nearby frequencies. This selectivity is more easily provided in the IF amplifier. On the other hand, it is necessary for this stage to reject signals that, even though they are not close in frequency to the desired signal, may cause problems in later stages. Sometimes the RF stage is tuned along with the local oscillator, but more often in current designs an RF stage uses relatively broadband filters so that the frequency band of interest is covered but spurious responses are excluded.

The RF stage is a Class A amplifier. It should have a good noise figure and a wide dynamic range. AGC can be used, but designers often prefer not to apply it to the RF stage because any alteration of the stage gain from the optimum level will degrade the noise figure.

A switch is sometimes provided to remove the RF stage from the signal path for strong signals to prevent overloading; such signals are applied directly to the mixer after going through the input filter. Another position on the same switch may add a few decibels of attenuation to prevent very strong signals from overloading the mixer. Of course, both removing the RF amplifier and adding attenuation have adverse effects on the receiver's noise performance, but for very strong signals it is less important to maintain the noise figure than to avoid overloading.

The Mixer and Local Oscillator

Any type of mixer circuit will work. Diode mixers are generally rejected as too noisy and too lossy, except in the simplest receivers. Either bipolar transistors or FETs can be used as mixers, but the latter are preferred because they create fewer intermodulation distortion components. The problem here, as in the RF amplifier, is not so much distortion of the modulating signal as the creation of spurious responses due to interactions between desired and interfering signals.

It is extremely important that the local oscillator be stable, as any frequency change will result in the receiver drifting away from the station to which it is tuned. *LC* oscillators must be carefully designed for stability. This is expensive, of course, and some designs have used various types of automatic frequency control (AFC), which reduce drift by feeding an error signal back to the local oscillator.

Another approach to local-oscillator stability is to use crystal control. The L.O. can be a simple crystal oscillator with a switch to change crystals for different channels. This becomes unwieldy with more than a few channels, so crystal-controlled frequency synthesizers have become very popular in newer designs. The use of a synthesizer also allows remote control, computer control, direct frequency or channel number entry via a keypad, and other similar conveniences. AFC is not necessary with a properly designed synthesizer.

Oscillator spectral purity is also important. Any frequency components at the oscillator output can mix with incoming signals and creating undesirable spurious responses. Noise generated by the local oscillator degrades the noise performance of the receiver and should be minimized. Some designs use a bandpass filter to remove spurious signal components and noise from the local oscillator signal before it is applied to the mixer.

Generally the intermediate frequency is chosen to be lower than the signal frequency to make it easier to provide gain and selectivity in the IF amplifier. The IF is the difference between the incoming signal frequency and the local oscillator frequency. The local-oscillator frequency can be higher than the signal frequency, in which case the receiver is said to use high-side injection, or lower than the signal frequency (low-side injection).

For high-side injection, we need

$$f_{IF} = f_{LO} - f_{SIG} \tag{9.7}$$

so

$$f_{LO} = f_{SIG} + f_{IF} \tag{9.8}$$

where

f_{IF} = intermediate frequency
f_{LO} = local-oscillator frequency
f_{SIG} = signal frequency

Similarly, for low-side injection we need

$$f_{IF} = f_{SIG} - f_{LO} \tag{9.9}$$
$$f_{LO} = f_{SIG} - f_{IF} \tag{9.10}$$

An example will show how this works.

EXAMPLE 9.7 ▼

A receiver tunes from 500 MHz to 600 MHz with an IF of 20 MHz. Calculate the range of local-oscillator frequencies required if the receiver uses:

(a) high-side injection

(b) low-side injection

SOLUTION

(a) For a signal frequency of 500 MHz we require

$$f_{LO} = f_{SIG} + f_{IF}$$
$$= 500 \text{ MHz} + 20 \text{ MHz}$$
$$= 520 \text{ MHz}$$

Similarly, a signal frequency of 600 MHz requires a local-oscillator frequency of 620 MHz.

(b) For a signal frequency of 500 MHz we need

$$f_{LO} = f_{SIG} - f_{IF}$$
$$= 500 \text{ MHz} - 20 \text{ MHz}$$
$$= 480 \text{ MHz}$$

Similarly, for a signal frequency of 600 MHz, the local oscillator must operate at 580 MHz.

▲

The above example shows that if either Equation (9.7) or (9.9) is satisfied, a signal will be produced at the intermediate frequency. For any given local-oscillator and IF frequencies then, there ought to be two signal frequencies that will mix with the local oscillator to produce a signal at the IF. An example should make this clear.

EXAMPLE 9.8 ▼

A receiver has its local oscillator set to 550 MHz and its IF amplifier designed to work at 20 MHz. Find two signal frequencies that can be received.

SOLUTION

First we use Equation (9.7):

$$f_{IF} = f_{LO} - f_{SIG}$$

$$f_{SIG} = f_{LO} - f_{IF}$$
$$= 550 \text{ MHz} - 20 \text{ MHz}$$
$$= 530 \text{ MHz}$$

Now try Equation (9.9):

$$f_{IF} = f_{SIG} - f_{LO}$$
$$f_{SIG} = f_{LO} + f_{IF}$$
$$= 550 \text{ MHz} + 20 \text{ MHz}$$
$$= 570 \text{ MHz}$$

Obviously, it is not a good idea to have the receiver tune to two frequencies at once. Which one is received depends on the bandpass filter in the front end; it must be designed to accept one and reject the other. The undesired signal is called the **image frequency**. You can see why by looking at Figure 9.12. Think of the local oscillator frequency as a mirror; the two possible signals are equidistant from it, on opposite sides. One of these is the desired frequency; the other is the image.

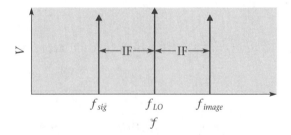

(a) High-side injection

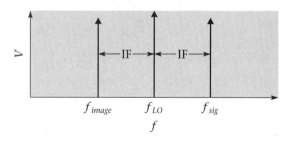

(b) Low-side injection

FIGURE 9.12 Images

IF Amplifier The IF amplifier accounts for most of the receiver's gain and selectivity. The classical method to provide this is to use several stages, to give sufficient gain, coupled by tuned transformers that provide the selectivity. In modern receivers, other types of bandpass filters (for example, crystal and ceramic filters) are more popular. The filter is often placed at the beginning of the IF chain, right after the mixer.

The IF amplifier of any receiver where the signal has an envelope must be linear to avoid distorting the signal envelope. This requires that the gain of the amplifier be reduced with strong signals to prevent overloading. The circuitry for this is called *automatic gain control* (*AGC*).

Receivers for constant-amplitude signals, such as FM, need not be linear; in fact, it is better if they incorporate limiting to avoid any effect on the demodulator due to amplitude variations.

It is important to understand the different functions of the bandpass filters in the front end and in the IF amplifier. Filtering before the mixer is required to remove image frequencies and any other signals that can mix with the local oscillator or with each other to produce a signal at the IF. Such signals cannot be removed after the mixer; they are at the intermediate frequency so they will be passed by the filters in the IF amplifier. The IF filters, on the other hand, should restrict the receiver bandwidth to that of the signal, thereby reducing noise and eliminating interfering signals that are close in frequency to the desired signal.

Demodulation

Demodulation, also called *detection,* is the inverse of modulation. Its purpose is to restore the original baseband signal. The type of circuit used depends on the modulation scheme. A few representative types will be examined here.

Analog Demodulators Full-carrier amplitude modulation can be demodulated with a very simple circuit, as shown in Figure 9.13 on page 346. This accounts for its early popularity. All that is necessary to rectify the signal is to remove half of the envelope and then filter out the intermediate-frequency component. A simple RC filter is sufficient for this.

Demodulation of SSBSC AM is more complex. The carrier must be reinserted at the receiver. Next a product detector is used. This is the same as the balanced modulator described earlier, since both are actually modulators. It can be shown that multiplying the SSB signal with the carrier results in the original baseband signal with some additional high-frequency components that are easily filtered out.

Analog FM requires a detector that converts frequency variations to voltage variations. Many circuits have been used for this, but two types account for the overwhelming majority of current FM detectors.

PLL Method The use of a phase-locked loop to demodulate FM signals is very straightforward. See Figure 9.14 on page 347, for a typical PLL detector. The incoming FM signal is used to control the frequency of the VCO. As the incoming frequency varies, the PLL generates a control voltage to change the VCO frequency, which will follow that of the incoming signal. This control voltage varies at the same rate as the frequency of the incoming signal, thus it can be used directly as the output of the circuit. Unlike the PLLs used in transmitter modulator circuits, this PLL must have a short time constant so that it can follow the modulation. The capture range of the PLL

FIGURE 9.13
AM detector

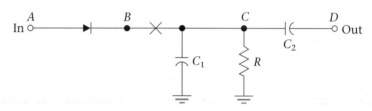

(a) Circuit

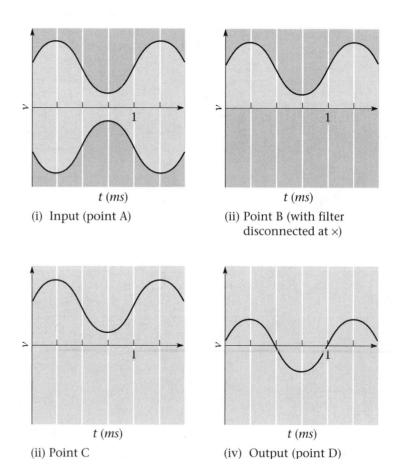

(i) Input (point A)

(ii) Point B (with filter
disconnected at ×)

(ii) Point C

(iv) Output (point D)

(b) Time domain

is not important, since the free-running frequency of the VCO will be set
equal to the signal's carrier frequency at the detector (that is, to the center of
the IF passband). The lock range must be at least twice the maximum deviation of the signal. If it is deliberately made wider, the detector will be able to

FIGURE 9.14
PLL FM detector

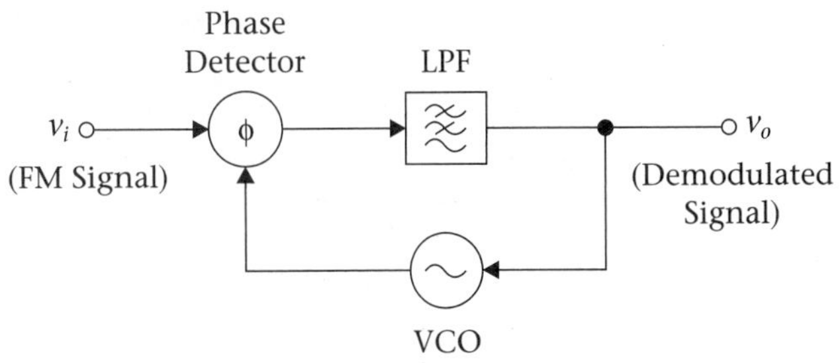

Phase Detector   LPF

v_i (FM Signal) ϕ

v_o (Demodulated Signal)

VCO

function in spite of a small amount of receiver mistuning or local oscillator drift. Amplitude variations of the input signal will not affect the operation of this detector, unless they are so great that it stops working altogether.

Quadrature Detector   Like the PLL detector, the *quadrature detector* is adapted to integrated circuitry. In the quadrature detector (see Figure 9.15), the incoming signal is applied to one input of a phase detector. The signal is also applied to a phase-shift network. This consists of a capacitor (C_1 in the figure) with high reactance at the carrier frequency, which causes a 90° phase shift (this is the origin of the term *quadrature*). The tuned circuit consisting of L_1 and C_2 is resonant at the carrier frequency. Therefore, it causes no phase shift at the carrier frequency but does provide a phase shift at other frequencies that will add to or subtract from the basic 90° shift caused by C_1.

The output of the phase-shift network is applied to the second input of the phase detector. When the input frequency changes, the angle of phase shift in the quadrature circuit varies, as the resonant circuit becomes inductive or capacitive. The output from the phase detector varies at the signal frequency but has an average value proportional to the amount the phase angle differs from 90°. Low-pass filtering the output will recover the modulation. Figure 9.15 shows this function, which is accomplished by a simple first-order filter consisting of R_2 and C_3. The cutoff frequency should be well

FIGURE 9.15
Quadrature FM detector

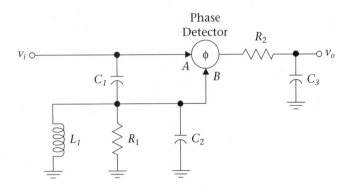

Phase Detector

above the highest modulating frequency and well below the receiver inter-mediate frequency.

The phase detector is the same as is used for phase-locked loops. It can be an analog multiplier (product detector) or a digital gate (either an AND or an exclusive-OR gate).

Digital Demodulators FSK, being very similar to analog FM, can be de-tected using similar means. QPSK, which is more common in wireless use, can be demodulated by the circuit shown in block-diagram form in Figure 9.16. Note that it is quite similar to the QPSK modulator shown in Figure 9.7.

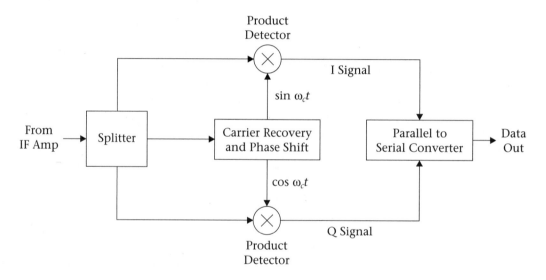

FIGURE 9.16 QPSK demodulator

In order to demodulate the QPSK signal correctly, the carrier must be re-covered from the original signal. This can be done using a phase-locked loop. The incoming signal is then applied to two multipliers, which are also fed the carrier signal but with a 90° phase difference between them. The re-sulting I and Q signals must be applied to a parallel-to-serial converter to restore the original serial bit stream.

Receiver Specifications This section defines and describes some of the specifications commonly used to describe and compare receivers. Receivers for specialized applica-tions often have additional specifications, but those listed below apply to most kinds of receiver.

Sensitivity The transmitted signal may have a power level, at the transmitting antenna, ranging from milliwatts to hundreds of kilowatts. However, the losses in the path from transmitter to receiver are so great that the power of the received signal is often measured in dBf, that is, decibels relative to one femtowatt (1 fW = 1 × 10⁻¹⁵ W). A great deal of amplification is needed to achieve a useful power output. To operate an ordinary loudspeaker, for instance, requires a power on the order of 1 W, or 150 dBf.

Because received signals are often quite weak, noise added by the receiver itself can be a problem. Most demodulators are inherently noisy and also operate best with fairly large signals (hundreds of millivolts), so some of the amplification must take place before demodulation.

The ability to receive weak signals with an acceptable signal-to-noise ratio is called **sensitivity**. It is expressed in terms of the voltage or power at the antenna terminals necessary to achieve a specified signal-to-noise ratio or some more easily measured equivalent. When digital modulation is used, the criterion may be bit-error rate rather than signal-to-noise ratio.

Selectivity In addition to noise generated within the receiver, there will be noise coming in with the signal, as well as interfering signals with frequencies different from that of the desired signal. All of these problems can be reduced by limiting the receiver bandwidth to that of the signal (including all its sidebands). The ability to discriminate against interfering signals is known as *selectivity*.

Selectivity can be expressed in various ways. The bandwidth of the receiver at two different levels of attenuation can be specified. The bandwidth at the points where the signal is 3 or 6 dB down is helpful in determining whether all the sidebands of the desired signal will be passed without attenuation. To indicate the receiver's effectiveness in rejecting interference, a bandwidth for much greater attenuation, for example 60 dB, should also be given.

The frequency-response curve for an ideal IF filter would have a square shape with no difference between its bandwidth at 6 dB and 60 dB down. The closer the two bandwidths are, the better the design. The ratio between these bandwidths is called the shape factor. That is,

$$ SF = \frac{B_{-60}}{B_{-6}} \tag{9.11} $$

where

SF = shape factor

B_{-60} = the bandwidth at 60 dB down from maximum

B_{-6} = the bandwidth at 6 dB down from maximum

The shape factor should be as close to one as possible. The following example shows the effect of changing the shape factor on the rejection of interfering signals.

EXAMPLE 9.9

Calculate the shape factors for the two IF response curves shown in Figure 9.17, and calculate the amount by which the interfering signal shown would be attenuated in each case.

FIGURE 9.17

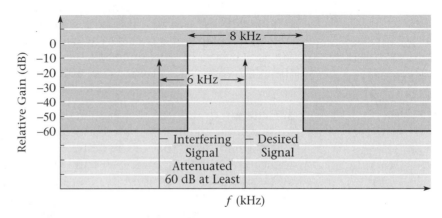

(a) Shape factor = 1   Bandwidth (6 dB) = 8 kHz

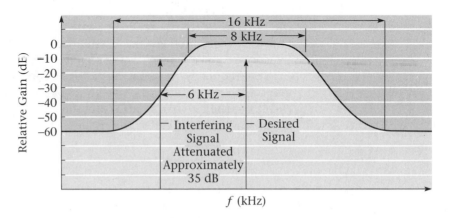

(b) Shape factor = 2   Bandwidth (6 dB) = 8 kHz

SOLUTION

Figure 9.17(a) shows an ideal filter. Since the −6 dB and −60 dB bandwidths are equal, the shape factor is 1. The interfering signal is attenuated by 60 dB.

In Figure 9.17(b), the −6 dB bandwidth is 8 kHz and the −60 dB bandwidth is 16 kHz, so the shape factor is 2. The interfering signal is attenuated approximately 35 dB compared to the desired signal.

Adjacent channel rejection is another way of specifying selectivity that is commonly used with channelized systems. It is defined as the number of decibels by which an adjacent channel signal must be stronger than the desired signal for the same receiver output.

Alternate channel rejection is also used in systems, such as FM broadcasting, where stations in the same locality are not assigned to adjacent channels. The alternate channel is two channels removed from the desired one. It is also known as the second adjacent channel. Figure 9.18 shows the relation between adjacent and alternate channels. For example, if 94.1 MHz were assigned to a station, the adjacent channel at 94.3 MHz would not be assigned in the same community, but the alternate channel at 94.5 MHz might be. Any signal at 94.3 MHz, then, will be relatively weak, and a receiver's ability to reject strong adjacent channel signals will be less important than its ability to reject the alternate channel (in this case, the one at 94.5 MHz).

FIGURE 9.18
Adjacent and alternate channels

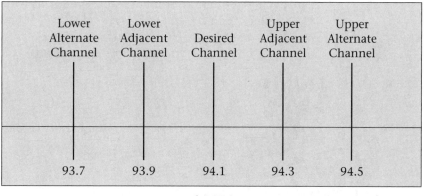

Lower Alternate Channel	Lower Adjacent Channel	Desired Channel	Upper Adjacent Channel	Upper Alternate Channel
93.7	93.9	94.1	94.3	94.5

f (MHz)

Distortion In addition to good sensitivity and selectivity, an ideal receiver would reproduce the original modulation exactly. A real receiver, however, subjects the signal to several types of distortion. They are the same types encountered in other systems: harmonic and intermodulation distortion, uneven frequency response, and phase distortion.

Harmonic distortion occurs when frequencies are generated which are multiples of those in the original modulating signal.

Intermodulation takes place when frequency components in the original signal mix in a nonlinear device, creating sum and difference frequencies. A type of intermodulation peculiar to receivers consists of mixing between the desired signal and an interfering one that is outside the IF passband of the receiver but within the passband of the RF stage and is therefore present in the mixer. This can result in interference from a strong local station that is not at all close in frequency to the desired signal. In addition, any non-linearities in the detector or any of the amplification stages can cause intermodulation between components of the original baseband signal.

EXAMPLE 9.10

A receiver is tuned to a station with a frequency of 100 MHz. A strong signal with a frequency of 200 MHz is also present at the amplifier. Explain how intermodulation between these two signals could cause interference.

SOLUTION

See Figure 9.19. Assume that the tuned circuit at the mixer input has insufficient attenuation at 200 MHz to block the interfering signal completely. (Remember that the interfering signal may be much stronger than the desired signal and therefore may still have similar strength to the desired signal, even after being attenuated by the input tuned circuit.) The two signals could mix in the mixer stage to give a difference frequency of 100 MHz, at the same frequency as the original signal. This interfering signal will then pass through the rest of the receiver in the same way as the desired signal.

FIGURE 9.19

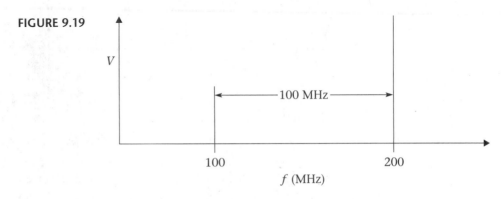

The frequency response of an AM receiver is related to its IF bandwidth. Restricting the bandwidth to less than the full width of the signal reduces the upper limit of response, since the sidebands corresponding to higher modulating frequencies are farther away from the carrier. For example, a receiver bandwidth of 10 kHz will allow a modulation frequency of 5 kHz to be reproduced. Reducing the bandwidth to 6 kHz reduces the maximum audio frequency that can be reproduced to 3 kHz. In addition, any unevenness within the IF passband will affect the audio frequency response.

The relation between frequency response and bandwidth is not so simple for FM receivers, since a single baseband frequency produces multiple sets of sidebands. To reduce distortion it is necessary for an FM receiver to have enough bandwidth to encompass all the sidebands with significant amplitude.

Phase distortion is slightly more difficult to understand than frequency response. Of course, the signal at the receiver output will not be in phase with the input to the transmitter: there will be some time delay, which can be translated into a phase shift that increases linearly with frequency. Phase distortion consists of irregular shifts in phase and is quite a common occurrence when signals pass through filters. Phase distortion is unimportant for analog voice communication, but it is crucial that it be minimized when digital PSK is used.

Dynamic Range   As mentioned above, a receiver must operate over a considerable range of signal strengths. The response to weak signals is usually limited by noise generated within the receiver. On the other hand, signals that are too strong will overload one or more stages, causing unacceptable levels of distortion. The ratio between these two signal levels, expressed in decibels, is the **dynamic range** of the receiver.

The above description is actually a bit simplistic. A well-designed AGC system can easily vary the gain of a receiver by 100 dB or so. Thus, almost any receiver can cope with very wide variations of signal strength, provided only one signal is present at a time. This range of signal strengths is sometimes referred to as dynamic range, but should properly be called AGC control range.

The hardest test of receiver dynamic range occurs when two signals with slightly different frequencies and very different power levels are applied to the antenna input simultaneously. This can cause overloading of the receiver input stage by the stronger signal, even though the receiver is not tuned to its frequency. The result can be *blocking* (also called *desensitization* or *desense*), which is a reduction in sensitivity to the desired signal. Intermodulation between the two signals is also possible.

EXAMPLE 9.11

A receiver has a sensitivity of 0.5 µV and a blocking dynamic range of 70 dB. What is the strongest signal that can be present along with a 0.5 µV signal without blocking taking place?

SOLUTION

Since both signal voltages are across the same impedance, the input imped-ance of the receiver, the general equation

$$10 \log \frac{P_1}{P_2} = 20 \log \frac{V_1}{V_2}$$

can be used. Here, let signal 1 be the stronger. Then

$$R = 10 \log \frac{P_1}{P_2}$$

where

$$R = \text{dynamic range in dB}$$

So

$$R = 20 \log \frac{V_1}{V_2}$$

Given that $R = 70$ dB, $V_2 = 0.5$ µV, we find

$$\frac{V_1}{V_2} = \text{antilog } \frac{R}{20}$$

$$V_1 = V_2 \text{ antilog } \frac{R}{20}$$

$$= 0.5 \, \mu V \text{ antilog } \frac{70}{20}$$

$$= 1581 \, \mu V$$

$$= 1.581 \, \text{mV}$$

Spurious Responses   The superheterodyne has many important advantages over simpler receivers, but it is not without its problems. In particular, it has a tendency to receive signals at frequencies to which it is not tuned and sometimes to generate signals internally, interfering with reception. Careful design can reduce these **spurious responses** almost to insignificance, but they will still be present.

Image frequencies have already been discussed. Image rejection is defined as the ratio of voltage gain at the input frequency to which the receiver is tuned to gain at the image frequency. Image rejection is usually expressed in decibels.

An image must be rejected prior to mixing. Once it has entered the IF chain, the image is indistinguishable from the desired signal and impossible to filter out. Image rejection is accomplished by filters before the mixer. Using a higher intermediate frequency can improve image rejection by placing the image farther away in frequency from the desired signal, where it can more easily be removed by tuned circuits before the mixer.

As well as images, superheterodyne receivers are subject to other problems. The local oscillator will have harmonics, for instance, and these can mix with incoming signals to produce spurious responses. The incoming signal may also have harmonics created by distortion in the RF stage. In fact, it is possible for a receiver to respond to any frequency given by the equation:

$$f_s = \left(\frac{m}{n}\right)f_{LO} \pm \frac{f_{IF}}{n} \tag{9.12}$$

where

$$f_s = \text{frequency of the spurious response}$$
$$f_{LO} = \text{local oscillator frequency}$$
$$f_{IF} = \text{intermediate frequency}$$
$$m, n = \text{any integers}$$

EXAMPLE 9.12

A receiver has an IF of 1.8 MHz using high-side injection. If it is tuned to a frequency of 10 MHz, calculate the frequencies which can cause an IF response, for values of m and n ranging up to 2.

SOLUTION

First, we find f_{LO}:

$$f_{LO} = f_{sig} + f_{IF}$$
$$= 10 \text{ MHz} + 1.8 \text{ MHz}$$
$$= 11.8 \text{ MHz}$$

Now the problem is easily solved using Equation (9.12) and a table of values. All frequencies in the table are in MHz.

m	n	$(m/n)f_{LO}$	f_{IF}/n	$(m/n)f_{LO} + (f_{IF}/n)$	$(m/n)f_{LO} - (f_{IF}/n)$
1	1	11.8	1.8	13.6	10
1	2	5.9	0.9	6.8	5.0
2	1	23.6	1.8	25.4	21.8
2	2	11.8	0.9	12.7	10.9

Rearranging the results in the last two columns in order of ascending frequency gives us the frequencies to which the receiver may respond. In MHz they are:

5.0, 6.8, 10.0, 10.9, 12.7, 13.6, 21.8, 25.4

In spite of these problems, however, the superheterodyne remains the preferred arrangement for almost all receiving applications. It proves to be easier to improve the design to reduce its problems than to go to a different system.

9.4 Transceivers

In wireless communication systems, base stations often use separate transmitters and receivers, but portable and mobile units generally combine the transmitter and receiver into one unit called a **transceiver.** This provides convenience, reduced size and weight, and sometimes allows costs to be reduced by using some components for both transmitting and receiving.

Half-Duplex Transceivers

Many wireless communication systems use transceivers that switch from transmit to receive, either using a push-to-talk switch or voice-activated switching. In this case, it is often possible to use many components for both transmit and receive functions. Consider the VHF-FM transceiver whose block diagram is shown in Figure 9.20, for example.

Here, the CPU and the frequency synthesizer that it controls are used for both transmit and receive. When transmitting, the synthesizer doubles as carrier oscillator and FM modulator. When receiving, the synthesizer functions as the first local oscillator. Some transceivers also share audio circuitry.

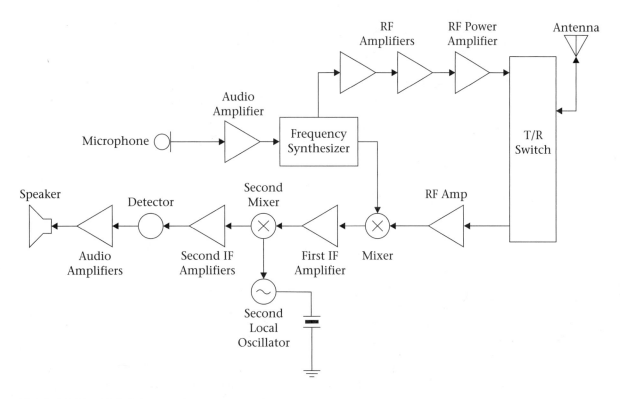

FIGURE 9.20 VHF-FM transceiver

Note that the receiver section of this transceiver uses *double conversion*; that is, there is a second local oscillator/mixer combination. The second local oscillator operates at a fixed frequency; the purpose of the second IF is to allow there to be a high first IF for good image rejection followed by a low second IF to simplify the design of a narrowband filter for good selectivity.

Half-duplex operation simplifies the connection of the transceiver to its antenna. A mechanical switch can be used, but as in the example, more often the switching is electronic. Since the antenna is never connected to both transmitter and receiver at the same time, there is no need for filters to separate transmitted and received signals.

Mobile and Portable Telephones When full-duplex operation is required as it is with cordless and cellular telephones, it is necessary to use two separate channels, one each for transmit and receive. They must be far enough apart in frequency that a filter can separate them, so that the transmitter signal does not overload, and possibly damage, the receiver. Figure 9.21 shows how this device, called a *duplexer,* fits into the system. Duplexers are also used at base stations to allow transmitters and receivers to use the same antenna simultaneously.

FIGURE 9.21
Use of duplexer
for full-duplex
communication

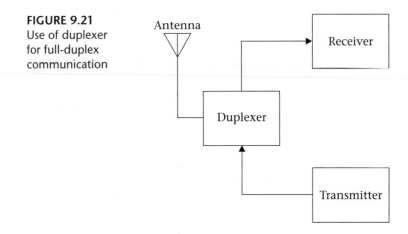

Summary The main points to remember from this chapter are:

- Transmitters are required to produce a modulated signal at the correct power level and frequency and with acceptable levels of spurious signals.
- Receivers perform the inverse operation to transmitters, restoring the original baseband signal with as little added noise and distortion as possible.
- The type of modulation and the point in the transmitter where modulation is introduced determine the type of amplifiers needed.
- Transmitter efficiency is an important consideration because of energy, size, and heat dissipation requirements.
- Most receivers use the superheterodyne principle, in which the signal is moved to a fixed intermediate frequency.
- Receiver design emphasizes sensitivity, which is the ability to receive weak signals, and selectivity, which is the ability to reject noise and interference.
- For a receiver, dynamic range is the ability to receive a weak signal in the presence of stronger signals.
- Transceivers often use some of the same circuitry for both transmitting and receiving.
- When a transmitter and receiver share the same antenna simultaneously, they must use different frequencies, and a filter called a duplexer must be used to separate the signals.

Equation List

$$f_0 = N f_{ref} \tag{9.1}$$

$$\eta = \frac{P_o}{P_s} \tag{9.6}$$

$$f_{IF} = f_{LO} - f_{SIG} \tag{9.7}$$

$$f_{LO} = f_{SIG} + f_{IF} \tag{9.8}$$

$$f_{IF} = f_{SIG} - f_{LO} \tag{9.9}$$

$$f_{LO} = f_{SIG} - f_{IF} \tag{9.10}$$

$$f_s = \left(\frac{m}{n}\right) f_{LO} \pm \frac{f_{IF}}{n} \tag{9.12}$$

Key Terms

driver   amplifier immediately preceding the power amplifier stage in a transmitter

dynamic range   ratio, usually expressed in decibels, between the strongest and weakest signals that can be present in a system

frequency multiplier   circuit whose output frequency is an integer multiple of its input frequency

frequency synthesizer   device to produce programmable output frequencies that are accurate and stable

image frequency   reception of a spurious frequency by a superheterodyne receiver, resulting from mixing of the unwanted signal with the local oscillator signal to give the intermediate frequency

mixer   nonlinear device designed to produce sum and difference frequencies when provided with two input signals

selectivity   ability of a receiver to discriminate against unwanted signals and noise

sensitivity   ability of a receiver to detect weak signals with a satisfactory signal-to-noise ratio

spurious responses   in a receiver, reception of frequencies other than that to which it is tuned

spurious signals   unwanted signals accidentally produced by a transmitter

superheterodyne principle   use of a mixer and local-oscillator combination to change the frequency of a signal

transceiver   a combination transmitter and receiver

variable-frequency oscillator (VFO)   an oscillator whose frequency can be changed easily, usually by means of a variable capacitor or inductor

voltage-controlled oscillator (VCO)   oscillator whose frequency can be changed by adjusting a control voltage

● Questions

1. What is frequency agility, and under what circumstances is it desirable?

2. Why is it necessary to suppress the emission of harmonics and other spurious signals by a transmitter?

3. What is meant by the overall efficiency of a transmitter?

4. Why is audio compression used with many transmitters?

5. What is meant by the duty cycle of a transmitter?

6. Is it possible to use Class C amplifiers to amplify an AM signal? Explain.

7. What advantages does the use of a frequency synthesizer for the oscillator stage of a transmitter have over:

   (a) a crystal-controlled oscillator?

   (b) an *LC* oscillator?

8. How is the power output of an SSB transmitter specified?

9. Why do FM transmitters use low-level modulation followed by Class C amplification?

10. How can the deviation of an FM signal be increased?

11. Explain the operation of a QPSK modulator.

12. Distinguish between low- and high-side injection of the local-oscillator signal in a receiver.

13. Explain how image-frequency signals are received in a superheterodyne receiver. How may these signals be rejected?

14. What are the main characteristics of a well-designed receiver RF amplifier stage?

15. Why is the stability of the local oscillator important?

16. In addition to images, what spurious responses are possible with superheterodyne receivers, and how are they caused?

17. State what is meant by the shape factor of a filter, and explain why a small value for the shape factor is better for the IF filter of a receiver.

18. What advantage is gained by using double conversion in a receiver?

19. Describe two modern types of FM detectors, and explain how each works. Use diagrams to help with the explanations.

20. Explain the operation of a QPSK demodulator.

● Problems

1. A frequency synthesizer of the type shown in Figure 9.2 is to generate signals at 5-kHz intervals from 10 to 12 MHz using a reference crystal oscillator operating at 10 MHz.

   (a) Find a suitable value for Q.

   (b) What is the range over which N has to be varied?

2. A frequency synthesizer of the type shown in Figure 9.2 has a crystal oscillator operating at 5 MHz, $Q = 100$, and N ranges from 1000 to 1200. Find the range of frequencies that can be generated and the minimum amount by which the output frequency can be varied.

3. Draw a block diagram for an SSBSC generator that will generate an USB signal with a (suppressed) carrier frequency of 9 MHz. Show the carrier-oscillator frequency and the filter center frequency and bandwidth.

4. Using the QPSK modulator shown in Figure 9.7, draw a series of vector diagrams showing the output for the following data input: 00100111.

5. Frequencies of 120 MHz and 100 MHz are applied to a mixer. Calculate the frequencies that appear at the output if the mixer is:

   (a) a balanced mixer.

   (b) a square-law mixer.

6. The frequency of a transmitter is guaranteed accurate to $\pm 0.0005\%$. What are the maximum and minimum frequencies at which it could actually be transmitting if it is set to transmit on a nominal carrier frequency of 472.05 MHz?

7. A transmitter has an output carrier power of 25 W. If spurious signals must be at a level of -70 dBc or less, what is the maximum power any of the spurious signals can have?

8. A transmitter is rated to supply 4 W of power to a 50-Ω load, while operating from a power supply that provides 13.8 V. The nominal supply current is 1 A. Calculate the overall efficiency of this transmitter.

9. A superheterodyne receiver is tuned to a frequency of 5 MHz when the local-oscillator frequency is 6.65 MHz.

   (a) What is the intermediate frequency?

   (b) Which type of injection is in use?

10. One receiver has a sensitivity of 1 μV and another a sensitivity of 10 dBf under the same measurement conditions. Both receivers have an input impedance of 50 Ω. Which receiver is more sensitive?

11. A receiver has a sensitivity of 0.3 μV, and the same receiver can handle a signal level of 75 mV without overloading. What is its AGC control range in dB?

12. The receiver of problem 11 has a blocking dynamic range of 80 dB. If the desired signal has a level of 10 μV, what is the maximum signal level that can be tolerated within the receiver passband?

13. A receiver uses low-side injection for the local oscillator and an IF of 1750 kHz. If the local oscillator is operating at 15.750 MHz,

    (a) to what frequency is the receiver tuned?

    (b) what is the image frequency?

14. An FM broadcast receiver with high-side injection and an IF of 10.7 MHz is tuned to a station at 91.5 MHz.

    (a) What is the local-oscillator frequency?

    (b) What is the image frequency?

    (c) Find four other spurious frequencies that could be picked up by this receiver.

15. A receiver's IF filter has a shape factor of 2.5 and a bandwidth, at the 6 dB down points, of 6 kHz. What is its bandwidth at 60 dB down?

16. What is the shape factor of the filter sketched in Figure 9.22?

FIGURE 9.22

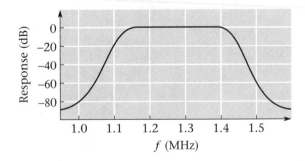

17. The block diagram of Figure 9.23 shows a double conversion receiver. It has two mixers, two local oscillators, and two intermediate frequencies. The idea is to have a high first IF for image rejection and a low second IF for gain and selectivity.

    (a) Does the first mixer use low-side or high-side injection?

    (b) What is the second IF frequency?

    (c) Suppose that the input signal frequency were changed to 17.000 MHz. What then would the frequencies of the two local oscillators be?

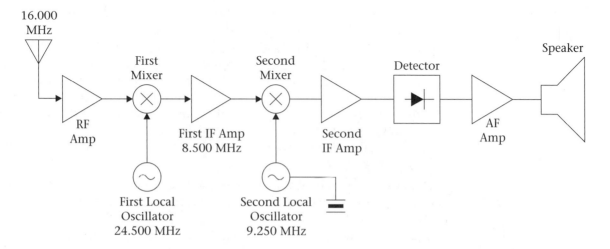

FIGURE 9.23

18. Suppose a receiver has the following gain structure (all stages operating at maximum gain):

RF amp:	12 dB gain
Mixer:	1 dB loss
IF amp:	76 dB gain
Detector:	3 dB loss
Audio amp:	35 dB gain

    (a) What is the total power gain of the receiver from antenna to speaker?

    (b) What would be the minimum signal required at the antenna in order to get a power of 0.5 W into the speaker? Express your answer in both watts and microvolts, assuming a 50-Ω input impedance.

(c) Now suppose a 100 mV signal is applied to the antenna. Calculate the output power. Is this a reasonable answer? Explain. What would actually happen in a real receiver?

19. (a) Draw the block diagram for an FM broadcast receiver. It is to have one RF stage and an IF frequency of 10.7 MHz. The local oscillator will operate above the signal frequency. Indicate on the diagram the frequency or frequencies at which each stage operates when the receiver is receiving a station at 94.5 MHz.

    (b) What is the image frequency of the receiver described above?

    (c) Give two ways in which the image rejection of the receiver could be improved.

20. An FM receiver is double conversion with a first IF of 10.7 MHz and a second IF of 455 kHz. High-side injection is used in both mixers. The first local oscillator is a VFO, while the second is crystal-controlled. There is one RF stage, one stage of IF amplification at the first IF, and three stages of combined IF amplification and limiting at the second IF. A quadrature detector is used as the demodulator. The receiver is tuned to a signal with a carrier frequency of 160 MHz. Draw a block diagram for this receiver.

21. Consider the transceiver whose block diagram is in Figure 9.20. Let the first I.F. be 16.9 MHz, with low-side injection of the first local-oscillator signal.

    (a) What frequencies would the synthesizer produce while transmitting and receiving, respectively, on marine channel 14 (156.7 MHz)?

    (b) Describe and explain the technique used for modulation in the transmitter section.

    (c) The transmitter section uses a maximum deviation of 5 kHz and operates with modulating frequencies in the range of approximately 300 Hz to 3 kHz. Suggest an appropriate receiver bandwidth.

Cellular Radio

Objectives

After studying this chapter, you should be able to:

- Describe the history of personal communication up to the beginning of digital cellular radio.

- Explain the operation and limitations of CB radio and cordless telephones.

- Explain the operation of and perform relevant calculations for North American analog cellular telephone systems.

- Explain the operation of and perform relevant calculations for North American digital cellular telephone systems.

10.1 Introduction

With this chapter we begin our discussion of several systems that are sometimes grouped together as **personal communication systems (PCS)**. Like many technical terms, this one has several meanings. Specifically, it is used for a particular variant of cellular radio which will be described in the next chapter. But more generally, it can be applied to any form of radio communication between individuals.

In this chapter, after some historical introduction, we look at the common North American cellular telephone system, known as the **Advanced Mobile Phone Service (AMPS)**. The network, the cell sites, and the portable and mobile telephones are described.

AMPS is still the most common cellular radio technology in North America, but a digital variation has become increasingly popular. We look at this system in this chapter as well. Newer *Personal Communication Systems (PCS)*, which use cellular techniques but at a different frequency range and with an assortment of digital modulation schemes, are the subject of the next chapter.

10.2 Historical Overview

The cellular radiotelephone system has its origins in much earlier systems. There has long been a need for portable and mobile communication. Three early concepts, two of which are still in wide use today, show aspects of what is needed. A brief look at each will show why cellular radio was created.

Citizens' Band Radio This is probably the earliest true personal communication system. Introduced in the United States in the 1960s, *citizens' band (CB) radio* enjoyed great popularity in the 1970s, followed by an almost equally steep decline as its limitations became better known.

CB radio was intended to do some of the same things that are envisaged by more recent personal communication systems. The relatively low frequency of 27 MHz made transceivers affordable when CB radio was introduced, and the absence of any test for a license made it easy for anyone to get involved. The transmitter power limit of four watts for full-carrier AM, or twelve watts peak envelope power for SSB, is designed to reduce interference by restricting the communication range to a few kilometers. The fact that most CB operation is between mobile units with low antenna heights and no repeaters also limits the effective range. The restricted range is necessary to limit interference since there are only 40 channels. This should not be a problem since CB radio is intended for local communication.

We can estimate the range of CB communication by making some assumptions based on typical equipment and propagation paths, using the techniques introduced in Chapter 7.

EXAMPLE 10.1

Two handheld CB transceivers are held 1 m above flat, level terrain. The transmitter power output is 4 W and the receiver sensitivity is 0.5 μV into 50 Ω. The transmitting and receiving antennas are both loaded vertical monopoles with a gain of 1 dBi. Determine whether the maximum communication range is limited by power or distance. Assume there is no interference and that free-space attenuation applies.

SOLUTION

First we should recognize that our answer is likely to be optimistic. Usually the terrain is not flat and there are reflections from buildings, vehicles, and so forth. Our results might be fairly accurate for transmission over water.

First let us find the possible line-of-sight range, which is limited by the distance to the radio horizon. In Chapter 7 we noted that

$$d = \sqrt{17h_t} + \sqrt{17h_r} \qquad (10.1)$$

where

$$d = \text{maximum distance in km}$$
$$h_t = \text{transmitting antenna height in m}$$
$$h_r = \text{receiving antenna height in m}$$

Applying this equation to our situation gives

$$d = \sqrt{17h_t} + \sqrt{17h_r}$$

$$= \sqrt{17 \times 1} + \sqrt{17 \times 1}$$

$$= 8.2 \text{ km}$$

This is the maximum distance regardless of power level. Now let us see what free-space distance would be possible with the given power level, antenna gains, and receiver sensitivity, ignoring the horizon. We'll ignore transmission line losses, which are probably negligible anyway since the antenna is mounted directly on the transceiver.

First we need to convert the required voltage at the receiver to a power level in dBm.

$$P = \frac{V^2}{R}$$

$$= \frac{(0.5 \times 10^{-6})^2}{50}$$

$$= 5 \text{ fW}$$

$$= -113 \text{ dBm}$$

We also need to express the 4 W transmitter power in dBm:

$$P_T = 4 \text{ W}$$
$$= 36 \text{ dBm}$$

The antenna gains increase the effective power by 2 dB. Our allowable path loss is then

$$L_{fs} = 36 \text{ dBm} - (-113 \text{ dBm}) + 2 \text{ dB} = 151 \text{ dB}$$

The path loss is given by

$$L_{fs} = 32.44 + 20 \log d + 20 \log f \qquad (10.2)$$

where

$$L_{fs} = \text{free-space loss in decibels}$$
$$d = \text{path length in km}$$
$$f = \text{frequency in MHz}$$

Here we know L_{fs} and f and we need to calculate d. Rearrange Equation (10.2):

$$20 \log d = L_{fs} - 32.44 - 20 \log f$$
$$= 151 - 32.44 - 20 \log 27$$
$$= 89.9$$
$$d = 313/8 \text{ km}$$

Therefore, this system is quite obviously limited by the distance to the radio horizon (and possibly by interference from other transmitters nearer to the receiver) and not by transmitter power. In this situation the power level could easily be reduced considerably with no effect on communication quality.

In a practical mobile situation the attenuation might actually be proportional to the fourth power of distance and would vary greatly depending on reflections. Still, range tends to be limited by the horizon, reflections and shadows, and interference, rather than power level.

Use of **half-duplex** (push-to-talk) operation for CB radio means only one channel is needed per conversation, and using AM (including its narrower bandwidth variant, SSB) keeps the required bandwidth less than would be needed for FM. The channels are spaced at 10-kHz intervals.

Selection among the 40 available channels is done by the operator, who is supposed to listen and manually switch to a clear channel before transmitting. Since anyone can listen in on any channel, there is no privacy.

The main disadvantages of CB radio are directly related to its simplicity and informality. Lack of privacy and co-channel interference are major problems. So is the lack of any connection to the wireline phone system or any access to repeaters for reliable, long-distance communication. The relatively high power level needed for communication without repeaters causes portable transceivers to be large and heavy. See Figure 10.1 for a typical example. The low frequency, which requires antennas to be large if they are to be efficient, is also a problem for portable transceivers.

FIGURE 10.1
Handheld CB transceiver
(Courtesy of Tandy Corporation)

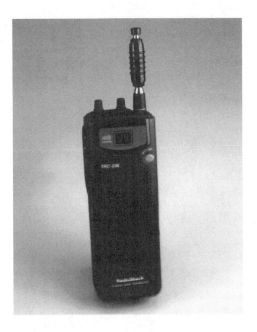

Some of these problems have been addressed with unlicensed FM transceivers currently being sold for the 46/49-MHz and 460-MHz bands. The latter frequency range is called the *Family Radio Service* (*FRS*). These transceivers are more compact but they still suffer from the other limitations mentioned earlier.

Cordless Telephones Most cordless phones are intended as simple wireless extensions to ordinary telephone service. For best results, a telephone, cordless or otherwise, should operate in full-duplex mode; that is, it should be capable of transmitting and receiving at the same time. Thus a cordless phone needs two radio channels, separated widely enough in frequency to avoid interference between them. Early designs had only a single channel for each direction, so the transmitter power levels and range had to be kept very small to minimize interference. Nonetheless, consumers found that a telephone that could be carried freely throughout a house and its grounds was very useful and cordless phones have been extremely popular. A typical modern example is shown in Figure 10.2.

FIGURE 10.2
Cordless phone

(Courtesy of Tandy Corporation)

Most current designs use analog FM in either the 43–49-MHz or 900-MHz bands. Older phones use AM at about 1.7 MHz for the handset, with FM at 49 MHz for the base unit. Some 900-MHz models use analog FM while others use digital spread-spectrum techniques, and there is at least one model (the Panasonic Gigarange™) that uses a 2.4-GHz spread-spectrum signal from base to handset, and 900 MHz from handset to base. Most cordless phones can automatically scan 10 or 25 channels, choosing a clear channel (assuming one exists) without any user involvement.

Table 10.1 shows the frequencies currently in use for 43–49-MHz cordless phones. Note that the base and portable frequencies must be widely separated, since both are in use simultaneously for **full-duplex** communication. Both the phone and base units require duplexers to separate the transmit and receive frequencies. Different manufacturers of 900-MHz phones use different channel frequencies and spacings, but the base and handset

TABLE 10.1   Cordless Telephone Frequencies: 43–49 MHz band

Channel	Base	Handset
1	43.720	48.760
2	43.740	48.840
3	43.820	48.860
4	43.840	48.920
5	43.920	49.000
6	43.960	49.080
7	44.120	49.100
8	44.160	49.160
9	44.180	49.200
10	44.200	49.240
11	44.320	49.280
12	44.360	49.360
13	44.400	49.400
14	44.460	49.480
15	44.480	49.500
16	46.610	49.670
17 (B)	46.630	49.845
18 (C)	46.670	49.860
19	46.710	49.770
20 (D)	46.730	49.875
21 (A)	46.770	49.830
22 (F)	46.830	49.890
23	46.870	49.930
24	46.930	49.990
25	46.970	49.970

The letters A through F denote channels that are also used for baby monitors.

frequencies are typically separated by about 20 MHz. The power level is deliberately set very low so that range is power-limited rather than extending to the radio horizon. It is surprising how low the power level for a cordless phone can be and still give reasonable results.

EXAMPLE 10.2 ▼

A cordless phone operating at 49 MHz is to have a range of 50 m. Assuming 0 dBi gain for the antennas and the same receiver sensitivity as in Example 10.1, what transmitter power is required?

SOLUTION

Obviously the distance to the horizon is not the limiting factor here. We can use Equation (10.2) to calculate the loss for a path length of 50 m:

$$L_{fs} = 32.44 + 20 \log d + 20 \log f$$
$$= 32.44 + 20 \log 0.05 + 20 \log 49$$
$$= 40.2 \text{ dB}$$

If the required signal strength at the receiver is −113 dBm as before, then the transmitter power must be at least

$$P_t = -113 \text{ dBm} + 40.2 \text{ dB} = -72.8 \text{ dBm} = 52.7 \text{ pW}$$

▲

In practice, the power levels are much higher to cope with fading due to reflections and absorption. Cordless phones in the 46/49-MHz band are restricted to an EIRP of about 30 μW, while 900 MHz digital phones can use about 16 mW EIRP.

Cordless telephones share much of the simplicity of CB radio. There are no license requirements, and there is no official coordination of frequencies. Users, or in most cases the phones themselves, simply try to choose a channel that is not in use. The newer cordless phones use digital access codes to prevent unauthorized persons from dialing the phone and possibly making unauthorized toll calls, but it is still not possible to use two nearby phones on the same channel at the same time. The use of FM does provide some protection from interference: due to the capture effect, the desired signal has only to be a few decibels stronger than the interfering signal in order to reduce interference to a reasonable level.

Privacy is not quite as nonexistent as with CB radio, since the newer phones automatically avoid occupied channels. This reduces accidental privacy violations, but anyone who wants to eavesdrop on an analog phone can do so by using a scanner, for instance. The digital phones offer much better privacy.

At this time, listening to others' cordless or cellular phone calls is illegal in the U.S.A., though not in Canada. (In Canada, listening to these conversations is legal, but divulging what you heard to another person is not.)

Because of the limited number of channels, cordless phones rely on extremely low transmitter power (microwatts to a few milliwatts depending on the band and the phone) to limit interference. Of course, this also limits their range. These phones certainly provide access to the wireline phone network, but in general, only from the customer's own premises or very nearby. The newer spread-spectrum phones do have more range—up to a kilometer or so under ideal conditions.

Despite interference problems and severely limited range, cordless phones have been and remain very popular with consumers. Various attempts have been made, particularly in Europe, to devise systems called **telepoints** that would enable users to take their cordless handsets to public places like malls and office buildings and use them there. However, recent developments in cellular and PCS systems have caused these ideas to lose favor. The cordless phone seems likely to remain popular in its current niche as a low-cost, wireless extension phone. Its low power allows it to have long battery life (weeks of standby, hours of talk time), and especially at 900 MHz, antennas are reasonably small and unobtrusive. Comparing the cordless phone shown in Figure 10.2 with the portable CB transceiver displayed earlier shows that cordless phones are a step in the right direction in terms of convenience.

Improved Mobile Telephone Service (IMTS)

The familiar cellular radiotelephone system has its origins in much earlier systems that used a few widely spaced repeaters. Wide coverage was obtained by using powerful base-station transmitters with antennas mounted as high as possible. The mobile transceivers likewise used relatively high power, on the order of 30 watts. Very similar systems are still widely used in dispatching systems, such as those for taxicabs and ambulances, for example.

The most common type of mobile telephone, from its introduction in the mid-1960s until the coming of cellular radio in the early 1980s (the first commercial cellular system became operational in Chicago in 1983), was known as the **Improved Mobile Telephone Service (IMTS)**. IMTS is a *trunked* system; that is, radio channels are assigned by the system to mobile users as needed, rather than having one channel, or pair of channels, permanently associated with each user. Narrowband FM technology is used. Two frequency ranges, at about 150 and 450 MHz were used for IMTS, with an earlier system called MTS operating at around 40 MHz. The three systems combined had only 33 available channels. A few IMTS systems are still in use, mainly in remote locations.

IMTS is capable of assigning channels automatically, by the rather simple means of transmitting a tone from the base station on unoccupied channels. The receiver in the mobile unit scans channels until it detects the tone.

IMTS is capable of full-duplex operation using two channels per telephone call. Direct dialing is also possible, so using a mobile phone is almost as simple as using an ordinary telephone at home.

The main problem with IMTS and similar systems is that whatever bandwidth is made available to a single repeater, is tied up for a radius of perhaps 50 km or even more, depending on the height of the antenna and the power of the transmitter at the base station. Any attempt to reuse frequencies within this radius is likely to result in harmful interference. Simple systems like this also suffer from fading and interference near the edges of their coverage areas. For instance, suppose two similar trunked systems with identical repeaters are located 50 km apart. Then, at a location midway between the two, a receiver would receive equally strong signals from each. Communication would be impossible if both repeaters used the same frequencies.

10.3   Introduction to the Advanced Mobile Phone System (AMPS)

As the demand for mobile telephony grew, it became obvious that another way had to be found to accommodate more users. More spectrum could be found, but only by going up in frequency, since all the spectrum near the existing mobile telephone allocations was already occupied. The 800-MHz region was already assigned to UHF television broadcasting, but more spectrum had been assigned to this service than was actually being used, probably because of the popularity of cable television for specialty channels. A band approximately 40-MHz wide (increased to about 50 MHz in 1986) was assigned to the new system.

To make the system more efficient, cellular radio technology, based on many repeaters with their range deliberately restricted, was introduced at the same time. See Chapter 7 for a description of the cellular principle. Cellular radio had to wait until enough computing power could be introduced into the system, at reasonable cost, to allow the system to keep track of mobiles as they moved from one cell site to the next and to allow mobile phones to change frequency and power level by remote control from the cell site.

When cellular telephony was introduced in North America, it was decided to allow two different companies, called carriers, to operate in any given region. One of these would be the local wireline telephone company (*telco*). The other would be an independent company called a **radio common carrier (RCC)**. Each carrier was assigned half the channels in each area in an attempt to encourage competition. (It would actually be more efficient in terms of spectrum usage to have only one provider: with the current system it is possible for one provider's channels to be fully loaded while the other still has available frequencies.)

Cellular radio goes a long way toward relieving the congestion described above by essentially reversing the conventional wisdom about radio systems using repeaters. Instead of trying to achieve long range by using high power, cellular repeaters are deliberately restricted in range by using low power. As discussed in Chapter 7, the high path loss associated with mobile propagation actually makes it easier to reduce interference in a cellular system. A reasonable elevation for the base-station antennas is still required to minimize radio shadows (behind buildings, for instance). Similarly, the mobile radios use low power (no more than 4 W ERP for mobiles and 600 mW or less for portable phones), and in fact, the mobile transmitter power is automatically limited by the system to the minimum required for reliable communication.

Instead of one repeater, there are many, located in a grid pattern like that shown in Figure 10.3. Each repeater is responsible for coverage in a small cell. As shown, the cells are hexagons, but of course in a real situation the antenna patterns will not achieve this precision—the cells are more likely to be approximately circular, with some overlap. All the cell sites in a region are connected by copper cable, fiber optics, or microwave link to a central office called a **mobile switching center (MSC)** or **mobile telephone switching office (MTSO)**, and the MSCs are themselves interconnected so that the system can keep track of its mobile phones. The cellular system is connected at a point of presence to the wireline network, so that cellular customers can speak to wireline customers.

Note that there is no provision for direct mobile-to-mobile radio communication. Even if two cell phones are in the same room, a call from one to

FIGURE 10.3
Cell boundaries (seven-cell repeating pattern)

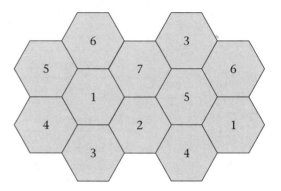

the other must go through a cell site and an MSC. Provided both portable phones are connected to the same network (A or B), there would be no need to go through the PSTN.

Since each transmitter operates at low power, it is possible to reuse frequencies over a relatively short distance. As we saw in Chapter 7, typical mobile propagation conditions allow for a repeating pattern of either seven or twelve cells; the available bandwidth is divided among these cells. The frequencies can then be reused in the next twelve or seven cells, with the lower number possible when directional antennas are used with three sectors using different frequencies per cell.

Cellular Carriers and Frequencies

In the current North American system, there are 395 duplex voice channels, each consisting of one channel in each direction for each of the two carriers. There are also 21 control channels for each carrier used to set up calls and administer the system. AMPS uses narrowband analog FM, with a maximum frequency deviation of 12 kHz and a channel spacing of 30 kHz.

Table 10.2 shows how these channels are divided between the two carriers: A represents the non-wireline carrier and B represents the wireline carrier. Note that the frequencies assigned to each carrier are not all contiguous because of the extra frequencies added to the system in 1986. Note also the rather large separation (45 MHz) between base and mobile transmit frequencies. This allows for simple duplexers to separate transmit and

TABLE 10.2 North American Cellular Radio Frequencies

Base Frequencies (forward channels)	Mobile Frequencies (reverse channels)	Type of Channel	Carrier
869.040–879.360	824.040–834.360	Voice	A
879.390–879.990	834.390–834.990	Control	A
880.020–880.620	835.020–835.620	Control	B
880.650–889.980	835.650–844.980	Voice	B
890.010–891.480	845.010–846.480	Voice	A*
891.510–893.970	846.510–848.970	Voice	B*

Table denotes transmit carrier frequencies. Mobile transmits 45 MHz below base.

A = non-wireline carrier (RCC)   B = wireline carrier (telco)
* = frequencies added in 1986

receive signals in the phones. The base transmits to the mobile on a **for-ward channel**, while transmissions from mobile to base use a **reverse channel**.

An individual cell doesn't use all these channels, of course. Each cell has only one-seventh or one-twelfth of the total number of channels assigned to a carrier, depending on the system. Contiguous frequencies are not used in order to reduce interference. With a seven-cell repeating pattern, transmitters in the same cell are generally separated by about seven channels or 210 kHz. Each cell in a seven-cell pattern also has three of the 21 control channels.

To further reduce receiver selectivity requirements, adjacent channels are not used in adjoining cells. Therefore, transmitters in adjacent cells are separated in frequency by at least 60 kHz.

Channel Allocation

The control channels are used, among other things, to allocate voice channels to phones. When a user dials a phone number on a mobile phone and presses the *Send* button, the phone scans all the control channel frequencies to find the strongest. This control channel should be associated with the closest cell site. The cell phone transmits on its corresponding control channel, and once the call has been set up, the cell site assigns it a clear voice channel, assuming one is available.

While the conversation continues, the cell sites adjacent to the one in use monitor the signal strength from the mobile. When the strength is greater in one of the adjacent cells, the system transfers the call to that cell. This procedure is called a **handoff**. Handoffs, of course, require a change in frequency for the mobile phone, under control of the system.

A similar procedure takes place for incoming calls. The mobile periodically identifies itself to the system whenever it is turned on, so the system usually has a good idea of its location. Paging signals are sent out on control channels and the mobile responds, enabling the system to locate it more precisely. In the early days of cell phones it often took some time, a minute or more, to find a mobile, but improved communication within the system has reduced this time to a few seconds in most cases. The phone is instructed to ring, and once it is answered, the system assigns it a voice channel. After that the system follows the phone as it moves from one cell to the next, as explained earlier.

Frequency Reuse

The reason for the complexity of the cellular system is, of course, frequency reuse. Once a mobile has moved out of a cell, the frequency pair it occupied is available for another conversation. By making cells smaller, frequencies can be reused at shorter distances. There is no theoretical limit to this, but there are practical limits. As cells become smaller, more cell sites are needed and handoffs occur more frequently, requiring more computing power and faster response both at the system level and in the individual mobile phone. Once the radius drops below about 0.5 km the handoffs occur so frequently that it is difficult to cope with a mobile moving at high speed. The flexibility of cell sizes allows for larger cells in less-developed areas and smaller cells in areas of the greatest traffic.

EXAMPLE 10.3

A vehicle travels through a cellular system at 100 kilometers per hour. Approximately how often will handoffs occur if the cell radius is:

(a) 10 km?

(b) 500 m?

SOLUTION

The reason for the word "approximately" in the problem statement is that we are not sure how the road crosses the cell boundaries. Let us assume for simplicity that the vehicle drives along a road leading directly from one cell site to the next. Thus, the vehicle will change cells each time it travels a distance equal to the diameter of a cell (twice the radius).

First convert the speed into meters per second.

$$v = \frac{100 \text{ km/hr} \times 1000 \text{ m/km}}{3600 \text{ s/hr}}$$

$$- 27.8 \text{ m/s}$$

Now we can find the time between handoffs.

(a) Let the diameter be $d = 20$ km

$$t = \frac{d}{v}$$

$$= \frac{20 \times 10^3 \text{ m}}{27.8 \text{ m/s}}$$

$$= 719 \text{ s}$$

$$= 12 \text{ min}$$

(b) This time, $d = 1$ km

$$t = \frac{d}{v}$$

$$= \frac{1 \times 10^3 \text{ m}}{27.8 \text{ m/s}}$$

$$= 36 \text{ s}$$

10.4 AMPS Control System

In this section we study in more detail the process by which the AMPS cellular system keeps track of phones and calls. We need to look at the functions of the control channels and also at the control information that is sent over the voice channels.

An effective control system has to do several things. It needs to keep track of mobile phones, knowing which ones are turned on and ready to receive a call and where they are. It needs to keep track of telephone numbers for authentication and billing, and it should have some way to detect and prevent fraudulent use. It must be able to set up calls, both from and to mobile phones and transfer those calls from cell to cell as required. It would be best if all this were transparent to the user, who should only have to dial the phone number or answer the phone, just as with a wireline phone at home. A more advanced system might also be able to send faxes and e-mail, surf the internet, and so forth, but let's look at the basics first!

First we need to understand the functions of the voice and control channels. You might assume that the voice channels are for talking and the control channels are for control signals, and that is mostly correct. However, there is a problem: cell phones contain only one receiver and one transmitter, so they can't receive both a voice channel and a control channel at the same time. Therefore, any control messages that have to be sent during a conversation must use the voice channel. Some of this is done using in-channel, out-of-band signaling (consisting of tones above the voice frequency range), and the rest is done with **blank-and-burst signaling**, during which the voice signal is muted for a short time (100 ms) while data is sent over the voice channel.

Digital signals on the control channel and those sent during blank-and-burst signaling on the voice channel are sent in a relatively simple way. They use FSK with 8 kHz deviation (16 kHz total frequency shift) and a channel bit rate of 10 kb/s. The data is transmitted using Manchester code. In order to reduce the likelihood of errors, the control channel sends each message five times and also uses Hamming error-correction codes. This increases the robustness of the control system but reduces the actual data throughput to 1200 b/s. There is no encryption in the AMPS system: all the data coding information is publicly available. This is a serious oversight that has been remedied in the newer PCS systems to be described in the next chapter.

Mobile and Base Identification

Each mobile unit has two unique numbers. The **mobile identification number (MIN)** is stored in the **number assignment module (NAM)** in the phone. The MIN is simply the 10-digit phone number for the mobile phone (area code plus 7-digit local number), translated according to a simple algorithm into a 34-bit binary number. The NAM has to be programmable, since it may be necessary to assign a different telephone number to the phone, but it is not supposed to be changeable by the user. In most cases, however, it can be changed from the keypad if the user knows the right procedure. (Check the internet—it took the author less than ten minutes to find the procedure for his own cell phone.)

Usually a cell phone is registered on either the A or B system and has one MIN. It can operate on the other system as a **roamer**, if necessary and if there is an agreement between the two systems to allow it. It is also possible for a phone to have two MINs so that it can be used on both A and B systems without roaming. In that case the user of the phone has two phone numbers (and two bills to pay).

The other identification number is an **electronic serial number (ESN)**, which is a unique 32-bit number assigned to the phone at the factory. It is not supposed to be changeable without rendering the phone inoperable, but in practice it is often stored in an EPROM (erasable programmable read-only memory chip) that can be reprogrammed or replaced by persons with the right equipment and knowledge. The combination of the MIN and the ESN enables the system to ensure proper billing and to check for fraudulent use (for instance, if a registered MIN appears with the wrong ESN the system will not allow the call to go through).

The mobile phone also has a number called the **station class mark (SCM)**, which identifies its maximum transmitter power level. There are three power classes corresponding to phones permanently installed in a vehicle, transportable "bag phones," and handheld phones. The maximum power levels, specified as ERP (effective radiated power with respect to a half-wave dipole) are as follows:

Class I (mobile):	+6 dBW (4 W)
Class II (transportable):	+2 dBW (1.6 W)
Class III (portable):	−2 dBW (600 mW)

Mobile transmitter power is controlled by the land station in 4 dB increments, with the lowest power level being −22 dBW (6.3 mW) ERP. The idea is to reduce interference by using as little power as possible. Mobile and transportable phones thus have better performance than portable phones only when propagation conditions are bad enough, or cells large enough, that the system needs to increase mobile power past the maximum for a portable phone. Using a portable phone inside a vehicle attenuates the signal considerably, so communication from a portable phone can sometimes be established in marginal areas by simply getting out of a car.

The cellular system has an identifying number called the **system identification number (SID)**. This enables the mobile phone to determine whether it is communicating with its home system or roaming. (Using a "foreign" system usually costs more and the user may disable this ability if desired.) In addition each cell site has a **digital color code (DCC)**. When the mobile detects a change in DCC without a change in frequency, it is an indication that co-channel interference is being received from another base station.

Turning on a Phone

When a cell phone is turned on, it identifies itself to the network. First it scans all the control channels for its designated system (A or B) and finds the strongest. It looks for the SID from the system to determine whether or not it is roaming. If it does not receive this information within three seconds, it tries the next strongest control channel. After receiving the system information, the mobile tunes to the strongest paging channel. Paging channels are control channels that carry information about calls that the system is trying to place to mobiles. If someone is calling the mobile, its number will be transmitted by the paging channel.

The control channel constantly updates the status of its associated reverse control channel (from mobile to system). Only the system transmits on the forward channel, but any mobile can transmit on the reverse channel. The system tells the mobiles when this channel is busy to reduce the chance of a *collision*, which occurs when two or more mobiles try to use the control channel at the same time. After checking that the reverse channel is free, the newly activated phone transmits its ESN and MIN to the land station so that the system knows the phone is ready for calls and in which cell the phone is located. If the mobile loses the signal and reacquires it or detects that it has moved to a different cell, it identifies itself again. In addition, the system may periodically poll its mobiles to see which are still active.

While turned on but otherwise idle, the mobile phone continues to periodically (at least once every 46.3 ms) check the control channel signal from the cell site. It has to verify that a signal is still available, that it is from the same system, and that there are no calls for the mobile phone.

Originating a Call   When the user of a mobile phone keys in a phone number and presses *Send,* the mobile unit transmits an origination message on the reverse control channel (after first checking that this channel is available). This message includes the mobile unit's MIN and ESN and the number it is calling. The cell site passes the information on to the mobile switching center for processing. Once authorization is complete, the cell site sends a message to the mobile on the forward control channel, telling it which voice channel to use for the call. It also sends the digital color code which identifies the cell site, and a **Control Mobile Attenuation Code (CMAC)**, which sets the power level to be used. This power level can be changed by the land station as needed during the call by means of a control message on the forward voice channel.

Now both stations switch from the control channel to a voice channel, but the audio is still muted on the phone. The cell site sends a control message on the forward voice channel confirming the channel. It then sends a **supervisory audio tone (SAT)** on the voice channel to the mobile phone. This is a continuous sine wave, with a frequency above the voice band. There are three possible frequencies: 5970, 6000, and 6030 Hz. The mobile relays the tone back to the cell site. Reception of this tone by the cell site confirms that the correct cell site and mobile are connected. The mobile sends a confirmation message on the reverse voice channel. After this handshaking, the call can begin.

During the call the SAT continues (it is filtered out before the audio reaches the speaker in the phone, of course). Reception of the wrong frequency tone by the base station indicates an interfering signal and interruption of the tone indicates that the connection has been lost, perhaps due to severe fading. If the tone is not resumed within five seconds, the call is terminated.

A 10-kHz *signaling tone* (*ST*) may also be transmitted on the voice channel during a call. It is used to signal handoffs to another cell and the termination of the call.

Receiving a Call   An incoming call is routed by the network to the cell where the mobile last identified itself. (If it has not identified itself to the network, it is assumed to be turned off and a recorded message to that effect is given to the caller.) The land station sends the MIN on the paging channel along with the voice channel number and power level to use. The mobile confirms this message and sends its ESN on the reverse control channel to be matched by the network with the MIN. This is to avoid fraudulent use. The base sends its information again on the forward voice channel along with the digital color code information, then the mobile confirms the information on the reverse voice channel. After this handshaking, the supervisory audio tone is transmitted on the voice channel and the conversation can begin.

Handoffs   The network monitors the received power from the mobile at adjacent cell sites during a call. When it detects that its strength is greater at an adjacent cell site than at the site with which it is communicating, it orders a handoff from one cell to the next. This always involves a change in channel, since to avoid co-channel interference the same channels are never used in adjacent cells. The order to do this is sent by the first cell site to the mobile on the forward voice channel using blank-and-burst signaling. The resulting 100 ms interruption in the conversation is barely perceptible. The voice channel must be used, because during a conversation the mobile is not monitoring any of the control channels. The mobile is given the new channel number, new attenuation code, and new SAT frequency. After confirmation on the reverse voice channel, the mobile switches to the new channel, which connects to the new cell site, and the conversation continues. There will probably be an audible disturbance while this occurs.

10.5   Security and Privacy

The AMPS system is not very private. Voices are transmitted using ordinary FM and conversations can be picked up with any FM receiver that will tune to the correct frequency. Base stations often repeat mobile transmissions, so quite often both sides of the conversation can be picked up with one receiver, just as with a cordless phone but from much greater distances (typically a few kilometers). The change in channels as a mobile is handed off does make it hard to follow conversations when the cell phone user is talking from a moving vehicle.

In 1988, in an attempt to increase cell phone privacy, the United States government banned the import or sale of scanners or other receivers that can tune to cellular frequencies. However, these are still legal in many countries (including Canada); there are millions of old scanners around and frequency converters are easy to build. It should therefore be assumed that AMPS voice transmissions are public. The transmission of confidential information, such as credit card numbers for instance, is not advisable with analog cell phones.

Stolen cell phones work only until the owner has the service cancelled. (There is a code to lock the phone, but most people don't bother, or they leave the password at the factory setting, such as "1234." Generally they do this because the password is supposed to be changeable only by the dealer, though in fact it can often be done from the keypad if one knows the correct method.) Even if the phone number is changed by a knowledgeable thief, the Electronic Serial Number will give away the fact that the phone is "hot." However, it is not impossible, at least on some phones, to change the ESN. It is also possible to acquire valid pairs of MIN and ESN numbers by monitoring the reverse control channels. There is no encryption and the exact specifications for the data fields are publicly available. It is just a matter of acquiring the hardware and software to decode a 10 kb/s FSK data stream—not a very formidable task. Once the numbers are available it is possible to "clone" a cell phone to emulate a valid phone. Calls made on the cloned phone are billed to the unfortunate legitimate subscriber.

Service providers do have some protection. For instance, if the network detects the "same" phone trying to make two calls at once or two calls in quick succession from widely separated locations, it will flag the occurrence and someone will investigate. As the networks become larger and better integrated, this type of fraud becomes a little more difficult.

Another fraud is to use a cloned or stolen phone on another network as a roamer. If the foreign network is not capable of checking the phone's home network in real time, it may accept the call. This is becoming less likely as networks become better connected with each other. In the meantime, networks are becoming less trusting (especially in the United States) and less likely to allow roaming without identification.

10.6 Cellular Telephone Specifications and Operation

In this section we look at the requirements for the mobile or portable phone itself and consider some examples of phone construction. See Figure 10.4 for a block diagram of a typical analog cell phone. Because the system is full-duplex, the transmitter and receiver must operate simultaneously with a single antenna. A duplexer is used to separate the two signals. The wide 45-MHz frequency separation between transmit and receive frequencies makes this relatively easy. The constant frequency separation also simplifies frequency synthesizer design.

Microprocessor control is necessary to allow the phone to switch channels and power levels by remote control from the base station. The processor and its associated memory are also useful for timing calls, storing passwords to unlock the phone, storing lists of frequently-called numbers, and so on. Some cell phones can enter a sleep state between calls to conserve battery life; they must emerge from this state to check the control channel at least once in each 46.3 ms time period.

Transmitter Power and Frequency

In the previous section we noted that cell phones come in three *station classes*. This term refers to the maximum power level produced. The actual transmitted power level is adjusted in 4 dB steps by signals from the cell site. The mobile transmitter must transmit at within 3 dB of the correct power

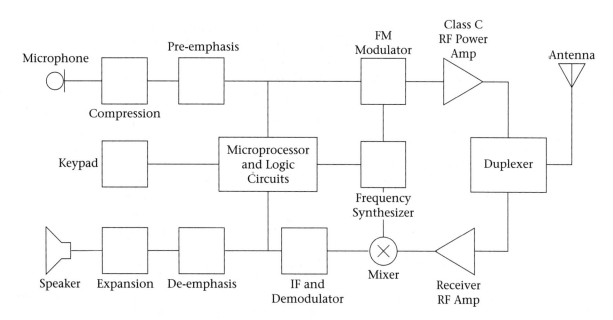

FIGURE 10.4 Block diagram of analog cell phone

level within 2 ms of turning on and must reduce its output to −60 dBm ERP or less within 2 ms of being turned off. The transmitted frequency must be within 1 kHz of the specified channel frequency.

The power levels for mobile, transportable, and portable phones are shown in Table 10.3. The abbreviation *MAC* refers to the *mobile attenuation*

TABLE 10.3   Power Levels for Mobile Phones (EIRP in dBW)

MAC	Class I	Class II	Class III
000	+6	+2	−2
001	+2	+2	−2
010	−2	−2	−2
011	−6	−6	−6
100	−10	−10	−10
101	−14	−14	−14
110	−18	−18	−18
111	−22	−22	−22

There is a range of 28 dB between maximum and minimum mobile power levels with a Class I phone.

code, which is transmitted from the base station to adjust the power of the mobile according to propagation conditions. Because FM and FSK are used, there is no need for linearity in the transmitter power amplifier, and Class C operation can be used for greater efficiency.

Transmitter Modulation

As mentioned earlier, voice transmission uses FM with a maximum deviation of 12 kHz each way from the carrier frequency. Data transmission uses FSK with 8-kHz deviation each way.

Companding with a ratio of 2:1 is used in voice transmission. That is, in the transmitter, a 2 dB change in audio level from the microphone causes only a 1 dB change in modulation level. The process is reversed in the receiver. The result is an improvement in signal-to-noise ratio for low-level audio signals.

As with almost all FM systems, pre-emphasis is used in the transmitter and de-emphasis in the receiver. This means, the higher audio signals are given more gain in the transmitter and correspondingly less gain in the receiver. In the cell phone system, all frequencies above 300 Hz are boosted at the transmitter, with a slope of 6 dB per octave. Figure 10.5 shows the pre-emphasis curve with the corresponding de-emphasis curve for the receiver.

FIGURE 10.5
Cellular radio
pre-emphasis and
de-emphasis

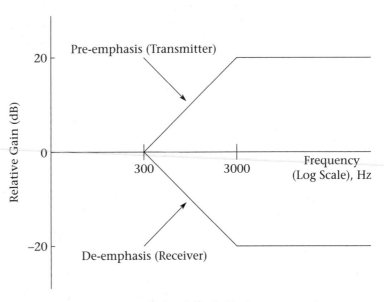

Note: This is an idealized (Bode Plot) representation.
Real filters have a slight curvature in the response
at both ends of the frequency range.

Mobile and Portable Antennas

Since the transmitted power of cellular phones is specified in terms of ERP, the use of more efficient antennas allows transmitter power to be reduced. This is especially important in the case of portable phones, since lower transmitter power leads to longer battery life. On the other hand, more efficient antennas tend to be larger.

Most portable cell phones use a quarter-wave monopole antenna. At 800 MHz, the length of this antenna is about 9.5 cm. The options are wider for mobile antennas. Many of these use a quarter-wave and a half-wave section, separated by an impedance-transforming coil. See Figure 10.6 for examples of typical portable and mobile antennas.

FIGURE 10.6
Portable and mobile antennas

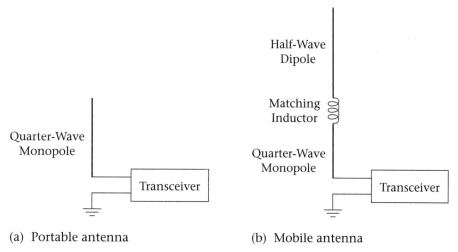

(a) Portable antenna   (b) Mobile antenna

10.7 Cell-Site Equipment

The radio transmitting equipment at the cell site operates at considerably higher power than do the mobile phones, but this power is shared among all the channels that are used at the site. Similarly, there must be receivers for each voice and control channel in use at the site, as well as extra receivers for monitoring the signal strength of mobiles in adjacent cells. Consequently, the cell site equipment is much more complex, bulky, and expensive than the individual cell phones. In addition, as already noted, cell sites often need directional antennas to facilitate the division of each cell into sectors. Transmit and receive antennas may be separate or combined at the cell site. Often two receive antennas and one transmit antenna are used per cell, or per sector, in a sectorized configuration. This allows for space diversity in the receiver; to counteract the effects of fading, the receiver monitors the signal from both antennas and chooses the stronger signal. The effects of fading are likely to be greater for one antenna than the other at any given moment. Figure 10.7 is a block diagram of the equipment in a typical cellular base station.

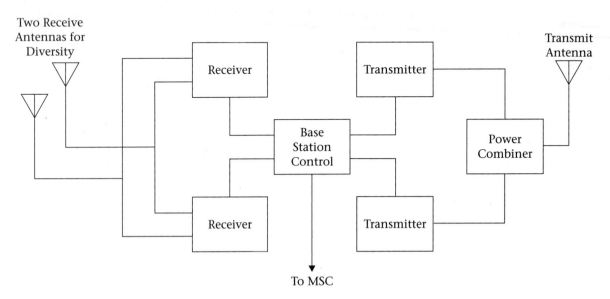

Note: Two transceivers shown for clarity. An
actual site would have many more.

FIGURE 10.7   Cellular radio base station

The combination of the mobile cellular phone and the cell-site radio equipment is known as the **air interface**. There is much more to cellular telephony than radio, however. The network must be organized and administered as a whole. This administration includes keeping track of phone locations, billing, setting up and handing off calls, and so on. The substantial computing resources required to do all this are, at least in part, responsible for the delay in the introduction of cellular telephony for many years after the idea was first proposed.

Figure 10.8 shows a typical cellular telephone system. Each cell has several radio transceivers (one per channel); usually one wideband power amplifier is used to provide the transmit power for all channels in a site (or sector, for sectorized systems). The site's radio equipment is operated by a **base station controller (BSC)**. The base station controller takes care of the air interface: assigning channels and power levels, transmitting signaling tones, and so on. The mobile switching centers (MSCs), also called mobile telephone switching offices (MTSOs), route calls along a private copper, fiber optic, or microwave network operated by the cellular service provider.

FIGURE 10.8
Cellular radio system

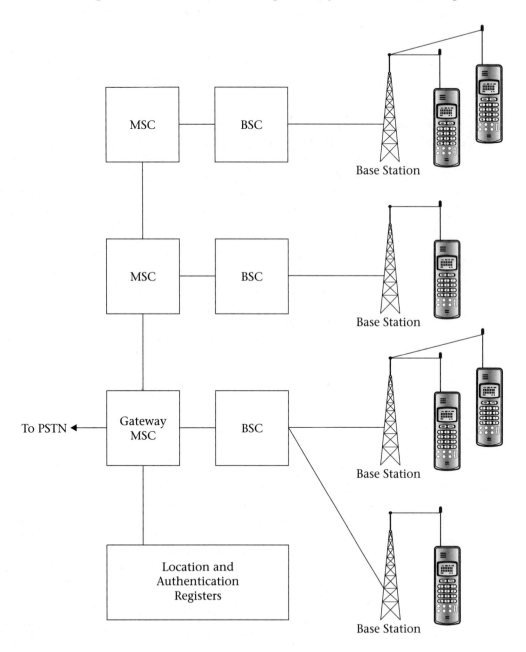

Their action is also required in authorizing calls, billing, initiating handoffs, and so on. Sometimes the BSC and MSC are combined. Associated with the MSCs are data banks where the locations of local and roaming mobiles are stored.

At certain points in the system, the cellular network is connected to the public switched telephone network (PSTN). These gateways allow calls to be made between landline and cellular phones, and between cell phones using different service providers. The cellular system communicates with the PSTN using Signaling System Seven, which was described in Chapter 5.

Traffic and Cell Splitting

The optimum size of a cell depends on the amount of traffic. Ideally, most of the available radio channels should be in use at peak periods, but situations where all channels are in use should be rare. If all channels in a cell are busy, it is impossible for anyone to place a call to or from that cell. The user has to hang up and try again later. This situation is called **call blocking** and is obviously undesirable. It causes revenue loss, and if it is frequent, unhappy customers may switch to the competing system. (This is always a possibility with the North American AMPS system since there are two competing systems in each area.) Call blocking takes place on the wireline network as well. For instance, long-distance trunks are sometimes unavailable during peak calling periods. This means that customers are used to blocking and will put up with a small percentage, perhaps one or two percent, of calls being blocked.

A more unpleasant situation occurs when a mobile phone moves into a cell that has all its channels busy. The attempt by the system to hand off the call to the new cell is frustrated by the lack of free channels, and the call must be terminated, or *dropped*. **Dropped calls** are very inconvenient and sometimes embarrassing, as the effect is the same as if one of the parties had hung up on the other. Call dropping is very rare on the wireline system, so customers are not as tolerant of this problem.

Since call blocking also occurs on the wireline network, which has been in operation for about a century, you might guess that someone has already studied the problem. You'd be correct: A. K. Erlang, a Swedish engineer, studied the problem using statistical analysis early in the twentieth century. He found, not surprisingly, that the more channels there were, the smaller the possibility of blocking for a given amount of traffic. Perhaps less obviously, he found that with more channels, the amount of possible traffic per channel increases for a given blocking probability. This phenomenon is called *trunking gain,* and it is the reason a two-provider system is theoretically less efficient than one using a single provider.

Trunking gain can perhaps be better understood by looking at an everyday situation: customers lining up to use tellers at a bank. Suppose there are two tellers and a separate line for each. Further suppose that the lines are assigned to customers on the basis of the type of accounts they have. Those with checking accounts use the first line, those with savings accounts use the second line.

Now suppose I arrive at the bank. My account is a checking account, but there are several people in line at that window. There is no line at the savings window, but I can't use that one. I am *blocked* and decide to try again later. Of course, after several frustrating attempts, I may notice that the competing bank across the street has shorter lines and change service providers. Now my bank changes its policy. There is only one line, and anyone can use the next available teller. The next time I arrive at the bank I have a much lower probability of being blocked. A similar logic applies to many situations: it is always more efficient to combine channels, and the gains are greater with more channels.

Phone traffic is defined in *erlangs* (E). One erlang is equivalent to one continuous phone conversation. Thus if 1000 customers use the phone ten percent of the time each, they generate 100 E of traffic on average. Mathematically,

$$T = NP \tag{10.3}$$

where

T = traffic in erlangs
N = number of customers
P = probability that a given customer is using the phone

We see immediately that traffic analysis does not coordinate channels directly with the number of customers. Real customers may use the phone continuously for an hour, then not at all for the rest of the day. If enough of them use the phone at the same time (afternoon rush hour, for example, is a peak in cell phone usage in cities), the peak traffic will be much greater than the average. Some cell phone owners use the phone continually for business during the working day; others use it mainly for emergencies and generate very little traffic. Also, the usage patterns may vary, in response to changing rates, for instance (lower rates generate more traffic). Evening and weekend use tends to be light for cell phones, just as it is for wireline long distance, and both wireline and cellular providers commonly provide monetary incentives to customers to shift some of their usage to those periods.

The most obvious way to avoid call blocking and call dropping is, of course, to provide more channels. However, the number of channels for the system is fixed, and so is the number per cell once the repeating pattern, usually seven or twelve cells, has been established. In a 12-cell repeating pattern, each provider has $395/12 \cong 33$ voice channel pairs per cell. According to

Erlang's theory, this number of channels will accommodate about 23 E of traffic for one percent blocking.

The amount of traffic can be increased at the expense of a larger blocking probability. For instance, with 33 channels, a traffic level of 24.6 E can be accommodated with a two percent blocking probability. It might seem at first glance that a 7-cell repeating pattern can allow more traffic, but this is illusory. The cells are each divided into three sectors, using different channels. Therefore, each sector has only $395/21 \simeq 19$ duplex voice channels. Using the 7-cell pattern saves money by reducing the number of cell sites needed, not by increasing the number of channels.

Table 10.4 shows traffic levels in erlangs for 19 and 33 channels, with various blocking probabilities. It also shows traffic levels for larger numbers of channels. We'll use these with digital cell phones shortly.

TABLE 10.4   Cellphone Traffic in Erlangs per Cell or Sector

Number of Channels	Blocking Probability		
	1%	2%	5%
19	11.2	12.3	14.3
33	22.9	24.6	27.7
55	42.4	44.9	49.5
97	81.2	85.1	92.2

EXAMPLE 10.4 ▼

A cellular telephone system uses a 12-cell repeating pattern. There are 120 cells in the system and 20,000 subscribers. Each subscriber uses the phone on average 30 minutes per day, but on average 10 of those minutes are used during the peak hour.

Calculate:

(a) the average and peak traffic in erlangs for the whole system

(b) the average and peak traffic in erlangs for one cell, assuming callers are evenly distributed over the system

(c) the approximate average call-blocking probability

(d) the approximate call-blocking probability during the peak hour

SOLUTION

(a) The average traffic is

$$T = 20{,}000 \times \frac{0.5}{24}$$

$$= 416 \text{ E}$$

The peak traffic is

$$T = 20{,}000 \times \frac{10}{60}$$

$$= 3333 \text{ E}$$

(b) The average traffic per cell is

$$t = \frac{416}{120}$$

$$= 3.47 \text{ E}$$

The peak traffic per cell is

$$t = \frac{3333}{120}$$

$$= 27.8 \text{ E}$$

(c) Use the line from Table 10.4 corresponding to 33 channels, since this is the number available in a 12-cell repeating system. For average traffic at 3.47 E, the blocking probability is much less than 1% (in fact it is negligible, since the average number of calls is much less than the number of channels). At peak periods, however, the blocking probability increases to just over 5%.

The other way to increase capacity is to increase the number of cells. The number of channels per cell remains the same as before, but since each cell covers a smaller area, with less potential traffic, the probabilities of call blocking and call dropping are reduced. The downside of this, of course, is that the expense of the system increases with the number of cell sites, and more frequent handoffs occur, increasing the system overhead.

This reduction of cell size to increase traffic is called *cell-splitting*. Cell-splitting allows the network to begin with large cells throughout, with the cell size decreasing in high-traffic areas as the traffic increases. Cellular telephone infrastructure is expensive, but it does not all have to be built at once.

Cell-splitting allows spectrum space to be used more efficiently, but it is not particularly cost-effective in terms of equipment. Doubling the number of cells doubles the system capacity with the same bandwidth allocation, but it also doubles the number of cell sites, roughly doubling the system cost.

Microcells, Picocells and Repeaters

Cell-splitting, as described in the preceding paragraphs, can be used to increase the capacity of a cellular system. At a certain point diminishing returns set in, however. Cell sites are expensive, and the increase in capacity does not justify the increase in cost for very small cells. Another problem is that real estate costs are highest in the areas where demand is greatest, and the use of small cells means less choice in cell-site location and thus higher costs for access to the sites. Finally, there is increased load on the switching system due to the increase in the number of handoffs with small cells.

In high-demand areas, **microcells** are often used to help relieve the congestion at relatively low cost. A microcell site is a very small unit that can mount on a streetlight pole. The microcell antenna is deliberately mounted lower than the tops of nearby buildings to limit its range. A typical microcell covers about 500 meters of a busy street, but has very little coverage on side streets. See Figure 10.9 for a typical unit.

FIGURE 10.9
Microcell site
(Courtesy of Bell Mobility)

Because microcells have such small, narrow patterns, it is difficult to obtain general coverage this way. Consequently the original larger cells (*macrocells*) are left in place, so that calls can be handed off between microcells and conventional cells as required. The microcells must use different frequencies than the overarching conventional cells, of course; this is accomplished by assigning to the microcells some of the channels that were formerly used by the macrocells.

A microcell is often under the control of a conventional cell site, with which it usually communicates by microwave radio. The microcell may itself be divided into several zones. Figure 10.10 illustrates how microcells and conventional cells (sometimes called *macrocells*) can work together.

FIGURE 10.10
Overlay of microcells and macrocells

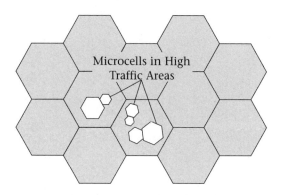

Microcells in High Traffic Areas

In order to save costs, many microcell sites are not true transceivers but are only amplifiers and frequency translators. Figure 10.11 on page 396 shows the idea. The main cell site up-converts the whole transmitted spectrum to microwave frequencies. The transmitter at the microcell site simply down-converts the block of frequencies to the cellular-radio band and amplifies it. No modulation or channel switching is required at the microcell site. Similarly, the microcell's receiver consists of a low-noise amplifier that amplifies the entire frequency range in use at that site plus an up-converter to convert the signal to microwave frequencies. All demodulation is handled at the main cell site.

Sometimes cellular radio signals are too weak for reliable use indoors. This is especially true in well-shielded areas like underground concourses. When reliable indoor reception is needed, sometimes very small cells called **picocells** are used. These are more common with the PCS systems to be described in the next chapter but are sometimes used with AMPS.

Indoor picocells can use the same frequencies as the outdoor cells in the same area if the attenuation of the structure is sufficient. This is the case in underground malls, for instance.

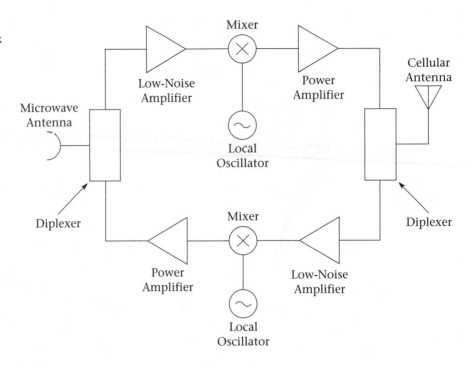

FIGURE 10.11
Microcell block diagram

Sometimes the problem is not excessive traffic, but a "hole" in the system coverage caused by propagation difficulties, such as a tall building or hill that casts a radio shadow. Obviously cell sites should be chosen to minimize these problems, but it is not always possible to eliminate them completely. In that case a *repeater* can be used, as shown in Figure 10.12. The repeater simply amplifies signals from the cell site and from the mobiles.

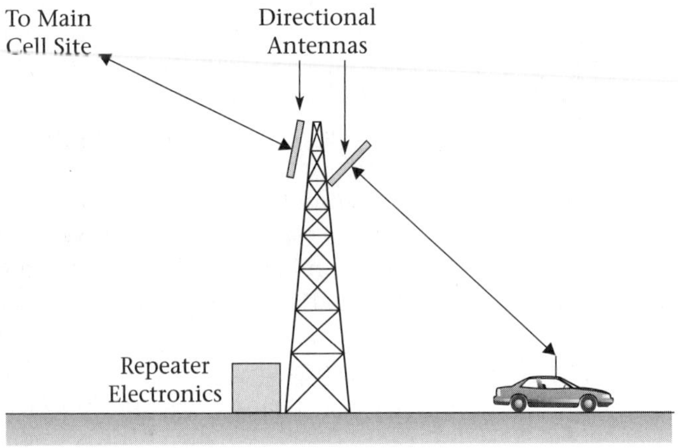

FIGURE 10.12
Cellular repeater

It can be connected to the main site by a microwave link, but often the repeater simply receives and transmits at the same frequencies, avoiding feedback by careful location of directional transmit and receive antennas.

10.8 Fax and Data Communication Using Cellular Phones

As can be seen from the foregoing, AMPS is a circuit-switched analog system designed for voice communication, as was the wireline system in its original form. Consequently, the most straightforward way to send data, as with the conventional wireline system, is to use modems at each end of the link. It is also possible to send packet-switched data using AMPS, as will be described shortly.

Cellular Modems The main differences between wireline phone service and analog cellular phones, for modem use, is that cell phone connections tend to be noisier and are subject to interruption during handoffs and fading. These interruptions, though brief in human terms (usually on the order of 100 ms to about 2 s), result in the loss of a considerable amount of data and possibly in a dropped connection. Consequently, the error-correction schemes on modems for cell phone use should be more robust than is necessary for wireline operation. Of course, this greater robustness results in slower data transmission. In addition, the modem must be set up not to require a dial tone before dialing. The dialing connections must be made separately to the phone: the situation is more complex than just plugging in a phone jack, as for a landline modem.

Many (but not all) modem cards for notebook and laptop computers will work with cell phones. Similarly, many, but not all, cell phones can be used with modems. A proprietary cable is required to connect modem to phone. It is also possible to find cell phone cards which plug into a notebook computer and enable it to send data via cellular radio.

Cellular modems are advertised as having speeds of up to 28.8 kb/s but actual speeds are usually 9600 b/s or less. Performance is improved by operating from a stationary vehicle, as this eliminates handoffs and reduces fading.

An error-correcting protocol called MNP10 is usually used with cellular connections. It must be used at both ends of the connection. MNP10 incorporates some special cellular enhancements. For instance, rather than trying to connect at the maximum speed and reducing speed if necessary, an MNP10 modem starts at 1200 b/s and gradually increases the speed. Multiple attempts are made to connect, and the packet size starts small and increases provided error rates remain low.

Another protocol, developed by AT&T, is called *Enhanced throughput cellular (ETC)*. It has the advantage of being able to work with a conventional wireline modem at the other end of the connection while retaining at least some of its benefits.

In order to allow communication between cellular modems and conventional wireline modems which may lack these more robust protocols, some cellular providers maintain a modem pool. This allows for conversion between cellular and wireline standards. Generally, the cellular user accesses this service by dialing a special prefix before dialing the telephone number of the landline modem. See Figure 10.13 for the idea.

FIGURE 10.13
Data transmission
by cellular radio

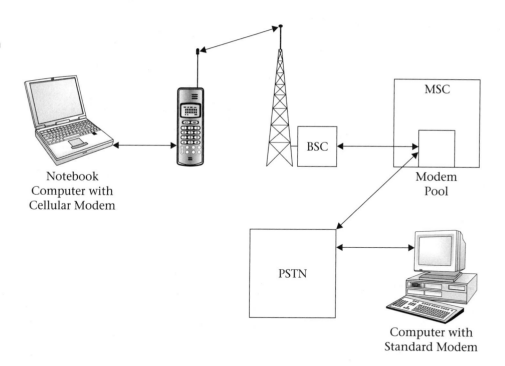

Facsimile transmission is also possible with cell phones. A fax modem and notebook computer can be used, or a conventional fax machine can be used with a special adapter. The adapter allows the cell phone to simulate a conventional two-wire telephone line with dial tone. Fax performance is much better from a stationary vehicle, since no special protocol is used.

Cellular Digital Packet Data (CDPD)

The previous section describes how data can be sent over a cellular voice channel in a manner similar to that used with landline telephony. That procedure is relatively simple and requires no prior arrangement with the cellular service provider, but it tends to be expensive, as full airtime costs, plus long-distance costs if applicable, must be paid for the entire time the call is connected.

Another way exists to send data over the AMPS cellular radio system. The **Cellular Digital Packet Data (CDPD)** system uses packet-switched data and tends to be less expensive than using a cellular modem, especially when data needs to be transmitted in short bursts. On the other hand, a separate account is required, and the cellular system has to be specially configured to use CDPD.

The principle behind CDPD is that at any given moment there are usually some voice channels in an AMPS system that are not in use. The CDPD system monitors the voice channels, using those that are idle to transmit data. When traffic is detected on the voice channel, the data transmissions cease within 40 ms. Since this is less than the setup time for a voice call, the voice customer is not affected. Users, once registered with the CDPD system, can transmit data as required without maintaining a continuous connection and tying up an expensive pair of voice channels.

The bit rate in the RF channel for CDPD is 19.2 kb/s, achieved using Gaussian minimum-shift keying (GMSK), a form of FSK. When overhead is taken into account, the maximum data rate is comparable to that obtained with a 14.4 kb/s modem—slow by current wireline standards, but not too bad for wireless. When the network is busy, the throughput is lower, as packets are stored and forwarded when a channel becomes available.

10.9 Digital Cellular Systems

Until recently, digital voice communication used more bandwidth than analog. The conventional wisdom was that digital was preferred where bandwidth was not an issue, because of its flexibility and immunity to the accumulation of noise, but that analog was better when bandwidth was constrained. For instance, you will recall that wireline telephony uses 64 kb/s for each one-way voice channel. In order to transmit this data rate in a 30-kHz channel, an elaborate modulation scheme would be needed. This would not be robust enough for the mobile radio environment with its noise and fading. Consequently, all first-generation cellular radio systems use analog modulation schemes. FM needs more bandwidth than AM and its variants, but it was found that FM's resistance to noise and interference more than made up for the additional required bandwidth.

Recent advances in data compression and voice coding, as discussed in Chapter 3, have reversed the conventional wisdom about bandwidth. It is now possible to transmit a digitized voice signal in less bandwidth than is required for an analog FM signal. This removes the last big obstacle to digital communication. Digital systems still tend to be more complex than analog, but large-scale integrated circuits have made it possible to build complex systems at low cost and in small packages. In fact, it appears likely that the analog cellular radio system just described will be the last of the great analog communication systems. Soon there will be no reason to use analog except for compatibility with legacy systems and perhaps for very low-cost communication devices.

Analog cellular radio can be seen as the first generation of wireless communication. Digital systems, which we begin to examine here and continue in the next chapter, are the second. The third generation will undoubtedly be digital. Its exact makeup is being debated at this time. A synopsis of the progress thus far can be found in Chapter 14.

Advantages of Digital Cellular Radio

The main incentive for converting cellular radio to a digital system was, as suggested earlier, to reduce the bandwidth requirements, allowing more voice channels in a given spectrum allotment. Other reasons also exist. Digital systems have more inherent privacy than analog, being harder to decode with common equipment. They also lend themselves to encryption, if required. Note that analog AMPS already uses digital signaling data. The fact that it is not encrypted is due to oversight: the system designers underestimated the ingenuity of hackers.

Digital communication systems can use error correction to make them less susceptible to noise and signal dropouts. They lend themselves to time- and code-division multiplexing schemes, which can be more flexible than the frequency-division multiplexing used in analog systems. Digital signals are easier to switch: in fact, most of the switching of analog telephone signals, including AMPS cellular telephony, is done digitally after analog-to-digital conversion.

The gradual conversion of the world's cellular radio systems to digital form has been accomplished differently in various parts of the world. In North America, which is the focus of this book, there existed one analog system with a very large installed base. Most of this infrastructure was not yet fully paid for, so operators were understandably reluctant to replace it. The requirement was for a digital system that would allow as much as possible of the analog equipment to be retained.

There were also millions of analog phones in consumers' hands. Many of these also were not yet paid for, especially since operators had given many analog phones to customers in exchange for service contracts. Perhaps you still have one of these (the author does). Making the analog phones obsolete would probably obligate the operators to provide new digital phones to their customers at no charge.

For these reasons, the system that evolved in North America uses the same radio frequencies, power levels, and channel bandwidths as AMPS. Some of the existing analog channels were converted to digital, leaving some analog channels in place in every cell for coexistence with analog equipment. The reduction of bandwidth requirements for digital communication allowed those analog channels that were converted to digital to combine three communication channels on one RF channel using TDMA, as described in the following section. The new system is so similar to AMPS that sometimes it is referred to as *digital AMPS* (*D-AMPS*).

The situation in Europe was completely different. Almost every country seemed to have a different analog cell phone scheme. Some had more than one! There were even several different frequency bands in use in different countries. The European Community made the radical decision to scrap the entire analog infrastructure and begin again with a common digital system. This system is called *GSM* (*Global System for Mobile*). It is not used in the cellular radio bands in North America but it is one of three systems in use for North American PCS, so GSM will be described in the next chapter.

Other countries varied in the type and extent of analog cell phone penetration. The result is a worldwide mixture of incompatible analog and digital formats, which will probably continue at least until the third generation of wireless telephony.

Conversion of AMPS to TDMA

The compatibility requirements outlined in the previous section constrained the development of a digital cellular radio system for North America. It was decided to combine three digital voice channels into one 30-kHz radio channel using TDMA. This is known as a full-rate TDMA system. The first such system went into operation in 1990. The specifications also encompass a half-rate system with six voice channels in one 30-kHz slot to be implemented at a later date when vocoder technology has improved.

The digital system would seem to be able to carry three times as much traffic as the analog system, but, due to trunking gain, the actual increase for a given level of blocking is greater. In the last two lines of Table 10.4, the traffic for various blocking probabilities is shown for both 7-cell sectorized and 12-cell repeating systems. The numbers of voice channels are calculated on the basis that one analog channel, for backward compatibility, is available in each cell or sector. Note that the traffic capacity is more than tripled.

At first, the AMPS control channels were left alone, and only the voice channels were digitized. This digital specification is known as IS-54B. A later

modification (1991), called IS-136, added high-speed digital control channels for better security and additional features. IS-136 is used both in the
800-MHz cellular radio band and the 1900-MHz PCS band.

TDMA Voice Channel   Figure 10.14 shows how the RF voice channel is divided in the TDMA
system. There are 25 frames per second so each frame is 1/25 s = 40 ms in
length. Each frame has 1944 bits so the total bit rate for the RF signal is 1944
× 25 = 48.6 kb/s. Phase-shift keying with four levels (π/4 QPSK) is used, so
there are two bits per symbol and the baud rate is 24.3 kbaud. Note that this
is a data rate of 48.6/30 = 1.6 b/s per hertz of bandwidth. This is quite conser-
vative, necessarily so because of the radio environment, which is subject to
noise, interference, and very deep fading.

FIGURE 10.14
TDMA frame

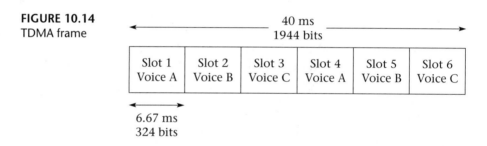

Each frame has six time slots lasting 40 ms/6 = 6.67 ms and containing
1944/6 = 324 bits each. For full-rate TDMA, each voice signal is assigned to
two time slots as shown. Six voice signals, occupying one slot each, can be
accommodated with half-rate TDMA. For the full-rate system, speech data
corresponding to 40 ms of real time is transmitted in 2 × 6.67 ms = 13.3 ms.
As with analog AMPS, the TDMA system uses separate RF channels for transmit and receive.

Speech encoding (see Chapter 3) is used to limit the bit rate to approximately 8 kb/s for each speech channel for the full-rate system. The full-rate
system allocates two noncontiguous time slots to each voice channel: slots 1
and 4 for the first, 2 and 5 for the second, and 3 and 6 for the third. In addition, the bits corresponding to each 20 ms of speech are divided among
two time slots. Interleaving the data bits in this way reduces the effect of
burst errors.

Overhead reduces the number of data bits available per time slot to 260.
The data rate available for each voice channel is 260 bits/20 ms = 13 kb/s.
The voice is actually encoded at 7.95 kb/s and the remaining bits are used for
error correction. The half-rate system will use 4 kb/s for voice coding.

The frames as just shown are similar for both forward (base to mobile) and reverse (mobile to base) channels, but the composition of the time slots differs. The frames are synchronized for the forward and reverse channels, but the timing is offset so that a frame starts 90 bits (1.85 ms) earlier at the mobile. A mobile transmits during two of the six time slots and receives on a different two slots. The remaining two time slots are idle: the phone may use these to check the signal strength in adjacent cells to assist in initiating a handoff. This technique allows the digital cell phone to have only one transmitter and one receiver, just as for AMPS. In fact the RF situation is a little simpler, since there is no need for the mobile to transmit and receive simultaneously. (Since digital cell phones have to work with the analog AMPS system as well, this unfortunately does not really simplify the design, as a duplexer is still needed for analog operation.)

Each slot contains 324 bits for both forward and reverse channels. The allocation of these bits is different for the forward and reverse links, however. See Figure 10.15 for an illustration of the differences.

In particular, the mobile needs time to turn its transmitter on for each transmit time slot, since to avoid interference it must be off when the mobile is not scheduled to transmit. Six bit periods (123 μs) are allocated to this. The base station transmitter is on all the time, since it uses all six time slots to

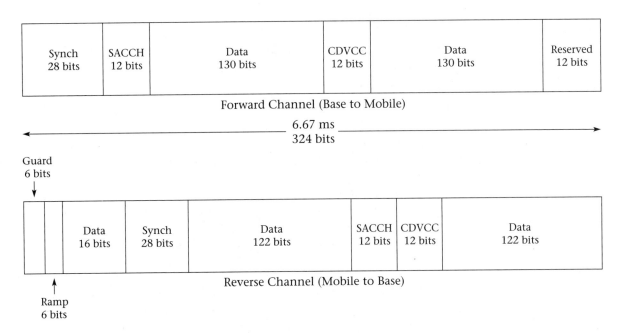

FIGURE 10.15 TDMA voice time slots

transmit to three mobiles on the same RF channel. The mobile also waits for a guard time (another six bits) to pass before transmitting. This is necessary because different amounts of propagation delay could cause mobile transmissions (on the same RF channel but in different time slots) to overlap when received at the base station.

EXAMPLE 10.5

Calculate the maximum distance between base and mobile that can be accommodated with a guard time of 123 μs.

SOLUTION

The signal from base to mobile is delayed by the propagation time between the two. This causes the mobile's synchronization to be off by that much time, and it will be late in starting its transmission. In addition, the signal is further delayed by the propagation time from mobile to base. Therefore, the 123 μs allowance must include the round-trip propagation time. Since radio waves propagate at the speed of light, the maximum total round-trip distance is:

$$d = ct$$
$$= 300 \times 10^6 \text{ m/s} \times 123 \times 10^{-6} \text{ s}$$
$$= 36.9 \text{ km}$$

The maximum one-way distance is half that, or about 18.5 km. For even larger cells, the mobile can instruct the base to advance its transmission by up to 30 bit periods (617 μs).

In addition to voice samples and their associated error correction, the digital traffic channels contain synchronizing, equalizer training, and control information.

The *Coded Digital Verification Color Code (CDVCC)* provides essentially the same information as the Supervisory Audio Tone (SAT) in AMPS. The **Slow Associated Control Channel (SACCH)** provides for control signal exchanges during calls and essentially replaces the blank-and-burst signaling in AMPS, though there are provisions to "steal" voice data bits for additional control information as required. These stolen bits form the **Fast Associated Control Channel (FACCH)**, which is used for urgent information such as handoff commands. The *coded digital locator (CDL)* field tells the mobile where to find digital control channels, if available.

TDMA Control Channels

As mentioned, there are two "flavors" of TDMA cellular radio in use. The earlier specification, called IS-54B, uses the same control channels and formats as AMPS. These are called *Analog Control Channels* (*ACCH*) because of their association with the analog system, but as noted earlier, they are actually digital, using FSK and a channel data rate of 10 kb/s.

The IS-136 specification incorporates separate control channels for the digital system. These are called *Digital Control Channels* (*DCCH*) to distinguish them from the older type. Digital control channels consist of pairs of slots on the same RF channels that are used for voice. The DCCH can be assigned to any RF channel; it does not have to be one of the 21 control channels used in the analog system. As with the voice channels, separate forward and reverse channels are needed. Normally there is one DCCH pair per cell, or per sector in a sectorized system. See Figure 10.16.

FIGURE 10.16
TDMA digital
control channel

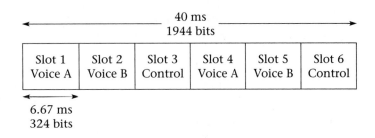

The total bit rate for a DCCH is one-third of the RF channel bit rate, or 44.6/3 = 14.9 kb/s, compared with 10 kb/s for an ACCH. This extra capacity makes the digital control channels useful for many added features, such as call display and short text messages. As with the analog system, digital control channels are used in setting up calls. They cannot be used during a call, since the single receiver in the mobile unit is otherwise occupied.

Privacy and Security in Digital Cellular Radio

Privacy is considerably improved in digital cellular radio compared to the analog system. Ordinary analog scanners can make no sense of the digitized voice signal. Even decoding it from digital to analog is not straightforward, due to the need for a vocoder. However, obviously vocoders are present in all digital cell phones, so a modified cell phone could do the job.

There is some encryption of the authorization information in the TDMA system, enough to make cell phone cloning and impersonation difficult. In general the level of security of the TDMA system is considerably better than with AMPS and is probably adequate for general use.

Dual-Mode Systems and Phones

One of the most important features of the TDMA digital cellular radio system is its backward compatibility with AMPS. Cell-site radio equipment does not need to be replaced, and provided that all cells keep some analog channels, neither do analog cell phones. Cell sites can incorporate digital channels as needed to cope with increased traffic.

Since not all cells are expected to have digital channels in the near future, it is necessary for digital cell phones to work with the analog system as well. While analog-only phones will continue to work throughout the system, digital-only phones would work only in major metropolitan areas. Therefore, all current TDMA digital cell phones are dual-mode: they attempt to make a digital connection first, then if that fails, revert to analog.

One difference in the RF component of TDMA cell phones is the addition of a new power class, Class IV. This is the same as the analog Class III, except for the addition of three new power levels at the bottom of the range. These levels of –26, –30, and –34 dBW ERP allow for better operation with microcells. In addition, lower power levels, coupled with the fact that the transmitter operates only one-third of the time, help to improve battery life in digital mode.

Figure 10.17 is a block diagram for a typical dual-mode TDMA cell phone. Note that a duplexer is required for analog operation. The transmitter power amplifier is linear, because QPSK requires this. Of course, a linear amplifier is also satisfactory for FM and FSK. The fact that the channel bandwidth and frequencies are the same for analog and digital systems simplifies the RF design. For instance, only one receiver IF filter is needed.

When we look at PCS systems in the next chapter, we'll see that the TDMA system just described is one of three systems used for PCS. This allows for the possibility of dual-mode, dual-band phones incorporating analog AMPS, as well as both digital cellular and TDMA PCS modes.

Data Communication with Digital Cellular Systems

Paradoxically, connecting data equipment like modems and fax machines to digital cell phones can be more complicated than with the ordinary analog system. Since voice is encoded using a vocoder, it is not satisfactory simply to insert modem tones instead of an analog voice signal. The vocoder is optimized to code voice and will make a mess of any other kind of input. In fact, even DTMF tones from the phone's keypad have to bypass the vocoder and be transmitted digitally in a time slot of their own.

Circuit-switched data communication can be accomplished with the digital system by inputting the data directly to the voice time slots without using the vocoder. The data rate is limited to 9600 b/s to allow for additional error correction and still fit within the 13 kb/s allocated for voice data. Of course, the system has to be told about this, so that the data can be output properly at the other end.

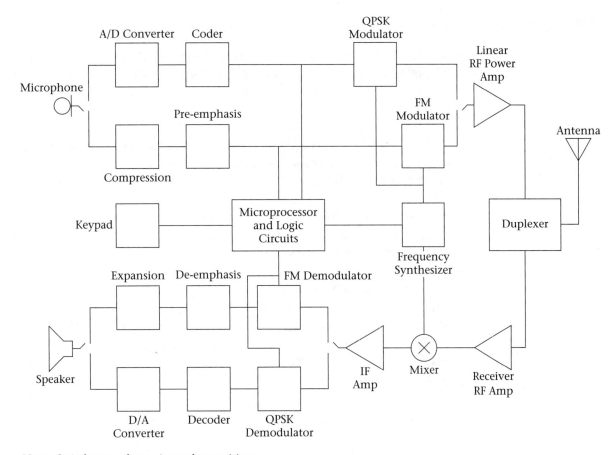

Note: Switches are shown in analog position.

FIGURE 10.17   Block diagram of dual-mode cell phone

In practice, since all 800-MHz digital cell phones and cell sites are also capable of analog operation, the usual way of sending data is to use an analog channel and a cellular modem as previously described. Similarly, packet-switched data can be sent via the analog system using CDPD as already described. Short data messages can be sent using the digital control channel.

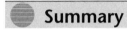

Summary

The main points to remember from this chapter are:

- Personal communication systems work best when they are portable, connected to the public switched telephone network, and widely available at reasonable cost.

- Early attempts at personal communication include CB radio, IMTS radio telephones, and cordless phones.

- Cellular radio systems are much more efficient users of spectrum than communication systems using a single repeater, but they are also much more complex and require computing power in the mobile phone.

- AMPS, the North American analog cell phone system, uses FM modulation with 30-kHz channels in the 800-MHz region of the spectrum.

- Cell-splitting, including the use of microcells and picocells, can greatly increase the capacity of a cellular system.

- Cell sizes have a lower limit due to the number and frequency of handoffs required for quickly moving vehicles.

- The North American digital cell phone standard builds on AMPS by splitting one analog channel into three digital channels using TDMA.

- Both circuit-switched and packet-switched data can be carried on the cellular radio system using special adapters. Performance is generally not as good as with the wireline system.

Equation List

$$d = \sqrt{17h_t} + \sqrt{17h_r} \qquad\qquad (10.1)$$

$$L_{fs} = 32.44 + 20 \log d + 20 \log f \qquad\qquad (10.2)$$

$$T = NP \qquad\qquad (10.3)$$

Key Terms

Advanced Mobile Phone Service (AMPS)   North American first-generation cellular radio standard using analog FM

air interface   in wireless communication, the radio equipment and the propagation path

base station controller (BSC)   in cellular and PCS systems, the electronics that control base station transmitters and receivers

blank-and-burst signaling in cellular communication, interrupting the voice channel to send control information

call blocking failure to connect a telephone call because of lack of system capacity

cellular digital packet data (CDPD) method of transmitting data on AMPS cellular telephone voice channels that are temporarily unused

control mobile attenuation code (CMAC) information sent by the base station in a cellular radio system to set the power level of the mobile transmitter

digital color code (DCC) signal transmitted by a cell site to identify that site to the mobile user

dropped call a telephone connection that is unintentionally terminated while in progress

electronic serial number (ESN) number assigned to a cell phone by the manufacturer as a security feature. It is transmitted with the phone's telephone number to authorize a call.

fast associated control channel (FACCH) in a digital cellular system or PCS, control information that is transmitted by "stealing" bits that are normally used for voice information

forward channel communication from a cell site or repeater to the mobile unit

full-duplex two-way communication in which both terminals can transmit simultaneously

half-duplex two-way communication in which only one station can transmit at a time

handoff transfer of a call in progress from one cell site to another

Improved Mobile Telephone Service (IMTS) a mobile telephone service, now obsolescent, using trunked channels but not cellular in nature

microcell in cellular radio, a small cell designed to cover a high-traffic area

mobile identification number (MIN) number that identifies a mobile phone in a cellular system; the mobile telephone number

mobile switching center (MSC) switching facility connecting cellular telephone base stations to each other and to the public switched telephone network

mobile telephone switching office (MTSO) see **mobile switching center (MSC)**

number assignment module (NAM) in a cellular phone, a memory location that stores the telephone number(s) to be used on the system

personal communication system (PCS) a cellular telephone system designed mainly for use with portable (hand-carried) telephones

picocells very small cells in a cellular radio system

radio common carrier (RCC) a company that acts as a carrier of radiotelephone signals

reverse channel communication channel from mobile station to base station

roamer a cellular customer using a network other than the subscriber's local cellular network

slow associated control channel (SACCH) in a digital cellular system or PCS, control information that is transmitted along with the voice

station class mark (SCM) code which describes the maximum power output of a cellular phone

supervisory audio tone (SAT) in the AMPS system, a sine wave above the voice frequency range, transmitted on the voice channel along with the voice, used by the base station to detect loss of signal

system identification number (SID) in the AMPS system, a number transmitted by the base station to identify the system operator

telepoint a very small cell used with some cordless phones to allow their use in public areas

● Questions

1. What usually limits the distance for CB communication?
2. What usually limits the distance from handset to base for cordless phones?
3. What advantage does the use of a higher frequency have for hand-held transceivers?
4. Why do cordless phones and cellular phones need two RF channels for one conversation?
5. What is a telepoint system?
6. Explain the concept of frequency reuse with reference to the IMTS and AMPS mobile telephone systems.
7. How do cellular telephone systems connect to the PSTN?
8. Explain why FM is better than AM for cellular radio, in spite of its wider bandwidth.

9. Why is it more efficient to use a single provider for cellular and PCS phone service? Why are multiple providers used in spite of this fact?

10. What happens when a mobile phone moves out of a cell? What is the name for this process?

11. What factors limit the maximum and minimum sizes of cells?

12. What are the functions of the control channels in the AMPS system?

13. What type of modulation and data rate are used on AMPS control channels?

14. How does the AMPS system attempt to prevent fraudulent billing of calls to the wrong customer? How effective are these techniques?

15. What are the MIN and ESN for a cell phone and what are the differences between them?

16. What are the differences between mobile phones, transportable phones, and portable phones from the point of view of the cellular system?

17. What are the SID and DCC and what is the difference between them?

18. When, why, and how are control signals sent over the voice channel?

19. How does the AMPS system keep track of which mobiles are turned on and where they are?

20. List and explain the steps in receiving a call to a mobile.

21. What is SAT and what is its function? Is it analogous to anything in a conventional local loop?

22. What is *blank-and-burst* signaling and why is it necessary?

23. What is a cloned phone? How does the AMPS system try to detect these phones?

24. What is the function of the *duplexer* in a cell phone?

25. Why is it unnecessary for cell phones and cellular base station transmitters to use linear RF power amplifiers in their transmitters? What is the advantage of using nonlinear amplifiers?

26. What is the range of transmit power adjustment (in dB) for each class of cell phone?

27. Under what circumstances would a Class I or Class II phone operate with higher transmit power than a Class III phone?

28. What is meant by pre-emphasis and how does it improve the signal-to-noise ratio of an FM cellular radio system?

29. What advantage does a typical mobile antenna have over a typical portable cell phone antenna? Given this advantage, why is the same type of antenna not used on portable phones?

30. What is the advantage of using two dedicated receive antennas per cell site or sector over a combined transmit/receive antenna?

31. What is the function of the base station controller?

32. What is meant by the term *air interface* in a cellular radio system?

33. What is the difference between call blocking and call dropping? Which is more objectionable?

34. Explain the concept of trunking gain.

35. What are microcells? How do they differ from macrocells and picocells?

36. What limits the size of a microcell?

37. Where are repeaters used in cellular systems?

38. What are the differences between cellular modems and ordinary wireline modems?

39. How can a fax machine be made to work with a cell phone?

40. What is the difference between circuit-switched and packet-switched data communication?

41. How does the CDPD system transmit packet-switched data over the normally circuit-switched AMPS cell phone system?

42. How does the North American digital cellular system increase system capacity?

43. What is meant by TDMA and FDMA? Show that the North American digital cellular system actually uses both TDMA and FDMA.

44. How do privacy and security in the North American digital cell phone system compare with these aspects of AMPS?

45. Digital wireline telephone systems use 64 kb/s per voice channel, yet digital cellular systems achieve almost equivalent voice quality using 8 kb/s. Explain briefly how this is possible.

46. State and compare the modulation schemes used by the control channel in analog AMPS and the digital control channel used in the TDMA digital cell phone system.

47. TDMA cell phones need a duplexer even though they transmit and receive in different time slots. Why?

48. What is the problem with connecting a cellular modem to a TDMA phone?

Problems

1. Look at Example 10.1.

   (a) Calculate the minimum transmitter power that would give line-of-sight communication at 27 MHz for the maximum distance set by the radio horizon. Assume free-space propagation.

   (b) Repeat the calculation for free-space propagation at 800 MHz. Account, physically, for the fact that more power is required at the higher frequency.

2. Repeat the calculation for Problem 1 assuming typical mobile propagation conditions and a frequency of 800 MHz. What can you conclude about the feasibility of portable communication without repeaters over this distance in an urban environment?

3. Now suppose we have a cellular system. The mobile is as described in Problem 2 but the base station has an antenna with a gain of 6 dBi located 50 m above average terrain. How much power is needed now? Is this reasonable?

4. Amateur radio clubs often operate repeaters, usually in frequency bands near 144 and 430 MHz. They do not have the same power restrictions as CB radio. Calculate the minimum repeater height and transmitted EIRP for a repeater that must communicate with mobile stations over a 50 km radius, maintaining a minimum field strength of 50 mV/m at the mobile receiver, which has an antenna height of 1.5 m. Assume free-space propagation conditions.

5. Suppose, that in an attempt to increase range, cordless telephone manufacturers double the power output of their transmitters. Assume, that in the cluttered in-house environment, power density is inversely proportional to the fourth power of distance.

   (a) Suppose the usable range, limited by signal strength without interference, was 50 m before the power increase. What is it after the increase?

   (b) What effect will the power increase have on interference from other portable telephones?

      (i) operating at the old (low) power level?

      (ii) operating at the new (high) power level?

   (c) What can you conclude from your answer about the effect of increasing power in an environment where transmission range is limited by

      (i) signal-to-noise ratio?

      (ii) signal-to-interference ratio?

6. (a) Calculate the length of a quarter-wave vertical antenna for each of the following services: CB radio at 27 MHz; cordless phone at 49 MHz; FRS at 460 MHz; cellular radio at 800 MHz; and PCS at 1800 MHz.

   (b) What can you conclude about the ease of designing portable transceivers for each of these services?

7. Suppose we are designing picocells for an indoor cellular system in an underground mall. If we assume people using the system are walking at a speed of 5 m/s or less, what is the smallest cell radius we could have if we wanted to have handoffs occur no more often than once every 45 seconds?

8. Suppose we are communicating on a cell phone using a digital modem at 9600 b/s. How many bits will be lost during

   (a) blank-and-burst signaling?

   (b) handoff, if it results in a 0.5 s loss of signal?

9. A cellular system is designed to operate reliably with traffic of 2000 E. If each customer on average uses the phone for three minutes during the busiest hour, how many customers can be accommodated, assuming an even distribution of customers?

10. If the number of channels were doubled in the previous question, would the maximum number of customers be double, more than double, or less than double that found above? Explain.

11. Suppose that a system's customers use their cell phones for an average of five minutes each during the busiest hour. The system has 10,000 customers, approximately evenly distributed over 100 square km. Half of the customers have dual-mode digital phones. The other half have analog-only phones. We require a maximum of 2% call blocking for those with digital phones at rush hour.

    (a) How many cells are required and with what radius, if we choose an analog system with a 12-cell repeating pattern? Assume hexagonal cells.

    (b) Suppose we try to improve service by changing to a digital system, leaving one analog channel per cell. What happens (qualitatively) to the blocking rate for customers with each type of phone?

    (c) How could we improve the situation in (b), as it applies to analog-only customers?

12. Suppose two cell phones are in the same room. Draw a diagram showing the route signals take if the cell phones are:

    (a) operating on the same system (A or B)

    (b) operating on different systems (one each on A and B)

13. Prepare a chart showing and explaining the steps in placing a call from a mobile.

14. Prepare a chart showing and explaining the steps in placing a call to a mobile.

15. (a) Assume that the range of a mobile phone is limited by its transmitter power. If a portable phone (Class III) can communicate over a distance of 5 km, what would be the maximum range for a permanently installed car phone (Class I) under the same circumstances. Assume free-space attenuation.

    (b) Repeat part (a), but assume that attenuation is proportional to the cube of distance. This is more typical of a mobile environment.

    (c) Examine the assumption made above. What other factors could limit communication range?

16. (a) By how much would a 2400-Hz baseband signal be boosted by the transmitter pre-emphasis circuit in a cell phone, compared with a frequency of 300 Hz?

    (b) Why is the boost due to pre-emphasis unlikely to cause over-modulation of the transmitter?

    (c) Compare the gain in the receiver for signals of frequencies 2400 Hz and 300 Hz after demodulation.

17. Suppose that the same type of vocoder used in full-rate TDMA cellular radio were to be used for wireline telephony. How many voice channels could be carried on a T-1 line?

18. A base and mobile are separated by 5 km. What is the propagation time for a signal traveling between them?

19. Express the minimum power output for a digital Class IV cell phone in mW. How does it compare with CB radio and cordless phones?

20. (a) Review the discussion of mobile propagation in Chapter 7. Find the path loss in a typical urban environment in the cell phone frequency range if the base station antenna has a height of 20 m and the distance is 5 km.

    (b) Find the power delivered to the base station receiver under the same circumstances if the base station antenna has a gain of 6 dBi and the feedline to the receiver has a loss of 2 dB.

    (c) Suppose the phone described in part (a) is used inside a vehicle, which causes a loss of 20 dB. What would be the effect on the answers for parts (a) and (b)?

Personal Communication Systems

11

Objectives

After studying this chapter, you should be able to:

- Compare cellular radio with PCS.

- Describe the operation of each of the personal communication systems used in North America.

- Compare the North American personal communication systems.

- Perform calculations of spectral efficiency for PCS.

- Calculate open-loop power and processing gain for CDMA PCS.

- Explain and compare the methods in which data can be carried on PCS.

11.1 Introduction

As pointed out in Chapter 10, the term *personal communication* is somewhat vague. In general, it refers to direct communication between people rather than between places. To understand the concept, consider Joe, a typical present-day North American. Joe has a home phone and an office phone. The home phone has an answering machine and the office phone has voice mail, but there is no connection between them. Joe also has a cell phone in his car. That system does not have voice mail, though it could if he wanted to pay extra. If you want to talk to Joe, you really call a place rather than a person. You decide if he's more likely to be at home, in the office, or in the car, and place your call accordingly. If he's in none of those places, you leave a message. Actually, you probably leave two messages, one at home and one at work.

Now consider Joan, a slightly more modern North American. Joan also has office and home phones, but her cell phone is a small portable phone which she usually carries with her. However, she has to pay airtime on the cell phone (but not on the others) so she only gives out that number to close friends and important business associates. Those people, trying to prevent Joan from having to pay airtime, usually try her home and office numbers first, then the cellular number. Joan is halfway toward a true PCS.

Move forward a little to the future. Ted has only one phone. It's small enough for him to slip into his shirt pocket, and he carries it with him all day. He has only one phone number. If you want to talk to Ted, you call that number. If he cannot take calls for a while, he turns off the phone and voice mail takes over. Similarly, if Ted wants to talk to someone, he uses his pocket phone—no matter where he is. Ted's phone number is no longer associated with a place; it is associated with a person.

It is possible to live like Ted today, but for most people it is tricky. How does the switchboard at Ted's company find his personal phone? How does he charge business calls to the company? How does he get a listing in the phone book? What happens when he is out of range of his service provider? All these problems are solvable, and there are many people who feel that, eventually, all phones will be wireless. Your pocket phone will connect to a wireless PBX at your office when you're at work, to a telepoint in your house when you're at home, to a cellular or PCS network when you're out, and to a satellite-based telephone system when you are at a remote location.

We are still some distance from true PCS, but the gap is closing. The systems described in this chapter are playing a major role in closing that gap. Perhaps it is appropriate, then, that the name for the larger concept is often applied to the systems described in this chapter. These so-called personal communication systems are derived from the cellular concepts introduced in Chapters 7 and 10, with enhancements to allow the phones to be smaller and lighter, to have improved battery life, and to have extra features not available in first-generation cellular systems.

There are three competing types of PCS in North America. This contrasts vividly with Europe, which simply extended its GSM digital cell phone system to a higher frequency range for PCS. We have already heard of two of the three North American personal communication systems. One is in fact the GSM system, which was mentioned, though not discussed, in Chapter 10. The second is IS-136, the North American digital cell phone standard, which has also been extended for use at PCS frequencies. The third is a direct-sequence spread-spectrum system developed in the United States by Qualcomm and known as IS-95, or by its tradename, CDMAone™. In the next section we look at some features that all these systems have in common and contrast them with conventional analog AMPS. Following that, we look at each of the three systems in more detail.

 ## 11.2 Differences Between Cellular Systems and PCS

Though based on the same cellular idea as the first-generation cell phone systems described in Chapter 10, PCS have significant differences which justify the use of a different term. You should realize, however, that many of the differences are transparent, or at least not immediately obvious, to the user. The systems described in this chapter are often called *second generation* personal communication systems; in other words, the analog cell phone system is really the first generation of PCS. The third generation, now being designed, will feature much wider bandwidth for high-speed data communication and it will be discussed in Chapter 14.

Frequency Range One of the reasons for establishing new PCS was that the cellular frequency bands were becoming crowded, especially in major metropolitan areas. There was no room for expansion in the 800-MHz band, so the new service was established in the 1900-MHz band (1800 MHz in Europe). This has advantages in terms of portable antenna size. A few years ago, electronics for this frequency range would have been prohibitively expensive, but advances in integrated circuit design have reduced the cost penalties.

In North America the broadband PCS band consists of 120 MHz in the 1900-MHz region. The term *broadband* here is relative. It refers to bandwidth sufficient for voice communication and distinguishes this service from such narrowband services as paging, which will be discussed later in this book. Sometimes the term *broadband communication* is used to refer to video and high-speed data; that is *not* the sense in which it is used here.

See Table 11.1 for the PCS band plan. Note that there are six frequency allocations, so up to six licenses can be awarded in any given area. There are three 30-MHz and three 10-MHz allocations. The reverse channel or **uplink** (mobile to base) is 80 MHz above the forward channel or **downlink** (base to mobile) frequency. Reverse and forward channel allocations are separated by a 20-MHz band, from 1910 to 1930 MHz, which is allocated for unlicensed services like short-range voice communication. In the United States the frequencies have been assigned by auction; in Canada licenses were allocated after public hearings. Some PCS carriers are established cellular providers with 800-MHz licenses; others are new to the field of wireless communication.

TABLE 11.1 Broadband PCS Band Plan

Allocation	Base Transmit (Forward Channel or Downlink)*	Mobile Transmit (Reverse Channel or Uplink)*
A	1850–1865	1930–1945
B	1870–1885	1950–1965
C	1895–1910	1975–1990
D	1865–1870	1945–1950
E	1885–1890	1965–1970
F	1890–1895	1970–1975

*Frequencies are in MHz

Smaller Cell Size Cellular telephony was originally conceived as a mobile radio system, with phones permanently mounted in vehicles. These phones use efficient external antennas on the roof of the vehicle and have a maximum ERP of 4 W. However, in recent years portable cell phones have outsold mobile phones by a considerable margin. This has implications for the system, as portable phones have lower power and are often in difficult locations—from a propagation point of view—such as inside vehicles and buildings. For these reasons, portable AMPS cell phones may not work reliably near the edges of the larger cells. PCS, on the other hand, were designed from the beginning with handheld phones in mind. At first it was thought that most PCS users would be on foot, but it is now quite obvious that subscribers expect to use the phone wherever they are: outdoors, indoors, in an underground shopping mall, or in their cars.

PCS cells are typically smaller than AMPS cells to accommodate more traffic and low-power handheld phones. They must hand off calls very quickly to handle users in moving cars.

All-Digital System

Because of technical constraints, all first-generation cellular systems are analog, though some progress has been made in converting them to digital technology. In fact, some providers have marketed 800-MHz digital cell phones as "PCS" systems.

Current digital technology is more efficient than analog FM in its use of bandwidth. It also allows lower power consumption in the portable phone and more advanced data communication and calling features. Security and privacy are inherently better with any digital system, since ordinary scanners cannot be used to intercept calls, and digital coding schemes can also incorporate encryption as required.

There is one major problem with North American digital PCS. Whereas first-generation cellular systems in North America all use the same analog technology, there are three incompatible digital systems in North America. This makes roaming more difficult with PCS than with cellular phones. Many providers and phone manufacturers have solved this problem by offering dual-band, dual-mode phones that are capable of both PCS and analog cellular operation. The solution is not ideal, because it results in phones that are larger and more expensive than they would otherwise have to be.

Extra Features

AMPS systems were designed with POTS (*plain old telephone service*) in mind. Even features commonly found on wireline phones, such as call display, present problems in AMPS. Digital systems allow a substantial amount of data transmission in their control channels, making all sorts of enhancements possible. In addition to obvious features like call display, digital systems can allow short printed messages, and even e-mail and limited web browsing are possible without additional modems and computers. The features available and the way they are implemented vary with the type of PCS, and we will look further into this later in this chapter.

Coverage

At least at present, the coverage for any PCS is much less universal than it is for AMPS cell phones. This will undoubtedly change in the future, as the systems acquire more customers and build more infrastructure. In the meantime PCS users have to pay more attention to local coverage areas than do analog cell phone users.

Rate Structure   One of the arguments for PCS is that they should be less expensive than analog cellular radio. The utilization of spectrum space is more efficient, for example. In practice, rates tend to be set by a combination of market forces. The analog systems have a head start in paying for their infrastructure and have been able to lower prices to match PCS in many cases.

11.3 IS-136 (TDMA) PCS

We looked at IS-136, the North American Digital Cellular standard, in Chapter 10. Most people just refer to it as *TDMA* (*time division multiple access*) when they are talking about PCS, though GSM is also a TDMA system. The most important difference between the 800-MHz and 1900-MHz versions of TDMA is that there are no analog control channels in the PCS bands. Rather than go over the ground already covered in Chapter 10, in this chapter we will take a closer look at the digital control channel and consider how enhanced services are provided. Much of this material is very similar for the GSM system, described next.

TDMA Digital   Recall from Chapter 10 that the digital control channel (DCCH) uses two of
Control Channel   the six time slots in a TMDA frame (slots 1 and 4, to be precise). Normally only one DCCH is required per cell or sector. Figure 11.1 shows how the time slot is divided up for both forward and reverse channels.

FIGURE 11.1
TDMA digital
control channel

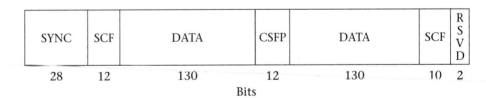

(a) Forward channel

(b) Reverse channel

Let us look at the forward channel first. The *SYNC (synchronizing)* bits have the same function as for the voice channels, allowing the mobile receiver to lock on the beginning of the transmission. The *SCF (shared channel feedback)* bits perform several functions. They provide acknowledgement of messages from mobiles and inform the mobiles of the status of the reverse control channel. Just as in analog AMPS, the forward channel is under the control of the base station, but many mobiles share a single reverse control channel. By monitoring the status of the reverse channel as reported by the base in the SCF field, the mobiles can reduce the possibility of collisions. These still occur occasionally; however, in that case, the message from the mobile will not be acknowledged by the base, and the mobile will try again after a random delay time.

The *CSFP (coded superframe phase)* bits identify the location of this time slot in a larger frame that extends over 16 TDMA frames or 32 blocks of control-channel data, representing a time period of 640 ms. Each block is designated as containing broadcast, paging, messaging, or access response information. Each of these types of data can be considered a separate logical channel of data, time-division multiplexed with the other types and with voice as well, since four of every six timeslots on the RF channel that carries the control channel are still used for voice. The number of control-channel blocks assigned to each type of use can be varied within limits. Table 11.2 on page 424 summarizes the logical channels, and a brief description of each is given below.

Two superframes comprise a *hyperframe*. The hyperframe structure allows data to be repeated. This means that a mobile receiver can check the signal strength on other channels, without missing data. It also provides redundancy: if the mobile misses some data because of a burst error, it gets a second chance.

The broadcast channel contains information intended for all mobiles. It is divided into two components. The *fast broadcast channel (F-BCCH)* is used to transmit system parameters to all the mobiles. These include the structure of the superframe itself, the system identification, and registration and access parameters. All of these must be communicated to the mobile before it can place a call, so all of this information is transmitted at the beginning of each superframe. The *extended broadcast channel (E-BCCH)* has less critical information, such as lists of the channels used in neighboring cells. This information can be transmitted over the course of several superframes.

The *short message service, paging, and access channel (SPACH)* is used for control messages to individual telephones and for short paging-type messages to be displayed on the phone. It is not necessary for every phone to monitor all these messages; the phone is told which block to monitor and can go into an idle or *sleep* mode the rest of the time while it waits for a call. This helps to extend battery life.

The reverse control channel is quite different from the forward channel. There is no broadcast information; there is only one logical channel called the *Random Access Channel* (*RACH*). This is used by the mobile to contact the base, for registration, authentication, and call setup. Normally the mobile will find out from the broadcast channel whether this channel

TABLE 11.2 Logical Channels in the Data Section of the TDMA Digital Control Channel

Name of Channel		Function	Time Slots per Superframe
Broadcast Channel (BCCH)	Fast Broadcast Channel (F-BCCH)	Urgent information for all mobiles, transmitted once per superframe, at beginning of superframe: • Superframe structure • System identification • Access parameters • Registration parameters	3–10
	Extended Broadcast Channel (E-BCCH)	Less urgent information for all mobiles (transmitted over several superframes): • Neighbor lists (control channel frequencies in nearby cells) • Regulatory configuration (spectrum allocation) • Mobile assisted-channel allocation (frequencies mobiles should monitor)	1–8
Reserved		As needed by system	0–7
Short Message Service, Paging, and Access Channel (SPACH)	Short Message Service Channel (SMSCH)	• Short message service • Remote phone programming	Remaining Slots
	Paging Channel (PCH)	Paging (ringing mobile phone)	
	Access Response Channel (ARCH)	Control messages to individual phones	

is clear; if that information is available, it can transmit at random. If there is no acknowledgement, it probably means a collision has occurred, and the mobile will try again after a short random delay. As with the reverse voice channel, time has to be allocated for ramping up the mobile transmitter power, and guard time is needed to avoid interference between mobiles at different distances from the base. Ramp time is shown as R and guard time as G, in Figure 11.1(b).

11.4 GSM

GSM is the system used in Europe and most of Asia for both cellular and PCS bands. It is not found at 800 MHz in North America, but it is used in the 1900-MHz PCS bands. Note that, even though the same modulation scheme is used, North American PCS phones will not work in Europe because the frequencies allocated to PCS are different—the European bands center around 1800 MHz.

GSM is another TDMA system but the details are different. GSM also has some unique features that make it arguably more sophisticated and versatile than IS-136. It is not compatible with existing IS-136 cell site equipment, but this is not an issue for the new PCS-only providers, as they have no legacy equipment with which to maintain compatibility.

GSM RF Channels and Time Slots

GSM channels are 200 kHz wide (compared with 30 kHz for IS-136 TDMA). The total bit rate for an RF channel is 270.833 kb/s; the modulation is a variant of FSK called *GMSK* (*Gaussian minimum shift keying*) using a frequency deviation of 67.708 kHz each way from the carrier frequency. GMSK was described in Chapter 4.

Voice channels are called *traffic channels* (*TCH*) in GSM. One RF channel is shared by eight voice transmissions using TDMA. In terms of spectral efficiency, GSM works out to 25 kHz per voice channel, compared to about 30 kHz for AMPS and about 10 kHz for TDMA. This is an approximate comparison as it ignores differences in control-channel overhead.

As in TDMA, the mobile transmitter operates only during its allotted time slot (one-eighth of the time, compared with one-third of the time in TDMA.) Other things being equal, a GSM phone should have longer battery life than a phone using either AMPS or TDMA.

Figure 11.2 on page 426 shows the structure of an RF channel and its division into time slots (called *bursts* in GSM).

Control information in GSM is on two logical channels called the *broadcast channel* (*BCCH*) and the *paging channel* (*PCH*). As with TDMA, it is unnecessary

FIGURE 11.2
GSM RF
channel

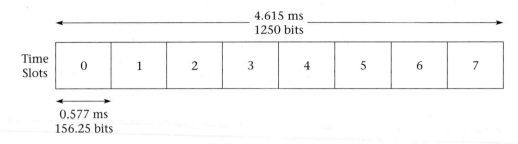

to use a whole RF channel for this. Instead, one of the eight time slots on one RF channel in each cell or sector is designated as a control channel. The broadcast information is transmitted first, followed by paging information. See Figure 11.3 for an illustration.

FIGURE 11.3
GSM control
channel

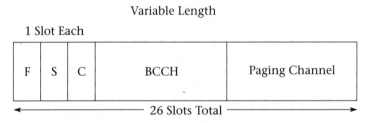

F = Frequency correction (sine-wave carrier)
S = Synchronizing
C = Contents list
BCCH = Broadcast control channel

The BCCH and PCH are forward channels only. The corresponding reverse channel is called the *random-access channel* (*RACH*) and is used by the mobiles to communicate with the base. Mobiles transmit on this channel whenever they have information; if a collision occurs, the mobile waits a random time and tries again. Transmissions are shorter than the duration of the slot to prevent interference caused by the propagation delay between mobile and base. The delay problem is avoided on the traffic channels, because the base instructs the mobile to advance or retard the timing of its transmissions to compensate for the changes in propagation delay as it moves about in the cell.

Just as with TDMA, it is also necessary to send control information on the traffic channels. This is because the mobile has only one receiver; it cannot count on receiving the broadcast channel during a call, because both channels may use the same time slot. Also as with TDMA, there are two control channels associated with the traffic channel. The *Slow Associated Control Channel* (*SACCH*) uses one of every 26 bursts on the voice channel. It is used to inform the base of power measurements made by the mobile of signal strength in adjacent cells. The *Fast Associated Control Channel (FACCH)* "steals" bits from the voice signal and is used for urgent messages from the base, such as handoff instructions.

Voice Transmission

Each voice transmission is coded at 13 kb/s. A linear predictive coder, which models the way sounds are produced in the human throat, mouth, and tongue, is used. Such coding allows the bit rate to be greatly reduced compared with straightforward PCM. In the future, it is planned to use more advanced voice coders (vocoders) to allow the bit rate to be reduced to 6.5 kb/s, doubling the capacity of the GSM system. Note the similarity with full- and half-rate TDMA, which code voice at 8 and 4 kb/s, respectively. See Chapter 3 for a discussion of vocoders.

The bits from the vocoder are grouped according to their importance, with the most significant bits getting the most error correction and the least significant bits getting none. Then the data is spread over several frames by **interleaving** it so that the loss of a frame due to noise or interference will have a less serious effect.

Each voice transmission is allocated one time slot per frame. A frame lasts 4.615 ms so each time slot is approximately 577 μs in duration. To allow time for transmitters to turn on and off, the useful portion of the time slot is 542.8 μs, which allows time for 147 bits. This gives a raw data rate of 31.8 kb/s per voice channel. The timing for mobile transmissions is critical so that each arrives at the base station in the correct time slot. Since the propagation time varies with the distance of the mobile from the base, the mobile has to advance its timing as it gets farther from the base. It does this by monitoring a timing signal sent from the base on a broadcast channel. Although the time slots used by a mobile for receiving and transmitting have the same number, they are actually separated in time by a period equal to three time slots (uplink lags downlink). This means that the mobile unit, unlike analog systems, does not have to receive and transmit at the same time. When neither receiving nor transmitting on the voice channels, the mobile monitors the broadcast channels of adjacent cell sites and reports their signal strengths to the network to help it determine when to order a handoff. See Figure 11.4 for the structure of a voice channel.

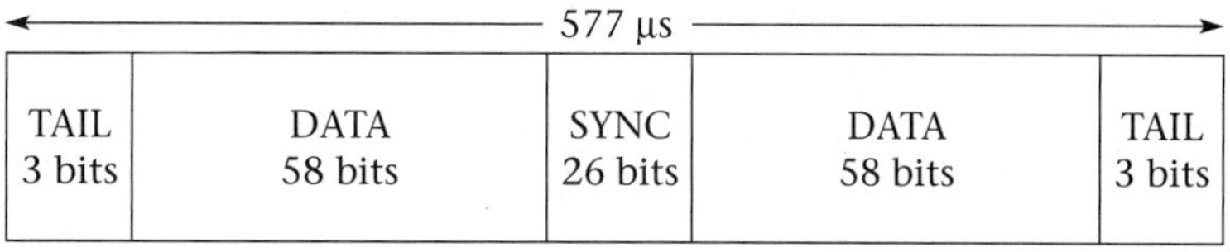

FIGURE 11.4
GSM voice channel

	577 μs			
TAIL 3 bits	DATA 58 bits	SYNC 26 bits	DATA 58 bits	TAIL 3 bits

Frequency Hopping in GSM

When multipath fading is a problem, the GSM system allows for frequency hopping, a type of spread-spectrum communication that was discussed in Chapter 4. This can often solve the problem, since multipath fading is highly frequency-dependent. All GSM mobiles are capable of frequency hopping, but only those cells that are located in areas of severe fading are designated as hopping cells. The system can hop only among the frequencies that are assigned to the cell, so there will be only a few hopping possibilities (on the order of three frequencies). Thus GSM is not really a true spread-spectrum system, but rather a TDM/FDM system with some spread-spectrum capability added on. This feature is unique to GSM; IS-136 TDMA has nothing like it.

Subscriber ID Module

The **subscriber ID module (SIM)** is unique to the GSM system. It is a **smart card** with eight kilobytes of memory that can be plugged into any GSM phone. SIMs come in two sizes: one is the size of a credit card, the other is about postage-stamp size. The SIM contains all subscriber information including telephone number (called the **International Mobile Subscriber Identification (IMSI)** in GSM), a list of networks and countries where the user is entitled to service, and other user-specified information such as memories and speed dial numbers. The card allows a subscriber to use any GSM phone, anywhere. For instance, since the PCS frequencies are different in Europe and North America, there is no point in a North American traveling in Europe with a PCS phone. If a traveler takes the SIM, however, it will work with any phone rented or purchased in Europe, as long as the subscriber has first contacted his or her North American GSM service provider to arrange for authorization.

The SIM also offers some protection against fraudulent use. A GSM phone is useless without a SIM; if the user removes the card when leaving the phone in a car, for example, the phone cannot be used unless the thief has a valid SIM. Unfortunately, the cards can be stolen too. The SIM can be set up to require the user to enter a *personal identification number (PIN)* whenever the phone is turned on to provide some security in case the card is lost or stolen.

Once a subscriber has a SIM, buying a new GSM phone is easy. No setup or programming by the dealer is required. Similarly, a user can have a permanently-installed mobile phone and a portable with the same phone number, provided that only one is used at a time. However, the TDMA system also makes purchasing a new phone fairly easy; it allows a phone to be activated and programmed over the air, using the control channel.

GSM Privacy and Security

The GSM SIM just discussed is only a part of the effort that has gone into securing this system. Both the data used in authorizing calls, such as the subscriber's identifying numbers, and the digitized voice signal itself, are usually encrypted. (It is possible to weaken or turn off the voice encryption, if a government requires it.) The security in GSM is better than in IS-136 and much better than in analog AMPS.

11.5 IS-95 CDMA PCS

This United States-designed system has an air interface that is radically different from either of the others, though its control and messaging structure is quite similar to GSM. CDMA is used to a limited extent on the 800-MHz band, but is much more common in the 1900-MHz PCS band. It uses code-division multiple access by means of direct-sequence spread-spectrum modulation. See Chapter 4 for an introduction to spread-spectrum radio and code-division multiple access (CDMA).

CDMA
Frequency Use

One CDMA RF channel has a bandwidth of 1.25 MHz, using a single carrier modulated by a 1.2288 Mb/s bitstream using QPSK. CDMA allows the use of all frequencies in all cells (not one-seventh or one-twelfth of the frequencies in each cell, as required by other systems). This gives a considerable increase in system capacity. Because of the spread-spectrum system, co-channel interference simply increases the background noise level, and a considerable amount of such interference can be tolerated. As with the other personal communication systems, base and mobile stations transmit on separate channels separated by 80 MHz.

Frequency diversity is inherent in any spread-spectrum system. This is especially beneficial in a mobile environment subject to multipath propagation. The GSM system discussed earlier can use a limited amount of frequency diversity by hopping among several (typically three) discrete channels. The CDMA system, on the other hand, uses the full 1.25-MHz bandwidth for all voice channels on a given RF channel. If a small portion of this spectrum suffers a deep fade due to reflections, the only effect will be a slight increase in the error rate, which should be compensated for by the error correction built into the coding of the voice and control signals.

Space diversity is also built into a spread-spectrum system. Other cellular systems and PCS typically employ two receiving antennas per cell or sector at the base station to provide some space diversity, but they use only one antenna at the mobile location. Multiple receiving antennas are also used with CDMA; but since all frequencies are used in all cells, it is possible to receive the mobile at two or more base stations. Similarly, a mobile can receive signals from more than one base station. Each can make a decision about the strongest signal and can, in fact, combine signals to obtain an even stronger one. Since there is no need for the mobile to change frequency on handoff, the CDMA system can use a *soft handoff,* in which a mobile communicates with two or more cells at the same time, rather than having to switch abruptly from one to another. This gives the ultimate in space diversity, with receiving antennas up to several kilometers apart. See Figure 11.5 for a comparison of soft and hard handoffs.

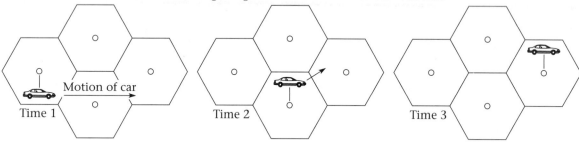

(a) Hard handoff

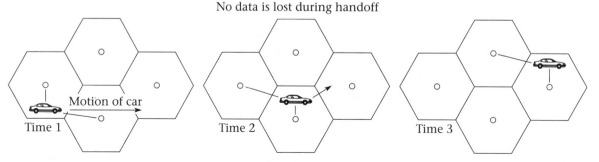

(b) Soft handoff

FIGURE 11.5 Hard and soft handoffs

CDMA Channels Each RF channel at a base station supports up to 64 orthogonal CDMA channels, using direct-sequence spread-spectrum, as follows:

- 1 pilot channel, which carries the phase reference for the other channels

- 1 sync channel, which carries accurate timing information (synchronized to the GPS satellite system) that allows mobiles to decode the other channels

- 7 paging channels, equivalent to the control and paging channels in TDMA and GSM
- 55 traffic channels

CDMA thus uses a bandwidth of 1.25 MHz for 55 voice channels, which works out to about 22.7 kHz per channel. This is similar to GSM and, at first glance, not as efficient as TDMA. However, the fact that all channels can be used in all sectors of all cells makes CDMA more efficient in terms of spectrum than any of the other systems. Since CDMA degrades gracefully with increasing traffic, it is difficult to arrive at a definite maximum for its capacity. Proponents of CDMA claim spectrum efficiencies ten to twenty times as great as for GSM; those using other systems dispute this and put the gain nearer two. Once there is a large body of data from all the PCS schemes, it will be easier to get at the truth.

Along with the other personal communication systems discussed in this chapter, the CDMA system also uses FDMA. Each PCS carrier has a spectrum allotment of either 5 MHz or 15 MHz in each direction (refer back to Table 11.1), so a cell site can have more than one RF channel.

Forward Channel   The forward and reverse channels are quite different in the CDMA system. Let us look at the forward channel first. We already know that sync, paging, and speech channels are combined on the same physical RF channel using CDMA. We learned in Chapter 4 that the direct-sequence form of CDMA is created by combining each of the baseband signals to be multiplexed with a *pseudo-random noise* (*PN*) sequence at a much higher data rate. Each of the signals to be multiplexed should use a different PN sequence. In fact, it can be shown that if the various sequences are mathematically *orthogonal,* the individual baseband signals can, at least in theory, be recovered exactly without any mutual interference. The math involved in proving this is beyond the scope of this text, but we should note that the number of possible orthogonal sequences is limited and depends on the length of the sequence. If the PN sequences are not orthogonal, CDMA is still possible, but there will be some mutual interference between the signals. The effect of this will be an increased noise level for all signals; eventually, as the number of non-orthogonal signals increases, the signal-to-noise ratio becomes too low and the bit-error rate too high for proper operation of the system. However, at no time do we hear audible crosstalk, as we do with two analog signals on the same frequency.

From the foregoing it would seem that using orthogonal PN sequences for CDMA is highly desirable, and this is what is done at the base station. A class of PN sequence called a **Walsh code** is used. The base station uses

64 orthogonal Walsh codes; each repeats after 64 bits. This allows for 64 independent logical channels per RF channel, as mentioned earlier. Walsh code 0 is used for the pilot channel to keep mobile receivers phase-aligned with the base station. This is a requirement for coherent demodulation, which is the only way to avoid interference among channels using the same carrier frequency.

In addition to the Walsh codes, two other codes are in use at a CDMA base station: a *short code* for synchronizing, and a *long code*, which is used for encryption of both voice and control-system data and is not used for spreading.

Figure 11.6 shows how the spreading works for one voice signal. The vocoder produces a voice signal with a maximum bit rate of 9.6 kb/s. Error-correction bits are added, and the samples are interleaved over time, just as they are for TDMA and GSM. This process increases the bit rate for one voice channel to 19.2 kb/s. Next the signal is exclusive-or'd with the long code. This code repeats only after $2^{42} - 1$ bits and is used for encryption, not spreading. The signal remains at 19.2 kb/s after this process.

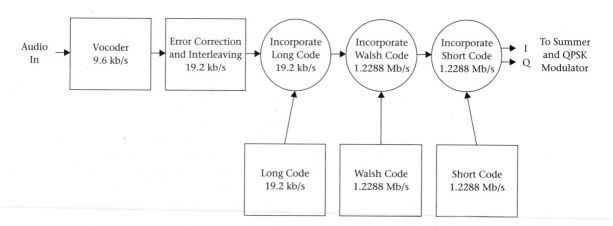

FIGURE 11.6 CDMA forward voice channel

Spreading occurs when the 19.2 kb/s baseband data stream is multiplied by one of the 64 Walsh codes. Each of the Walsh codes has a bit rate of 1.2288 Mb/s. The multiplication works as follows: when the data bit is zero, the Walsh code bits are transmitted unchanged; when the data bit is one, all Walsh code bits are inverted. The output bit stream is at 1.2288 Mb/s, which is 64 times as great a data rate as for the baseband signal at 19.2 kb/s. Therefore, the transmitted signal bandwidth is 64 times as great as it would be for the original signal, assuming the same modulation scheme for each.

Recall from Chapter 4 that the processing gain can be found as follows:

$$G_p = \frac{B_{RF}}{B_{BB}} \qquad\qquad (11.1)$$

where

G_p = processing gain

B_{RF} = RF (transmitted) bandwidth

B_{BB} = baseband (before spreading) bandwidth

Here,

$$G_p = \frac{B_{RF}}{B_{BB}}$$

$$= \frac{1.2288 \times 10^6}{19.2 \times 10^3}$$

$$= 64$$

In decibels, this is

$$G_p \text{ (dB)} = 10 \log 64$$
$$= 18.06 \text{ dB}$$

If we consider that the error-correction codes are a form of spreading as well, since they increase the data rate, the total spreading becomes

$$G_p = \frac{B_{RF}}{B_{BB}}$$

$$= \frac{1.2288 \times 10^6}{9.6 \times 10^3}$$

$$= 128$$

$$= 21.1 \text{ dB}$$

A signal-to-noise ratio of about 7 dB is required at the receiver output for a reasonable bit-error rate. This means that the signal-to-noise ratio in the RF channel can be about −14 dB for satisfactory operation; that is, the signal power can be 14 dB less than the noise power. This takes a little getting used to, but is typical of spread-spectrum systems.

The 64 orthogonal channels are transmitted on one RF carrier by summing them, as in Figure 11.7, and using QPSK to modulate them on a single carrier.

FIGURE 11.7
Multiplexing of
CDMA channels

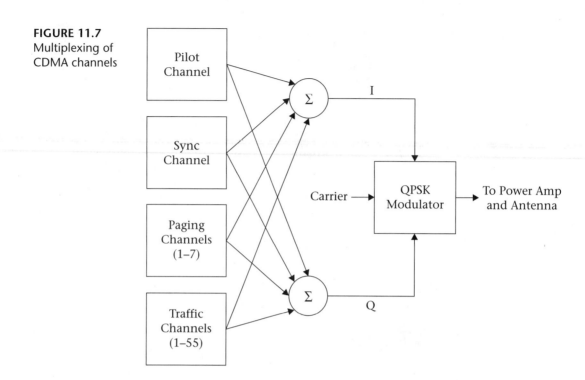

Reverse Channel

The mobile units cannot use truly orthogonal channels because they lack a phase-coherent pilot channel. Each mobile would need its own pilot channel, which would use too much bandwidth. Therefore, they use a more robust error-control system. It outputs data at three times the input data rate. Follow Figure 11.8 to see what happens to the signal.

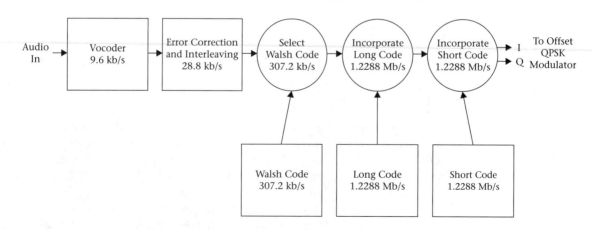

FIGURE 11.8   CDMA reverse voice channel

$$f_2 = 3 \times 9.6 \text{ kb/s}$$
$$= 28.8 \text{ kb/s}$$

The 28.8 kb/s signal is combined with one of the 64 Walsh codes and a long code to reach the full data rate of 1.2288 Mb/s. However, the purpose of each of these codes is different on the reverse channel. Here the long code is used to distinguish one mobile from another, as each uses a unique (though not necessarily orthogonal) long code. The Walsh codes are used to help the base station decode the message in the presence of interference. Each block of six information bits (64 different possible combinations) is associated with one of the 64 Walsh codes, and that code, rather than the actual data bits, is transmitted. Since each Walsh code is 64 bits long, this in itself does some spreading of the signal: the bit rate is increased by a factor of 64/6. The Walsh code mapping thus increases the data rate as follows:

$$f_3 = 28.8 \text{ kb/s} \times \frac{64}{6}$$
$$= 307.2 \text{ kb/s}$$

The long code is now multiplied with the data stream to produce a reverse channel bit rate of 1.2288 Mb/s, the same as for the forward channel. Each mobile transmits at the same rate to produce the spread-spectrum signal received at the base.

The modulation scheme is also slightly different on the reverse and forward channels. Both use a form of quadrature phase-shift keying (QPSK). The base station uses conventional QPSK. With this system the transmitter power has to go through zero during certain transitions. See Figure 11.9(a) on page 436.

The mobiles delay the quadrature signal by one-half a bit period to produce offset QPSK, which has the advantage that the transmitter power never goes through zero, though the amplitude does change somewhat. Linear amplifiers are still required in the mobile transmitter, but the linearity requirements are not as strict for offset QPSK as they are for conventional QPSK. See Figure 11.9(b).

Offset QPSK would have no advantage for the base station because a single transmitter is used for all the multiplexed signals. The summing of a large number of signals would result in a signal that still went through the zero-amplitude point at the origin.

Voice Coding

CDMA uses a variable rate vocoder. Four different bit rates are possible: 9600, 4800, 2400 and 1200 b/s. The full rate of 9600 b/s is used when the user is talking. During pauses, the bit rate is reduced to 1200 b/s. The other two rates are also in the specifications but are seldom used.

FIGURE 11.9
Standard and offset
QPSK

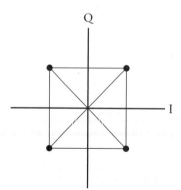

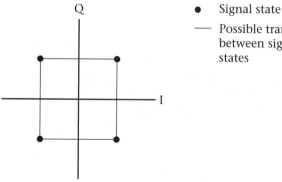

(a) Standard QPSK. Note that signal can move through the
origin (zero amplitude point).

(b) Offset QPSK. ½-bit delay in Q channel means that signal moves
along I, then along Q axis. It never goes to zero amplitude.

For many years it has been realized that each user typically talks less
than fifty percent of the time during a conversation. Theoretically, the bandwidth allocated to that customer can be reassigned during the pauses while
the other person is talking. However, until CDMA PCS came along there
were at least two problems with this. The first was that it does not sound natural to have the voice channel go dead when someone stops talking. It
sounds as if the phone has been disconnected. The reason is that there is
always background noise, even in a quiet room. The CDMA system transmits
this noise, but codes it at a lower rate (1200 b/s) because it is not important
that it be rendered accurately.

The other problem, with either FDMA or TDMA, was finding a use for the
vacated channel or time slot. The slot is usually available for only a few seconds or less, and the amount of time is not known in advance. In CDMA the

reduced amount of information to be sent can be translated directly into reduced interference to other transmissions on the same frequency, which automatically increases the capacity of the system. The way in which this is
done is different for the forward and reverse channels.

On the forward channel, data bits are repeated when the coder is running at less than the maximum rate of 9.6 kb/s. For instance, if the coder
operates at 1.2 kb/s, as it does during pauses in speech, each block of data is
transmitted eight times. Because the error rate at the receiver depends on
the energy per received data bit, the power in the transmit channel can be
reduced under these circumstances.

The mobile transmitter handles this situation differently. Rather than
reduce power, it simply transmits only one-eighth of the time, reducing
interference and increasing battery life.

Mobile Power Control

Controlling the power of the mobile stations is even more important with
CDMA than with other schemes. The power received at the base station from
all mobiles must be equal, within 1 dB, for the system to work properly. The
power level is first set approximately by the mobile, and then tightly controlled by the base. When first turned on, the mobile measures the received
power from the base, assumes that the losses on the forward and reverse
channels are equal, and sets the transmitter power accordingly. This is called
open-loop power setting. The mobile usually works with the equation:

$$P_T = -76 \text{ dB} - P_R \qquad (11.2)$$

where

P_T = transmitted power in dBm

P_R = received power in dBm

EXAMPLE 11.1

A CDMA mobile measures the signal strength from the base as –100 dBm.
What should the mobile transmitter power be set to as a first approximation?

SOLUTION

From Equation (11.2):

$$
\begin{aligned}
P_T &= -76 \text{ dB} - P_R \\
&= -76 \text{ dB} - (-100 \text{ dBm}) \\
&= 24 \text{ dBm} \\
&\approx 250 \text{ mW}
\end{aligned}
$$

The mobile begins by transmitting at the power determined by Equation (11.2) and increases power if it does not receive acknowledgement from the base. This could happen if a substantial amount of the received power at the mobile is actually from adjacent cells. We should also remember, that just as for the other systems, the forward and reverse channels are at different frequencies, so the amount of fading may be different.

Once a call is established, the open-loop power setting is adjusted in 1 dB increments every 1.25 ms by commands from the base station, to keep the received power from all mobiles at the same level. This *closed-loop* power control is required; for CDMA to work properly, all the received signals must have equal power. Otherwise the system suffers from the **near/far effect**, in which the weaker signals are lost in the noise created by the stronger ones. Careful power control has the added benefit of reducing battery drain in the portable unit, as the transmitted power is always the minimum required for proper operation of the system.

Rake Receivers and Soft Handoffs

One of the advantages of the CDMA system is that multipath interference can be reduced by combining direct and reflected signals in the receiver. The receivers used are called **rake receivers**; the reason can be seen in the diagram in Figure 11.10, which somewhat resembles a rake with several teeth for the reception of signals having different amounts of delay.

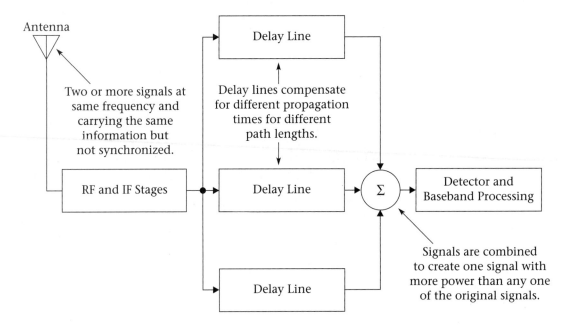

FIGURE 11.10 Rake Receiver

The mobile unit can combine three RF signals, delaying two of them to match the third. One of these signals can be assumed to be the base station in the current cell. The other two may be reflections or neighboring base stations. The base-station receiver can combine four signals: the direct signal from the mobile and three reflections.

In addition, two base stations may receive a signal from the same mobile. The base stations each send their signals to the MSC, which uses the higher-quality signal. Decisions about quality are made on a frame-by-frame basis every 20 ms. It is possible to have two base stations communicating with the same mobile indefinitely in what is referred to as a **soft handoff**. This avoids the dropping of calls that sometimes occurs when a handoff is unsuccessful in other systems, perhaps because there are no available channels in the new cell. The disadvantage is a considerably increased load on the base stations and the switching network.

CDMA Security

CDMA offers excellent security. A casual listener with a scanner will hear only noise on a CDMA channel. In order to decode a call it is not only necessary to have a spread-spectrum receiver, but also to have the correct despreading code. Since this so-called "long code" is $2^{42} - 1$ bits long before it repeats and is newly generated for each call, the chances of eavesdropping are small. Identification is done using private-key encryption, as for GSM.

11.6 Comparison of Modulation Schemes

All of the North American PCS have advantages. TDMA is compatible with much existing North American cell-site equipment. GSM has a long history and a large installed base, which tends to lead to lower prices. It also has more advanced features than IS-136. CDMA is the most sophisticated technically, offers the best security, and makes the best use of system bandwidth, at least in theory. All three are in wide use in the United States and Canada. Table 11.3 on page 440 compares the three systems under several headings.

Compatibility Issues: Multi-Mode Phones

From Table 11.3 there appears to be an obvious compatibility problem in PCS. The three systems have only their frequency range in common; none of the systems is compatible with either of the others. Consequently, roaming in the PCS band is possible only among providers that use the same system. At this writing, one or more of these systems is available in most populous areas, but not all areas have all three. Eventually the problem may disappear, as PCS coverage becomes ubiquitous with all three systems, but in the meantime, there is substantially less roaming capability with PCS than

TABLE 11.3   Comparison of North American PCS

Property	IS-136 TDMA	GSM	IS-95 CDMAone
RF Channel Width	30 kHz	200 kHz	1.25 MHz
Voice/Data Channels per RF Channel	3	8	64, including 1 sync and 7 control channels
Multiplexing Type	TDMA	TDMA plus limited frequency-hopping	Direct sequence spread-spectrum
Voice Coding Rate	8 kb/s full rate 4 kb/s half rate	13 kb/s full rate 6.5 kb/s half rate	Variable 9600 b/s max. 1200 b/s min.
Bit Rate for RF Channel	48.6 kb/s	270.833 kb/s	1.2288 Mb/s
Control Channel	2 time slots of 6 in one 30-kHz RF channel	1 time slot of 8 in one 200-kHz RF channel	7 of 64 orthogonal codes in one 1.25-MHz RF channel

with 800-MHz analog AMPS, which still has by far the widest distribution of any system in North America.

There is an obvious, though rather unwieldy, solution to the compatibility problem. This is to manufacture dual-band, dual-mode phones, which work with analog, 800-MHz AMPS as well as with one of the 1900-MHz personal communication systems. Dual-mode phones are currently available for all three of the PCS. Those PCS providers who do not also have an 800-MHz license often form alliances with a cellular provider to allow seamless roaming with only one monthly bill. Figure 11.11 shows examples of dual-mode phones.

FIGURE 11.11
Dual-mode phones
(Courtesy of Nokia, Inc.)

11.7 Data Communication with PCS

When we studied the TDMA cellular system, we observed that data communication can actually be more complex with a digital than with an analog system. This is because vocoders will not work properly with modems, so that the classic technique of connecting a modem to an analog voice channel does not work. We saw that at 800 MHz it is common for a digital phone to revert to analog mode for circuit-switched data communication and to use the CDPD system for packet-switched data.

The above techniques are still possible with a dual-mode PCS phone, but each of the three personal communication systems has developed its own techniques for data communication. This can be expected to become more important as a new generation of "smart phones" incorporating larger displays, (some even including web browsers) and portable computers incorporating RF communication modules are introduced.

At present the most popular use for PCS data seems to be short paging-type messages, followed by electronic mail. Worldwide web access is gaining in importance, but is currently limited by slow connection speeds and the limited graphics capability of PCS phone displays. Let us see how data transmission is handled with each of the three PCS.

TDMA Data Communication

The TDMA PCS standard allows for short messages and packet-switched data to be sent on the digital control channels (DCCH) or the digital traffic channels (DTC). Circuit-switched data is possible on the digital traffic channels.

The digital control and traffic channels support two main types of packet-switched data communication. A format called *cellular messaging teleservice (CMT)* is employed for a **short messaging service (SMS)**. This allows for brief paging-type messages and short e-mail messages (up to 239 characters), which can be read on the phone's display and entered using the keypad. For longer messages and extended services like web browsing, the *Generic UDP Transport Service (GUTS)* protocol is used. The acronym-within-an-acronym *UDP* stands for *User Datagram Protocol.*

Both of these services require extra equipment in the PCS network to translate between wireline protocols and those used with the radio link. With GUTS, the user connects to a network server that relays messages to and from the internet. The CMT system also requires the servers in the PCS network to assemble messages and interconnect with other services such as the user's e-mail service. See Figure 11.12 on page 442 for an illustration of packet-switched PCS data.

Circuit-switched data communication is accomplished on the digital traffic channels. The vocoder is bypassed and data is coded and sent directly

FIGURE 11.12
PCS packet-
switched data

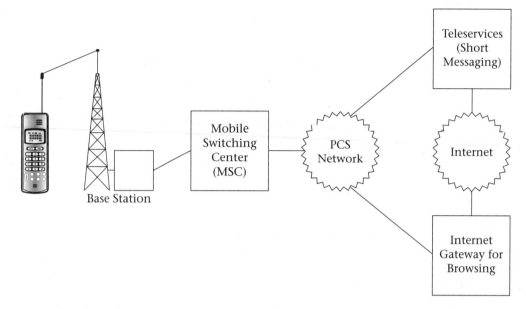

over a traffic channel at up to 9.6 kb/s. It is also possible to combine two or three traffic channels to send data at higher rates. This is done by using four or six of the time slots on a single RF channel, rather than the usual two.

Circuit-switched data is more expensive for the network to carry, as it uses at least one traffic channel full-time. On the other hand, it allows the user more freedom as to the type of data and the network called. For instance, a user could dial directly into a company mainframe computer. All that the PCS network has to provide is an interface to convert the protocols used on the air interface to an ordinary wireline modem standard for communication with the PSTN. At the mobile phone, a serial-port interface is typically provided for plugging in a computer. There is no modem in the phone, but it is set up to appear like a standard wireline modem to the computer. Figure 11.13 shows how circuit-switched data works.

GSM Data Communication

The types of data communication possible with GSM are similar to those used with TDMA. Short messages are available (up to 160 characters) using either the control or traffic channels, depending on whether the phone is in use for a voice call at the time. Circuit-switched data (including fax) can be accommodated at up to 9600 b/s using a traffic channel, just as for TDMA. A device especially designed to take advantage of GSM data communication, the Nokia 9000il Communicator, is shown in Figure 11.14.

FIGURE 11.13
PCS circuit-switched data

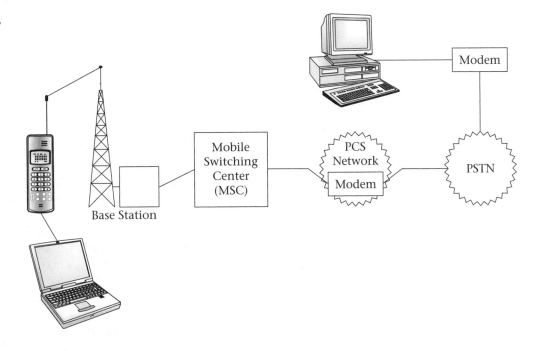

FIGURE 11.14
Nokia communicator
(Courtesy of Nokia, Inc.)

CDMA Data Communication There are some differences between CDMA and the other two systems in terms of data communication. Like the others, CDMA offers short messages via control channels. Its circuit-switched data capability using a single traffic channel is much greater, though, at 14.4 kb/s.

Wireless Web Browsing

Any of the PCS schemes just described can be used to access content on the World Wide Web. There are three major problems with all of them however: the data rate is low, even in comparison with ordinary telephone modems; the on-board computing power is low compared with a personal computer; and the handheld devices have very small, low-resolution displays. Many of these displays are not suitable for graphics. A typical web page would take a long time to load and when loaded would be almost, if not completely, unusable.

Third-generation wireless systems, which are described in Chapter 14, will help to solve the first problem, and perhaps make a start on the second. The third is more intractable: large displays and pocket-sized devices are simply not compatible. Therefore, even with third-generation systems, there will be a need for a means to display web pages on the small screens of PCS devices.

Until recently there have been many proprietary standards for displaying web content on wireless devices. Each worked only with a small number of specially created sites. Many of the major wireless manufacturers, including Ericsson, Nokia, and Motorola, have now combined to create a set of de facto standards for creating this content, known as the *Wireless Application Protocol* (*WAP*). The idea is to include a small program called a microbrowser in the wireless device, with most of the required computing done on network servers. These servers have access to specially modified pages on web sites and can also attempt to translate conventional sites so that they can be used by wireless devices. The pages have minimal graphics and condensed text so that they can be used with portable devices.

WAP is compatible with all of the current (second generation) systems and will be compatible with all third-generation systems as well. As more sites begin to provide pages compatible with WAP, the web should become quite accessible to portable wireless devices.

11.8 Testing Cellular Systems and PCS

We saw in Chapter 7 that all calculations of signal strength in a mobile environment are necessarily approximate, as there are too many variables for even a computer analysis to be accurate. Nonetheless, these predictions are generally accurate enough to locate cell sites and repeaters. Once these have been built, it is necessary to go into the field with a receiver to verify that the signal strength is satisfactory. The transmitter used for preliminary tests is often a portable model that puts out a carrier only. This simplifies the measurements, particularly when the system will eventually be CDMA with all channels active in all cells. Figure 11.15 shows a typical portable transmitter and receiver suitable for such testing.

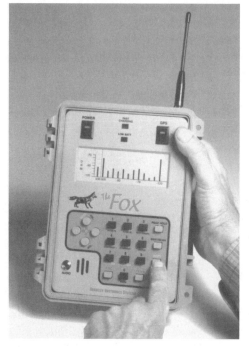

(a) Transmitter

(b) Receiver

FIGURE 11.15 Test transmitter and receiver
(Courtesy of Berkeley Varitronics Systems)

In addition to the RF equipment, it is desirable to have the test vehicle equipped with a global positioning system (GPS) receiver and a computer. It is then possible to plot signal strengths on a map and then to draw signal contour lines. In this way, any problem areas can be identified.

Once the system is in use, signal measurements can still be taken, but the process is more complex. This is especially true for CDMA, because all cells transmit on all frequencies with low power spread over a wide bandwidth. A conventional receiver would not be able to distinguish signals from noise, or one cell from another. A specialized receiver capable of extracting the pilot channel from a CDMA transmission is required. Each cell transmits a pilot with a slightly different delay added to the short code, and thus it can be identified if an accurate timing reference is available. Again, GPS is used by the CDMA system for this function. Figure 11.16 shows a specialized instrument designed for this purpose.

Testing of individual phones presents a problem, especially since it must be done quickly and economically; otherwise it will be less costly to

FIGURE 11.16
CDMA test receiver

(Courtesy of Berkeley
Varitronics Systems)

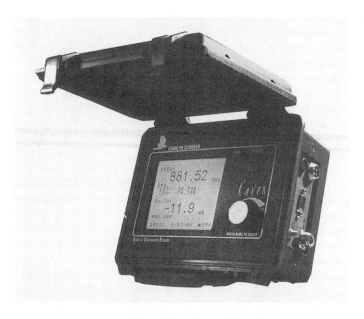

replace the phone than to repair it. Specialized and quite costly equipment is required. Figure 11.17 shows a unit that can be used to test all the functions of a PCS phone.

FIGURE 11.17
Mobile phone tester

(Courtesy of Agilent
Technologies, Inc.)

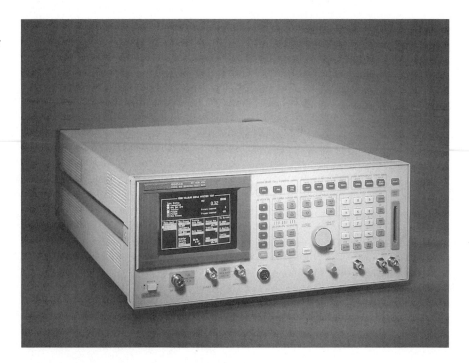

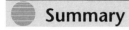 **Summary**   The main points to remember from this chapter are:

- The eventual goal of personal communication systems is to allow each individual to have one personal phone and phone number which will take the place of home, office, car, and portable phones.
- Current PCS resemble cellular radio systems except that they operate at a higher frequency and are completely digital.
- The PCS frequency range is divided in such a way that there can be up to six service providers in a given region.
- There are three incompatible PCS in current North American use: IS-136 TDMA, GSM1900, and IS-95 CDMAone. TDMA is also widely used at 800 MHz, and there is some use of CDMA in the 800-MHz band.
- All three PCS are digital. IS-136 and GSM1900 use TDMA, but with different modulation schemes and protocols. CDMAone uses a direct-sequence spread-spectrum system.
- Dual-mode, dual-band phones allow roaming on the analog cellular system when the correct type of PCS service is unavailable.
- In addition to voice, all systems have provisions for short written messages and also for circuit-switched data on their traffic channels.

Equation List

$$G_p = \frac{B_{RF}}{B_{BB}} \qquad\qquad (11.1)$$

$$P_T = -76 \text{ dB} - P_R \qquad\qquad (11.2)$$

Key Terms

downlink   signal path from a base station or a satellite to a mobile station or a ground station

interleaving   changing the time order of digital information before transmission to reduce the effect of burst errors in the channel

International Mobile Subscriber Identification (IMSI)   in the GSM system, a telephone number that is unique to a given user, worldwide

near/far effect   in a spread-spectrum system, the tendency for stronger signals to interfere with the reception of weaker signals

rake receiver   a radio receiver that is capable of combining several received signals with different time delays into one composite signal

short messaging service (SMS) transmission of brief text messages, such as pages or e-mail, by cellular radio or PCS

smart card a card with an embedded integrated circuit, that can be used for functions such as storing subscriber information for a PCS system

soft handoff in a cellular or PCS system, connecting a mobile to two or more base stations simultaneously

subscriber ID module (SIM) in the GSM system, a "smart card" containing all user information, which is inserted into the phone before use

uplink transmission channel to a satellite or base station

Walsh code class of orthogonal spreading codes used in CDMA communication

● Questions

1. The systems described in this chapter are sometimes called second-generation PCS. What was the first generation? What is expected from the third generation?

2. What frequency range is used for North American PCS? How were frequencies assigned to service providers?

3. How is interference between base and mobile transmissions prevented in PCS?

4. Give two other terms for the RF channel from base to mobile.

5. Give two other terms for the RF channel from mobile to base.

6. How is it possible for a digital system to use less bandwidth than a traditional analog FM system?

7. What other advantages do digital systems have over analog?

8. Do current PCS implementations have any disadvantages compared with analog 800-MHz cellular radio?

9. How does IS-136 PCS TDMA differ from the TDMA used at 800 MHz?

10. What are the meaning and function of the SCF bits in the TDMA digital control channel?

11. What happens if two mobiles transmit simultaneously on the reverse control channel in the TDMA system?

12. What is meant by a logical channel, and how is the control channel divided into logical channels in the TDMA system?

13. What is the meaning of a superframe in TDMA? What information must be transmitted completely in one superframe?

14. What is a hyperframe in TDMA, and what is its function?

15. What is sleep mode for a cell phone, and what is the advantage of being able to use a sleep mode?

16. Why won't a North American TDMA phone work in Europe?

17. Compare the number of TDMA channels on one RF channel for the TDMA and GSM systems.

18. Compare the RF channel bandwidths for the TDMA and GSM systems. Which uses bandwidth more efficiently?

19. Why do GSM and TDMA sometimes have to use traffic channels for control information?

20. How does GSM handle handoff requests from base to mobile?

21. What does a GSM mobile do, during a call, in those time slots when it is neither transmitting or receiving voice information?

22. How does frequency hopping improve the reliability of a communication system in the presence of multipath interference?

23. Which of the three North American PCS can use frequency hopping to improve its performance?

24. Describe the GSM SIM and explain how it makes traveling to other countries easier for GSM users than for those with other types of PCS phones.

25. How does GSM prevent people from using the system without authorization?

26. Why won't a North American GSM phone work in Europe?

27. How would a North American GSM user obtain service on a trip to Europe?

28. Explain the difference between frequency-hopping and direct-sequence spread-spectrum systems. Which of these is used with the CDMAone personal communication system?

29. How does CDMA improve the frequency reuse situation compared with TDMA and GSM?

30. Explain what is meant by a soft handoff. What advantages does it have?

31. Why are soft handoffs not possible with TDMA or GSM?

32. What is the use of the Walsh codes for each of the forward and reverse links of the CDMA system?

33. What is the use of the long code in each of the forward and reverse links in the CDMA system?

34. How does the CDMA system take advantage of pauses in speech to increase system capacity? Explain how it is done in both forward and reverse channels.

35. Why does QPSK require linear amplifiers while GMSK does not? Which PCS use each of these modulation schemes?

36. What advantage does offset QPSK have over normal QPSK? Where is offset QPSK used in PCS?

37. Why is control of the mobile transmitter power more important in CDMA than in the other PCS?

38. Explain the difference between open-loop and closed-loop power control.

39. What is a rake receiver, and what benefits does it provide?

40. Explain the difference in the way PCS handle short messages and longer data transmissions.

41. What is the difference between circuit-switched and packet-switched data transmissions?

42. Which of the personal communication systems has the highest speed for circuit-switched data?

43. Why is it difficult to measure signal strength in a CDMA PCS?

● Problems

1. (a) Calculate the length of a quarter-wave monopole antenna for the center of the cellular radio band (860 MHz) and the center of the PCS band (1910 MHz).

   (b) Assuming equal transmitter power applied to both of the above antennas, would there be any difference in the power density generated at a receiving antenna? Assume free-space propagation. Explain.

   (c) If the above antennas are used as receiving antennas, which of the two would produce a stronger signal at the receiver input, and by how many decibels? Assume equal power density at the antenna. Explain.

2. Compare the spectral efficiency of the TDMA and GSM PCS and analog AMPS in terms of kHz used per voice channel. Proceed as follows:

   (a) Take a simplistic approach to get a rough estimate. Find the width of each RF channel and the number of voice channels per RF channel.

Ignore control channels. Calculate spectrum use in kHz per channel.

(b) Refine your estimate by considering the proportion of the overall bandwidth used by control channels in each system and therefore unavailable for voice. Assume a 12-cell repeating pattern with one control channel for each cell in TDMA and GSM. Also assume the use of one of the larger PCS frequency blocks. Check Chapter 10 to refresh your memory about AMPS control channels.

(c) Compare CDMA with the others. First find the bandwidth per voice channel for one cell, then take into account the gain in capacity that results from the ability to use all channels in all cells to arrive at an equivalent bandwidth per voice channel.

3. (a) Suppose that the SMSCH in a TDMA system occupies five slots in each superframe. What is the total bit rate available for short messages?

(b) Suppose that five users in the cell or sector have messages addressed to them at the same time. What is the bit rate available for each user?

4. As we saw earlier in this book, ordinary digital wireline telephony uses a bit rate of 64 kb/s for each voice channel. Find the ratio by which the data has been compressed in each of the digital systems discussed. Assume full-rate vocoders for TDMA and GSM.

5. Compare TDMA, GSM, and CDMA by finding for each (take note that there are two different modulation schemes in use):

(a) the RF channel rate in bits per second per hertz of bandwidth

(b) the RF channel rate in symbols per second per hertz of bandwidth

6. (a) What is the difference in propagation time for signals from two mobiles, one 150 m from the base, and the other 2 km away?

(b) How many bit periods in the RF channel correspond to this difference, for each of the three systems discussed in this chapter?

7. (a) Calculate the time for one complete repetition of the Walsh code on the forward channel.

(b) Repeat (a) for the long code. (Now do you see why it's called the *long* code?)

8. (a) A CDMA mobile measures the signal strength from the base as −85 dBm. What should the mobile transmitter power be set to as a first approximation?

(b) Once the connection is made with the above power level, the base station needs the mobile to change its power to +5 dBm. How long will it take to make this change?

9. A rake receiver in a mobile receives a direct signal from a base 1 km away and a reflected signal from a building 0.5 km behind the mobile. It also receives a signal from another base station 3 km away. See Figure 11.18 for the situation. Calculate the amount of time delay each "finger" of the receiver needs to apply.

FIGURE 11.18

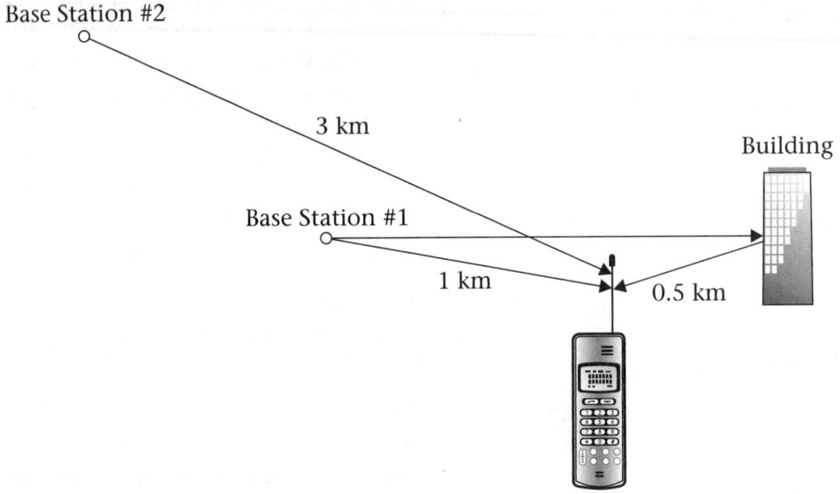

Base Station #2

3 km

Building

Base Station #1

1 km

0.5 km

12

Satellite-Based Wireless Systems

Objectives

After studying this chapter, you should be able to:

- Calculate velocity and period for artificial satellites in circular orbits.

- Distinguish between low-, medium-, and geostationary earth orbits, and explain the advantages and disadvantages of each for communication.

- Explain how elliptical orbits can be used in communication satellites.

- Explain the need for satellite tracking, and describe how it is done.

- Describe the types of satellite transponders used for wireless communication.

- Perform signal, noise, and signal-to-noise ratio calculations with satellite links.

- Describe several current and projected projects in the field of wireless communication by satellite, and discuss the merits of each.

12.1 Introduction

Cellular and personal communication systems work very well in populous areas, but they do require extensive infrastructure. Such systems are impractical for use in remote areas or at sea. Until quite recently the only means of portable communication in such areas was high-frequency (HF) radio. Over-the-horizon propagation is possible at frequencies in the range from about 3–30 MHz, due to reflection from the ionosphere; but HF propagation is unreliable, noisy, and suitable only for low-bandwidth applications. The antenna sizes required for wavelengths in the 10 to 100 meter range are awkward for mobile and all but impossible for handheld use.

A better alternative is to use satellites to relay signals from mobile or portable users to the public switched telephone network. Problems of cost, time delay, and equipment size exist, however, and will be addressed as we proceed. It seems unlikely that satellite wireless communication systems will ever be as popular as terrestrial systems, but they will fill an important role where terrestrial systems are not practical.

12.2 Satellite Orbits

We looked very briefly at satellite orbits in Chapter 7. In theory, a satellite can orbit at any altitude, but air resistance makes satellites impractical below about 300 km. A satellite can have any elliptical orbit but most (not all) of those used for communication have orbits that are at least approximately circular.

Satellite orbits can circle the equator (**equatorial orbit**); they can pass over both poles in a **polar orbit**, which has a 90° angle with respect to an equatorial orbit; or they can have any angle between these. Figure 12.1 shows some examples of possible circular orbits.

Orbital Calculations Satellites are held in orbit by their momentum. Gravity continually bends a satellite's path toward the earth, but the satellite's momentum is sufficient to prevent its trajectory from reaching the earth. This phenomenon is commonly called centrifugal force, though technically there is no such thing.

Any satellite orbiting the earth must satisfy this equation:

$$v = \sqrt{\frac{4 \times 10^{11}}{(d + 6400)}}$$

(12.1)

FIGURE 12.1
Circular orbits

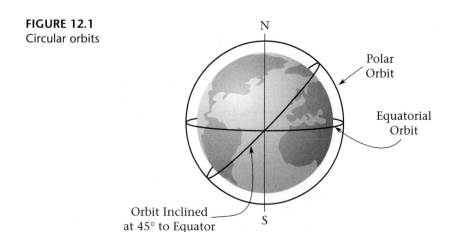

where

v = velocity in meters per second
d = distance above earth's surface in km

Several important things can be seen from Equation (12.1). First, the far-ther a satellite is from the surface of the earth, the slower it travels. Second, since a satellite that is farther from the earth obviously has farther to go to complete an orbit, and since it also travels more slowly than one closer to earth, then the orbital period of a distant satellite must be longer than the period of one closer to the earth. An example will illustrate this.

EXAMPLE 12.1

Find the velocity and the orbital period of a satellite in a circular orbit

(a) 500 km above the earth's surface

(b) 36,000 km above the earth's surface

SOLUTION

(a) From Equation (12.1):

$$v = \sqrt{\frac{4 \times 10^{11}}{(d + 6400)}}$$

$$= \sqrt{\frac{4 \times 10^{11}}{(500 + 6400)}}$$

$$= 7.6 \text{ km/s}$$

The circumference of the orbit can be found from its radius, which is that of the earth (6400 km) plus the distance of the satellite from the earth. In this case the total distance is

$$r = 6400 \text{ km} + 500 \text{ km} = 6900 \text{ km}$$

and the circumference of the orbit is

$$C = 2\pi r$$
$$= 2\pi \times 6900 \text{ km}$$
$$= 43.4 \text{ Mm}$$

The period of the orbit can be found by dividing the circumference by the orbital velocity,

$$T = \frac{C}{v}$$

$$= \frac{43.4 \times 10^6 \text{ m}}{7.6 \times 10^3 \text{ m/s}}$$

$$= 5.71 \times 10^3 \text{ s}$$

$$= 1.6 \text{ hours}$$

(b) Again, start with Equation (12.1):

$$v = \sqrt{\frac{4 \times 10^{11}}{(d + 6400)}}$$

$$= \sqrt{\frac{4 \times 10^{11}}{(36,000 + 6400)}}$$

$$= 3.07 \text{ km/s}$$

Note that the speed is less than before. The new radius is

$$r = 6400 \text{ km} + 36,000 \text{ km}$$
$$= 42.4 \text{ Mm}$$

The circumference of the orbit is, as before,

$$C = 2\pi r$$
$$= 2\pi \times 42.4 \text{ Mm}$$
$$= 266.4 \text{ Mm}$$

Now we can find the period of the orbit.

$$T = \frac{C}{v}$$

$$= \frac{266.4 \times 10^6 \text{ m}}{3.07 \times 10^3 \text{ m/s}}$$

$$= 86.8 \times 10^3 \text{ s}$$

$$\cong 24 \text{ hours}$$

Geostationary Orbit

The satellite in Example 12.1(b) has a particularly interesting orbit. It is, at least approximately, **geosynchronous**; that is, the satellite orbits the earth in the same amount of time it takes the earth to rotate once on its axis. If the orbit is circular and above the equator and the satellite travels in the same direction as the earth's rotation, it will also be **geostationary**; that is, it will appear stationary from the ground, because it rotates at the same rate and in the same direction as the earth. Though it is theoretically possible for a satellite to be geosynchronous without being geostationary by orbiting the earth in the direction opposite to the earth's rotation, in practice this is never done, and the two terms are used interchangeably. The orbit for a geostationary satellite has a radius more than five times as large as that of the earth. See Figure 12.2 for an illustration of the geostationary orbit.

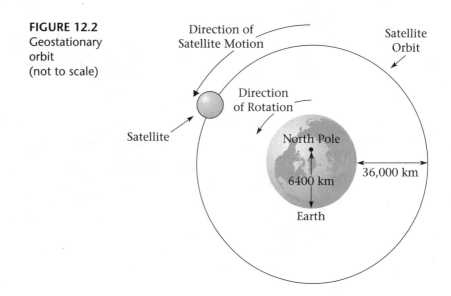

FIGURE 12.2
Geostationary orbit
(not to scale)

Elliptical Orbits   Technically all satellite orbits are elliptical. A circle is a special case of ellipse where the maximum distance from the earth (**apogee**) is equal to the minimum distance (**perigee**).

When a satellite has an elliptical orbit in which the earth is at one focus, it spends more time in that part of the orbit that takes it farthest from earth. This can be seen from Kepler's second law, which states that such a satellite sweeps out equal areas in space in equal times. See Figure 12.3 for the idea.

FIGURE 12.3
Elliptical orbit

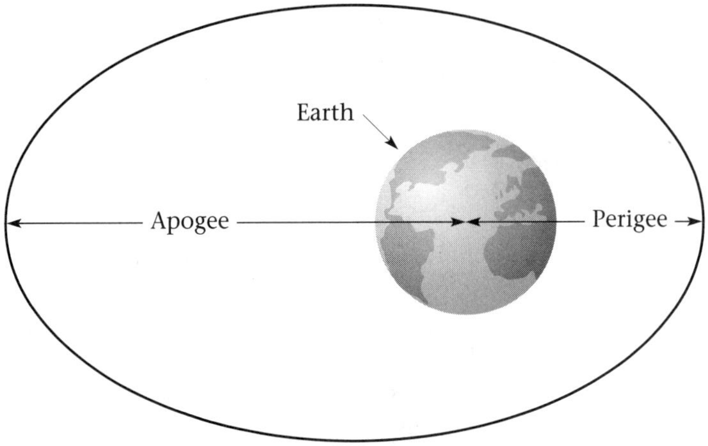

Such orbits can be used to advantage. The Russian Molniya series of communication satellites, for instance, uses highly elliptical polar orbits arranged so that the satellites spend about 11 hours of a 12-hour orbit over the northern hemisphere. This facilitates communication in regions near the North Pole. Geostationary satellites, which must be in a circular orbit above the equator, are near the horizon in these regions. The proposed Ellipso satellite system, described later, will also use elliptical orbits.

12.3   Use of Satellites for Communication

The traditional way to communicate with a satellite in non-geostationary orbit is to use a movable, directional antenna and point it at the satellite. The approximate *azimuth* (angle in the horizontal plane) and *elevation* (vertical angle measured from the horizon) can be calculated from orbital data, and fine position adjustments can be made for the strongest signal. The antenna must then follow, or **track**, the position of the satellite as it moves. At some point, of course, the satellite goes below the horizon and communication is lost.

There are two obvious problems with this for mobile and portable communication. The first is that rapidly changing the orientation of a highly directional antenna on a moving vehicle is difficult. This can be done on ships where changes in direction are relatively slow and there is room for a dish antenna, but it is very difficult on cars and just about impossible for a portable, hand-carried unit. Moving vehicles require antennas that either have a broad beam or are steered electronically using phased elements.

The disappearance of satellites over the horizon is another serious problem. If real-time voice communication is to be possible, at least one satellite must be visible at all times. Two solutions to the problem are explored below.

Using Satellites in Geostationary Orbit

There is an orbit in which a satellite appears to be stationary above a particular spot at the equator. This is very useful, because such a satellite never goes below the horizon and the antenna position with respect to the earth, once found, never needs to be changed. In fact this orbit is so convenient that until recently most communication satellites used it. It is still the method of choice for point-to-point service and broadcasting.

The geostationary orbit does leave something to be desired for portable and mobile use, however. Being able to point the antenna always in the same direction is not as useful on a car or with a handheld device as it is for a fixed installation. The orientation of the antenna with respect to the earth is stable, but the orientation of the antenna with respect to the vehicle must vary rapidly as the vehicle changes direction.

Another problem is the extreme path loss at the distance needed for a geostationary satellite. We have seen that if we require a geostationary orbit the altitude of 36,000 km, though greater than we would like, is set by the laws of physics. The large path loss usually means that a high-gain directional antenna is needed, and this is problematic for portable/mobile operation.

One of the main reasons for using satellites for wireless communication is to provide coverage in remote areas, which are out of reach of terrestrial cellular systems. Many of these areas are in high latitudes (northern Canada and Alaska, for example). Unfortunately, as the earth station moves toward the poles, the geostationary satellites are closer to the horizon. This makes it harder to achieve a direct line of sight to the satellite, which is free from obstructions such as buildings or hills. Thus the geostationary satellites, which must be above the equator, become less useful in extreme northern or southern latitudes. Another glance at Figure 12.2 should make this clear.

Yet another problem, though less important than those already mentioned, is propagation time. We saw in Chapter 7 that the propagation delay is about 0.25 s for a round trip to a geostationary satellite. This delay, while not fatal, is annoying in real-time conversations. It also causes delays in data transmission whenever a protocol requires acknowledgment from the receiving station before the transmission can continue.

Geostationary Satellite Beams and Footprints

A geostationary satellite can be "seen" from almost half the earth's surface. Therefore, three such satellites should be sufficient to cover the entire earth, except for the polar regions, with some overlap. A satellite designed for such wide coverage has an antenna with a relatively large beamwidth called a **hemispheric beam**. Because of its wide beam, such an antenna necessarily has relatively low gain.

Many geostationary satellites are not intended to cover an entire hemisphere. They use much more directional antennas, producing **spot beams** to cover populated regions. Such antennas have higher gain and can deliver a stronger signal on earth for the same transmitter power. Similarly, when used for receiving, such antennas can achieve a better signal-to-noise ratio at the satellite for a given transmitter power at the ground. This is very important in portable and mobile communication where ground-station transmitter power and antenna gain are strictly limited.

It is certainly possible for a satellite to have a combination of hemispheric and spot beams in order to provide basic service over a wide area and the ability to use smaller earth stations in particular areas.

When designing receiving installations on the ground, it is the EIRP of a satellite and not simply its transmitter power that is important. EIRP depends on the gain of the transmitting antenna in a particular direction, as well as the transmitter power, so it is different at different points on earth. Satellite operators publish maps showing the **footprint** of each geostationary satellite on the earth. These show the effective EIRP (in dBW) for the satellite at each point on the earth where reception is possible. See Figure 12.4 for an example of a satellite footprint.

Use of Low- and Medium-Earth Orbits

Geostationary satellites are usable for wireless communication, but as we have seen, there are reasons to prefer satellites with lower orbits. The two main problems with such satellites are (1) their position in space is not fixed with respect to a ground station and (2) the annoying tendency of such satellites to disappear below the horizon. (Another smaller problem is the Doppler effect, which causes frequencies to change. Transmitted frequencies are shifted higher as the satellite approaches a point on the ground and lower as the satellite recedes.)

The first problem is less important than it might seem. Shorter range results in much less propagation loss and removes the requirement for highly directional antennas. This makes antenna tracking less critical. In any case, if the antenna is mounted on a moving vehicle or person, the direction to the satellite is constantly changing, even with geostationary satellites.

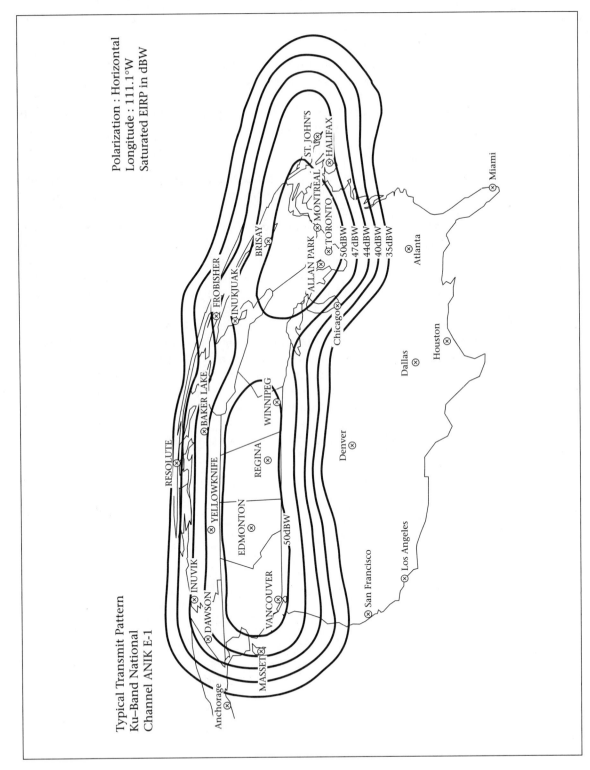

Polarization : Horizontal
Longitude : 111.1°W
Saturated EIRP in dBW

Typical Transmit Pattern
Ku-Band National
Channel ANIK E-1

FIGURE 12.4 Footprint for Anik E-1 satellite   (Courtesy of Telesat Canada)

When real-time communication is required, the only way to address the second problem is to use a **constellation** containing more than one satellite. The closer the satellites are to earth, the more of them are needed in order to have at least one satellite visible from a given point at all times. This can make the system quite complex and expensive, even allowing for the fact that it is less expensive to put satellites in lower orbits than in geostationary orbit.

The Doppler effect requires careful receiver design for both satellite and ground stations, so that the receiver can lock onto an incoming signal and track its frequency changes.

Satellite orbits are usually divided into three ranges. **Low-earth-orbit (LEO) satellites** range from about 300 to 1500 km above the earth. **Medium-earth-orbit (MEO) satellites** are about 8,000 to 20,000 km in altitude. The gap between LEO and MEO orbits is there to avoid the lower of the two Van Allen radiation belts that surround the earth; this radiation can damage satellites. These radiation belts extend from 1,500 to 5,000, and from 13,000 to 20,000 km above the earth's surface. MEO satellites are typically near the lower end of the MEO range to avoid the upper Van Allen belt. Finally, of course, there is the geostationary earth orbit (GEO) already mentioned. It would, of course, be possible to orbit satellites at still greater distances from the earth, but since the GEO is already farther from Earth than we would like, there is no point in using more distant satellites for communicating between points on Earth. Figure 12.5 shows the LEO, MEO, and GEO ranges, along with the Van Allen belts.

FIGURE 12.5
LEO, MEO and
GEO orbits
(not to scale)

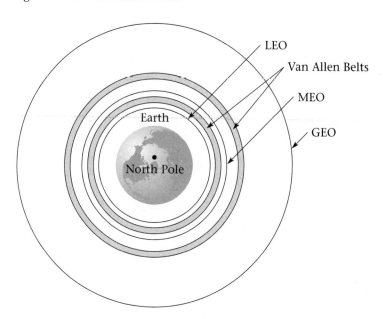

12.4 Satellites and Transponders

The satellite as a spacecraft, with its attendant guidance systems and positioning jets, is outside the scope of this book. We are concerned, however, with the satellite as a radio repeater. We need to know what the satellite looks like from the ground.

The traditional way to build a communication satellite is to design it as a frequency-shifting repeater or collection of repeaters (for some reason, a repeater on a satellite is called a **transponder**.) One satellite may have many transponders, each of which has a block diagram that looks like Figure 12.6. A range of frequencies is received from the ground via the *uplink,* amplified and shifted in frequency, and retransmitted on the *downlink*. No signal processing other than amplification and frequency shifting is done on the satellite. For obvious reasons, this type of transponder has what is known as the **bent-pipe configuration**.

FIGURE 12.6
Bent-pipe
transponder

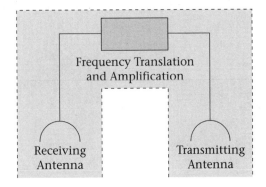

This transponder design is extremely versatile. A bent-pipe transponder can be used for anything from broadcast television using analog wideband FM to digital telephony using either time-division or frequency-division multiplexing or both. When used for one wideband FM signal, the satellite power amplifier can be operated in a saturated mode, much like a Class C amplifier, for greatest efficiency. If required to handle amplitude-varying signals, the amplifier can be "backed off" to a lower-power linear mode by remote control from the ground. Such a transponder is ready for just about any signal, or combination of signals, that will fit into its bandwidth.

It is also possible to design satellite transponders for specific applications. Some are designed to store digital information and retransmit it at a later time. Using this **store-and-forward** technique, data can be communicated using a low-earth orbit satellite that may not be visible to the transmitting and receiving stations at the same time. Also, satellites can be designed to communicate with each other. Such **crosslinks** can improve the efficiency of communication between earth stations, as indicated in Figure 12.7.

FIGURE 12.7
Crosslinks

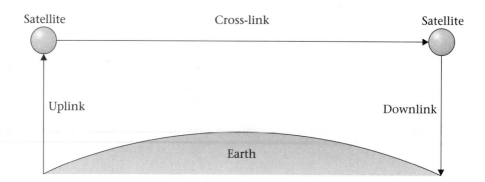

It is possible to turn a satellite transponder into a switching center so that a signal from a transmitter on the ground can be relayed to one of a variety of ground- or satellite-based receivers as required. However, the gain in efficiency is counterbalanced by an increase in satellite cost and complexity. It should be remembered that it is much easier to build and maintain complex equipment on the ground than in space. When we look at practical examples of wireless satellite communication later in this chapter, we'll see examples of both philosophies. Figure 12.8 shows a satellite with beam switching.

FIGURE 12.8
Beam switching

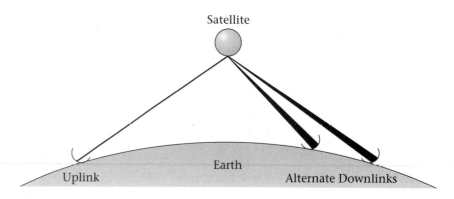

12.5 Signal and Noise Calculations

Because of the very weak received signals, satellite systems require low-noise receivers and relatively high-gain antennas for both the satellite and the earth station, especially when geostationary satellites are used. This is a different situation from the terrestrial systems we studied earlier, where interference or distance to the horizon was more likely than thermal noise to limit the range. This is, therefore, a good place to consider in more detail the factors that determine signal-to-noise ratio.

The signal strength is determined by the factors noted in Chapter 7. The important parameters are the EIRP of the transmitting installation, the distance to the receiver, and the gain of the receiving antenna, less any losses between antenna and receiver. Free-space propagation can usually be assumed in satellite communication, though for frequencies much above 10 GHz, losses in the atmosphere should be taken into account.

Noise Temperature

As discussed in Chapter 1, the noise level is determined by the bandwidth of the system and its equivalent noise temperature. To review,

$$P_N = kTB \tag{12.2}$$

where

P_N = noise power in watts
k = Boltzmann's constant, 1.38×10^{-23} joules/kelvin (J/K)
T = system noise temperature in kelvins (K)
B = noise power bandwidth in hertz

The system noise temperature is not necessarily the actual temperature at which it operates as measured with a thermometer. It depends on the contribution of the various electronic components, as well as any noise picked up by the antenna or contributed by the feedline from antenna to receiver.

The total noise temperature of a system can be found by adding together the noise temperatures of its various components, provided all are referenced to the same point. Usually the receiver input is used as a reference point. Receiver noise temperatures are specified at their input. Sometimes, *noise figure,* rather than noise temperature is given. The noise figure is a measure of how much an electronic system degrades the signal-to-noise ratio of a signal at its input; that is,

$$NF = \frac{(S/N)_i}{(S/N)_o} \tag{12.3}$$

where

$(S/N)_i$ = signal-to-noise ratio at the input
$(S/N)_o$ = signal-to-noise ratio at the output

The noise figure, expressed this way, is a dimensionless ratio. In practice, it is always specified in dB, where

$$NF_{dB} = 10 \log NF \tag{12.4}$$

It is simple to convert noise figure to equivalent noise temperature using the equation:

$$T_{eq} = 290(NF - 1) \tag{12.5}$$

where

T_{eq} = equivalent noise temperature in kelvins

NF = noise figure as a ratio (not in dB)

EXAMPLE 12.2

A receiver has an equivalent noise figure of 2 dB. Calculate its equivalent noise temperature.

SOLUTION

First convert the noise figure to a ratio.

$$NF = \text{antilog}\left(\frac{NF_{dB}}{10}\right)$$

$$= \text{antilog}\left(\frac{2}{10}\right)$$

$$= 1.58$$

Now use Equation (12.5) to find the equivalent noise temperature:

$$T_{eq} = 290(NF - 1)$$

$$= 290(1.58 - 1)$$

$$= 170 \text{ K}$$

To find the equivalent noise temperature of the receiving installation, we need to add the antenna noise temperature, as modified by the feedline, to the receiver equivalent noise temperature. That is,

$$T = T_{eq} + T_a \tag{12.6}$$

where

T = system noise temperature in K

T_{eq} = receiver equivalent noise temperature in K

T_a = antenna noise temperature in K

The antenna noise temperature results from thermal noise picked up from objects in the beam of the antenna. This depends on the angle of elevation of the antenna: when the antenna beam includes the ground, the noise level increases because of radiation from the ground itself. Luckily, this is seldom the case with ground stations used with geostationary satellites, except in very high latitudes where the satellite is just above the horizon. This means that noise entering the antenna originates mainly from extraterrestrial sources (stars, for instance) and from the atmosphere. Occasionally the sun passes through the main lobe of the antenna pattern; the sun is a very powerful noise source and makes communication impossible for the few minutes it takes to pass through the antenna beam. Otherwise the sky noise temperature for an earth-station receiving antenna is quite low, typically 20 K or less. The situation is very different for less directional antennas such as those used with portable phones in LEO systems. The noise temperature of these antennas may be about the same as the ambient temperature because of noise picked up from the surroundings; signals are much stronger at these antennas as well.

Losses in the antenna system contribute to its noise temperature. The noise temperature of an antenna system, at the far end of a feedline, is given by

$$T_a = \frac{(L - 1)290 + T_{sky}}{L} \tag{12.7}$$

where

T_a = effective noise temperature of antenna and feedline, referenced to receiver antenna input, in kelvins

L = loss in feedline and antenna as a ratio of input to output power (not in decibels)

T_{sky} = effective sky temperature, in kelvins

EXAMPLE 12.3 ▼

An earth station for use with a geostationary satellite has a dish antenna which sees a sky temperature of 25 K. It is connected to the receiver with a feedline having 1 dB loss. The receiver equivalent noise temperature is 15 K. Calculate the noise temperature for the system.

SOLUTION

First convert the feedline loss to a power ratio:

$$L = \text{antilog}\left(\frac{L_{dB}}{10}\right)$$

$$= \text{antilog}\left(\frac{1}{10}\right)$$

$$= 1.26$$

Now use Equation (12.7) to find the antenna noise temperature:

$$T_a = \frac{(L-1)290 + T_{sky}}{L}$$

$$= \frac{0.26 \times 290 + 25}{1.26}$$

$$= 80 \text{ K}$$

Now add this to the receiver equivalent noise temperature to find the system temperature, as in Equation (12.6).

$$T = T_{eq} + T_a$$

$$= 15 + 80$$

$$= 95 \text{ K}$$

Signal-to-Noise Ratio Once the system temperature is known, it is easy to calculate the noise power in any given bandwidth.

EXAMPLE 12.4

Calculate the noise power for the system in the previous example, if the bandwidth is 2 MHz.

SOLUTION

From Equation (12.2),

$$P_N = kTB$$

$$= 1.38 \times 10^{-23} \times 95 \times 2 \times 10^6$$

$$= 2.62 \text{ fW}$$

$$= -116 \text{ dBm}$$

Now, if we know the signal strength, we can easily find the signal-to-noise ratio. Usually the satellite power is specified as an EIRP in dBW, so all we need is the path loss and antenna gain, less any feedline losses, to find the received signal strength. We'll review free-space path loss after we look at another way to specify the noise performance of a receiving installation.

G/T A figure of merit called G/T has evolved to measure the combination of antenna gain and equivalent noise temperature for a receiving installation. G/T is defined as:

$$G/T(\text{dB}) = G_R(\text{dBi}) - 10 \log (T_a + T_{eq}) \qquad (12.8)$$

where

$G/T(\text{dB})$	=	figure of merit for the receiving system
$G_R(\text{dBi})$	=	receiving antenna gain in dBi
T_a	=	the noise temperature of the antenna
T_{eq}	=	the equivalent noise temperature of the receiver

The gain and noise temperatures should all be taken at the same reference point. The gain required is the antenna gain less any losses up to the reference point. As before, the reference point is usually the receiver input. See Figure 12.9. In a satellite receiver, the first stage is often located separately from the rest of the receiver and very close to the antenna. This stage is called the *low-noise amplifier* (*LNA*).

FIGURE 12.9
Reference plane for
G/T calculations

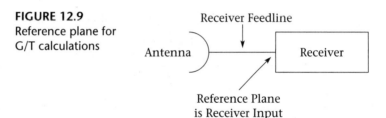

The antenna noise temperature is found as before.

EXAMPLE 12.5

A receiving antenna has a gain of 40 dBi and looks at a sky with a noise temperature of 15 K. The loss between the antenna and the LNA input, due to the feed horn, is 0.4 dB, and the LNA has a noise temperature of 40 K. Calculate G/T.

SOLUTION

First, we find G in dB. This is simply the antenna gain less any losses up to the reference point.

$$G = 40 \text{ dBi} - 0.4 \text{ dB}$$
$$= 39.6 \text{ dBi}$$

The calculation of T is a little more difficult. We have to convert the feedhorn loss into a ratio, as follows:

$$L = \text{antilog}\left(\frac{0.4}{10}\right) = 1.096$$

Substituting into Equation (12.7), we get

$$T_a = \frac{(L-1)290 + T_{sky}}{L}$$

$$= \frac{(1.096-1)290 + 15}{1.096}$$

$$= 39 \text{ K}$$

The receiver noise temperature is given with respect to the chosen reference point, so it can be used directly. Therefore,

$$G/T(\text{dB}) = G_R(\text{dBi}) - 10 \log (T_a + T_{eq})$$

$$= 39.6 - 10 \log (39 + 40)$$

$$= 20.6 \text{ dB}$$

Carrier-to-Noise Ratio Using G/T

Once G/T has been found, it can be used to help calculate the carrier-to-noise ratio for a system. We can use the following equation:

$$\frac{C}{N}(\text{dB}) = EIRP(\text{dBW}) - L_{fs}(\text{dB}) + G/T - k(\text{dBW}) \qquad (12.9)$$

where

$$
\begin{aligned}
C/N(\text{dB}) &= \text{carrier-to-noise ratio in decibels} \\
EIRP(\text{dBW}) &= \text{effective isotropic radiated power in dBW} \\
L_{fs}(\text{dB}) &= \text{free space loss in decibels} \\
G/T &= \text{figure of merit as given in Equation (12.8)} \\
k(\text{dBW}) &= \text{Boltzmann's constant expressed in dBW} \\
&\quad (-228.6 \text{ dBW})
\end{aligned}
$$

The free-space loss for this application is as found in Chapter 7, that is

$$L_{fs} = 32.44 + 20 \log d + 20 \log f \qquad (12.10)$$

where

L_{fs} = free space loss in decibels

d = path length in km

f = frequency in MHz

EXAMPLE 12.6

The receiving installation whose G/T was found in Example 12.2 is used as a ground terminal to receive a signal from a satellite at a distance of 38,000 km. The satellite has a transmitter power of 50 watts and an antenna gain of 30 dBi. Assume losses between the satellite transmitter and its antenna are negligible. The frequency is 12 GHz. Calculate the carrier-to-noise ratio at the receiver.

SOLUTION

The earth station was found to have $G/T = 20.6$ dB. The satellite transmitter power, in dBW, is

$$P_T(\text{dBW}) = 10 \log 50 = 17 \text{ dBW}$$

The EIRP in dBW is just the transmitter power in dBW plus the antenna gain in dBi, less any feedline losses, which are negligible here. Here, we have

$$EIRP(\text{dBW}) = 17 \text{ dBW} + 30 \text{ dBi} = 47 \text{ dBW}$$

Next we find $L_{fs}(\text{dB})$. Using Equation (12.10),

$$
\begin{aligned}
L_{fs} &= 32.44 + 20 \log d + 20 \log f \\
&= 32.44 + 20 \log 38{,}000 + 20 \log 12{,}000 \\
&= 205.6 \text{ dB}
\end{aligned}
$$

Now we can find C/N. From Equation (12.9),

$$
\begin{aligned}
C/N(\text{dB}) &= EIRP(\text{dBW}) - L_{fs}(\text{dB}) + G/T - k(\text{dBW}) \\
&= 47 \text{ dBW} - 205.6 \text{ dB} + 20.6 \text{ dB} + 228.6 \text{ dBW} \\
&= 90.6 \text{ dB}
\end{aligned}
$$

12.6   Systems Using Geostationary Satellites

In spite of their disadvantages, the relative simplicity of geostationary systems has made them attractive for the first generation of mobile systems. Global coverage can be achieved with only three GEO satellites, and all of North and South America can be covered with one. Let us look first, then, at two of the pioneers in satellite wireless communication. Both of them are still very much alive, and both use geostationary satellites.

Inmarsat   *Inmarsat (International Maritime Satellite Organization)* was established in 1979 as an intergovernmental treaty organization (it has since been privatized) and began service in 1982. It is now into its third generation of satellites, known as Inmarsat-3. Originally Inmarsat's mandate was to provide voice and data services to ships at sea, supplementing and eventually partly supplanting high-frequency radio. Since then its services have expanded to include land and aeronautical mobile communication.

Inmarsat uses a total of nine GEO satellites (four are in service, the rest are spares or leased out), covering the whole world except for the polar regions. Each satellite has a hemispherical beam and five spot beams with a maximum EIRP of 48 dBW in the spot beams. Power and bandwidth can be dynamically allocated among the beams. The satellites operate in the L-band (1.5/1.6 GHz).

Inmarsat offers several different types of telephone services. Inmarsat-A is the original analog telephone service. Inmarsat-B is a more modern digital service with similar capabilities. Both of these are intended mainly for communication with ships and need dish antennas with diameters of about 80 cm.

The Inmarsat mini-M service is more closely related to the theme of this book. Using the satellite spot beams, it is designed mainly for operation on land and in coastal waters. A typical portable transceiver, including antenna, is about the size of a laptop computer. It cannot be carried about when in use, but must be set up where its antenna can be aimed at one of Inmarsat's geostationary satellites. See Figure 12.10.

Handheld Inmarsat terminals are available, but only for text messaging services. Inmarsat is a major partner in ICO, a proposed MEO system for use with handheld phones.

MSAT (Mobile Satellite)   This joint Canadian and United States project uses one GEO satellite to provide coverage for North and Central America, the Caribbean and Hawaii (via a spot beam), and the surrounding coastal waters. See Figure 12.11 for the satellite footprint.

FIGURE 12.10
Inmarsat portable
terminal
(Courtesy of Inmarsat)

FIGURE 12.11
MSAT footprint

MSAT's satellite is about ten times more powerful than those used by Inmarsat and has an EIRP of at least 57.3 dBW in its coverage area so antennas can be smaller. Mobile terminals use a reasonably compact roof-mounted antenna, and portable terminals are about the size of a notebook computer and have a lid-mounted antenna. See Figure 12.12(a) for a typical mobile installation. The geostationary system does not allow for portable telephones that can be carried when in use.

The use of a single geostationary satellite allows MSAT to use a relatively simple network. Only one ground station with an 11 m dish antenna is needed. All calls are relayed through the satellite to the single ground station and from there to the PSTN. See Figure 12.12(b).

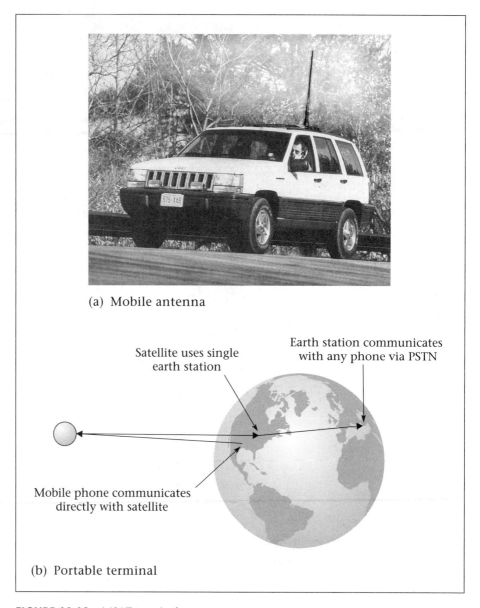

(a) Mobile antenna

Satellite uses single
earth station

Earth station communicates
with any phone via PSTN

Mobile phone communicates
directly with satellite

(b) Portable terminal

FIGURE 12.12 MSAT terminals

Comparison of Geostationary Systems

Table 12.1 shows a comparison of some of the important features of Inmarsat-3 and MSAT. Both use bent-pipe transponders connecting mobile users to a gateway that is in turn connected to the PSTN. Figure 12.13 shows the relatively simple network that results.

TABLE 12.1 Geostationary Satellite Systems

System	Inmarsat-3	MSAT
Date of operation	1996–1998	1996
Major uses and coverage	Voice, data, especially for ships Worldwide except polar regions	Voice, data, mainly for land mobile Western hemisphere
Number of satellites in active service	4	1
Uplink frequency (mobile to satellite), GHz	1.6	1.6315–1.6005
Downlink frequency (satellite to mobile), GHz	1.5	1.53–1.559
Uplink frequency (gateway to satellite), GHz	6.4	10.75–10.95
Downlink frequency (satellite to gateway), GHz	3.6	13–13.15, 13.2–13.25
Satellite EIRP, dBW	48	57.3

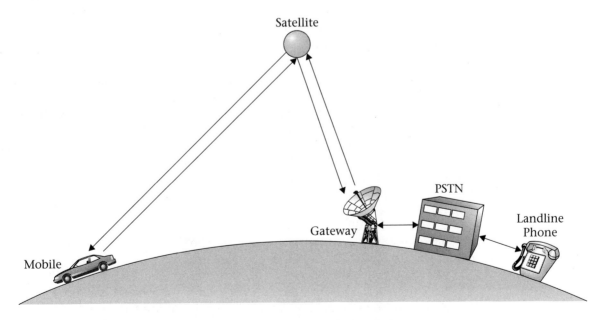

FIGURE 12.13 Mobile telephone network using geostationary satellites

12.7 Systems Using Low-Earth-Orbit Satellites

LEO satellite systems are very attractive, especially for use with handheld portable phones. The short distance to the satellite allows transmitter power and antenna gain requirements to be relaxed. This permits the use of portable phones that are only somewhat larger than a conventional cell phone. On the other hand, such a system requires many satellites (on the order of 40 to 70) and a complex network.

LEO systems are the most complex and expensive wireless communication systems yet devised. Several proposed systems have been cancelled before becoming operational due to a shortage of funds. As of April 2000, one LEO wireless telephony system (Globalstar) was operating, and another (Iridium) had been forced to discontinue service after only a year in operation, due to bankruptcy. If their technical advantages can be coupled with the necessary financial resources, LEO systems look like the ultimate in satellite telephony.

Iridium

Iridium began service in November 1998—and applied for bankruptcy protection in August 1999. As of April 2000, many attempts to restructure and continue to operate the network had failed, all commercial operation had ceased, and it seemed very likely that the satellites would be destroyed by deliberately taking them out of orbit. Nonetheless, a brief look at the Iridium system will give useful insights into the possibilities of LEO satellite telephony. Technically, the system was a success, and if the economic obstacles could be overcome, a similar system could be built in the future.

The Iridium system comprises 66 LEO satellites in a complex constellation, such that at least one satellite is visible from any location on earth at all times. The satellites are crosslinked; telephone calls can traverse the network from one satellite to another before being relayed to a ground station. This means that it is not necessary for every satellite to be in range of a ground station at all times and reduces the number of ground stations required. See Figure 12.14 for an illustration of this idea.

Iridium uses digital modulation, with a combination of FDMA and TDMA to assign channels.

Because Iridium's satellites are powerful and close to the ground, portable phones are usable with this system. The phone weighs about half a kilogram and has a fairly awkward helical antenna which must be positioned vertically and with a clear view of the sky for reliable operation. For operation where terrestrial cellular radio is available, a *cellular cassette* with its own antenna can be installed in the back of the phone. Although the Iridium satellite system was available nearly everywhere (in a few countries local

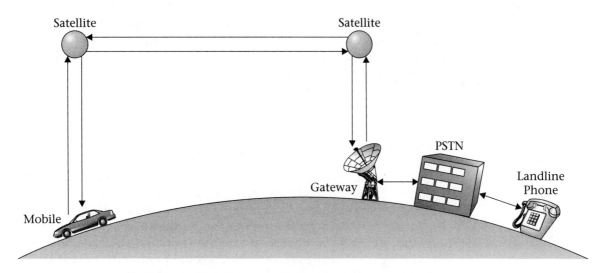

FIGURE 12.14 Mobile telephone network using LEO satellite and crosslinks

authorities forbade it), users generally preferred to use their local cellular system where possible to save on airtime costs. Airtime for Iridium ran several dollars per minute, compared to a few cents per minute for most terrestrial cellular systems. The high cost of airtime, and of the phones themselves, may have been contributing factors to Iridiums's financial problems.

Globalstar The Globalstar system began commercial operation in 1999, and by April 2000 service was available in more than 100 countries.

This system is slightly less ambitious than Iridium. It uses a constellation of 48 LEO satellites (plus four spares) at an altitude of 1414 km. The satellites use simple "bent pipe" transponders but have high power (about 1 kW per satellite). The satellites are in eight orbital planes of six satellites each, inclined at 52 degrees with respect to the equator. This allows the system to provide service from 70 degrees North latitude to 70 degrees South, which includes most of the Earth except for the polar regions.

CDMA is used, allowing a ground user to access two or more satellites simultaneously, provided they are in view, and utilizes the soft handoff techniques introduced for CDMA PCS services.

Because there is no switching on the satellites, communication is possible only when at least one satellite is visible simultaneously from the mobile phone and a ground station. This will require at least 38 ground stations, called *gateways*, for worldwide coverage. By April 2000 there were 18 gateways in operation.

Like Iridium, the Globalstar system is usable with handheld phones that resemble cell phones, but are larger and heavier. Some dual-mode phones are already available, so that Globalstar users can access lower-cost terrestrial cellular radio where it is available.

Teledesic

The Teledesic system, still under development, is the most ambitious of the proposed LEO systems. When operational it is expected to use 288 satellites plus spares, orbiting at an altitude of 1375 km. It is intended to be a high-speed data service, designed more for fixed terminals in homes and businesses than for mobile use. The frequency band it uses is much higher than the other LEO services: 28.6–29.1 GHz for the uplink and 18.8–19.3 GHz for the downlink. Atmospheric losses are high at these frequencies. It is necessary for earth stations to use satellites at a high angle, because signals pass through less of the atmosphere en route to and from these satellites than when satellites at lower elevation angles are used.

The main application for Teledesic is expected to be high-speed commercial data, in competition with fiber optics, and for individual use, high-speed internet access. Here Teledesic will have competition from cable modems, telephone lines using ADSL, and existing satellite access using geostationary satellites. Standard terminals are expected to support data rates from 16 kb/s (for a single voice channel) to 2.048 Mb/s (for high-speed data).

Table 12.2 provides a comparison of the three "Big LEO" systems just described.

"Little LEOs"

In addition to the huge projects sometimes referred to as "big LEOs," there are a number of more modest schemes, both existing and proposed, that exist only to provide low-data-rate digital services such as paging, short messaging, and vehicle tracking for trucking companies. These operations are called "little LEOs" because the systems are simpler and smaller. It is not necessary that there be a satellite in view of all stations at all times for these services because messages can be stored briefly and forwarded when a satellite becomes available (typically within a few minutes). Here are some examples of "little LEO" systems.

ORBCOMM The ORBCOMM system went into operation in 1998. It uses 28 LEO satellites and plans to increase this to 48. The system is used for short messaging, e-mail, and vehicle tracking. Unlike the big LEOs, little LEOs typically use frequencies in the VHF range to communicate with customer earth stations. ORBCOMM's uplink from mobile to satellite is at 148–150.05 MHz, with downlink at 137–138 MHz.

TABLE 12.2 "Big LEO" Systems

System	Iridium	Globalstar	Teledesic
Date of operation	1998	1999	2000
Major uses	Voice, paging, low-speed data	Voice, paging, low-speed data	High-speed data, voice
Number of satellites	66	48	288
Satellite altitude (km)	780	1414	1375
Uplink frequency (mobile to satellite), GHz	1.616–1.6265	2.4835–2.500	28.6–29.1
Downlink frequency (satellite to mobile), GHz	1.616–1.6265 (same range as uplink, TDM used)	1.610–1.6265	18.8–19.3
Uplink frequency (gateway to satellite), GHz	29.1–29.3	5.025–5.225	27.6–28.4
Downlink frequency (satellite to gateway), GHz	19.4–19.6	6.875–7.075	17.8–18.6
Intersatellite crosslink frequency, GHz	23.18–23.38	(No crosslinks)	65–71

LEO One LEO one is a proposed "little LEO" system with a similar structure to that of ORBCOMM. It will be designed to use 48 satellites at an altitude of 950 km for paging and short messaging. It is expected to be operational in 2002.

E-Sat E-Sat, expected to begin this year (2000), is an interesting "special case" LEO system. Using only six satellites, orbiting at an altitude of 1260 km, and using CDMA, it is intended to concentrate on the niche market of remote meter reading, especially in rural areas. Since meter readings are not urgent, store-and-forward technology is very appropriate for this type of system.

Table 12.3 summarizes the specifications of the little LEO satellite systems discussed in this chapter.

TABLE 12.3 "Little LEO" Systems

System	ORBCOMM	LEO One	E-Sat
Date of operation	1998	2002	2000
Major uses	Paging, short messaging/ e-mail, vehicle location	Paging, short messaging/ e-mail, vehicle location	Remote meter reading
Number of satellites	28	48	6
Satellite altitude (km)	785	950	1260
Uplink frequency (mobile to satellite), MHz	148–150.05	148–150.05	148–148.905
Downlink frequency (satellite to mobile), MHz	137–138, 400	137–138	137.0725–137.9725
Uplink frequency (gateway to satellite), MHz	149.61	148–150.05	148–148.905
Downlink frequency (satellite to gateway), MHz	137–138	400.15–401	137.0725–137.9725

12.8 Systems Using Medium-Earth-Orbit Satellites

Satellites in medium earth orbit are a compromise between the LEO and GEO systems. More satellites are needed than for GEO (on the order of 6 to 20 for real-time communication), but fewer than for LEO. Delay and propagation loss are much less than for GEO, but greater than LEO. Portable phones are possible with MEO systems, but they must be heavier and bulkier than for LEO systems.

The main advantage of using MEO rather than LEO satellites is financial. These systems promise rates for airtime that are at least on the same order of those for terrestrial cellular systems, unlike LEO systems.

At this printing no MEO systems are up and running, but some proposed systems appear likely to become operational in the near future.

Ellipso   Ellipso uses an interesting combination of elliptical and circular orbits. Its constellation is based on the fact that there is far more land mass and far greater population in high northern latitudes than in similar southern latitudes. A glance at any globe will confirm that most of the world's land mass is north of 40° south latitude.

The Ellipso system is designed to take advantage of this asymmetry. It will initially include six satellites, later increasing to ten, in a circular orbit about 8000 km above the equator for worldwide coverage at low latitudes. These are to be complemented by eight active satellites (plus two spares) in inclined elliptical orbits, designed so that they spend most of their time above the northern hemisphere. The elliptical orbits have a maximum height of approximately 7800 km and a minimum height of approximately 520 km. The elliptical-orbit satellites will have an orbital period of about three hours. See Figure 12.15.

FIGURE 12.15
Ellipso system orbits

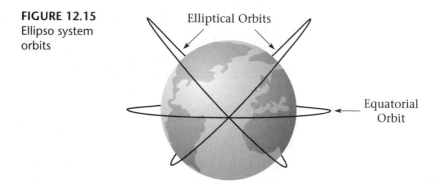

Ellipso's coverage is scheduled to be phased in gradually, beginning with the deployment of the circular-orbit satellites in 2002. These satellites can provide coverage to about 80% of the world's population.

The main focus of the Ellipso system is expected to be voice communication using portable and mobile terminals that are projected to be roughly the size of conventional cell phones but are likely to be considerably larger. CDMA is used, with the uplinks of all satellites receiving on the same frequency bands. This allows use of the soft-handoff feature of CDMA and avoids any necessity for portable phones to switch frequencies. The satellites will relay signals directly to ground-station gateways, with no on-satellite processing or intersatellite links.

ICO   *ICO* stands for *Intermediate Circular Orbit*. The plan, initiated by Inmarsat but since spun off and privatized, is to launch ten operational satellites and two spares in two orthogonal planes at an altitude of 10,355 km, each at a 45°

angle to the equator, thus providing global coverage. In fact, most of the time a given location will be in view of two satellites to provide diversity in case the view of one is blocked (by a tall building, for instance). Each satellite would be able to support 4,500 telephone calls using TDMA technology. The satellites are to use a simple bent-pipe configuration. The frequency ranges are projected to be similar to those for Ellipso, in the area of 1.6 GHz for satellite-to-mobile links.

Summary

The main points to remember from this chapter are:

- Satellites are especially useful for telecommunication in remote areas where terrestrial cellular systems are prohibitively expensive or impossible to build.

- The orbital period of a satellite in a circular orbit depends on its distance from the earth, with satellites farther from the earth having a longer period.

- Satellite orbits are classified according to their distance from earth, as low earth orbit (LEO), medium earth orbit (MEO), and geostationary earth orbit (GEO). Geostationary satellites appear stationary at a point above the equator.

- Systems using satellites in lower orbits have lower path loss and shorter propagation times, but require more satellites for real-time coverage.

- Satellites can act as simple repeaters or can contain elaborate switching systems to route calls. They can also store data for later forwarding.

- Current satellite systems for mobile communication use either GEO or LEO satellites. Only the latter systems allow handheld transceivers. There are MEO systems in the planning and construction stages.

Equation List

$$v = \sqrt{\frac{4 \times 10^{11}}{(d + 6400)}} \tag{12.1}$$

$$P_N = kTB \tag{12.2}$$

$$NF = \frac{(S/N)_i}{(S/N)_o} \tag{12.3}$$

$$NF_{dB} = 10 \log NF \tag{12.4}$$

$$T_{eq} = 290(NF - 1) \tag{12.5}$$

$$T = T_{eq} + T_a \tag{12.6}$$

$$T_a = \frac{(L - 1)290 + T_{sky}}{L} \tag{12.7}$$

$$G/T(\text{dB}) = G_R(\text{dBi}) - 10 \log (T_a + T_{eq}) \tag{12.8}$$

$$\frac{C}{N}(\text{dB}) = EIRP(\text{dBW}) - L_{fs}(\text{dB}) + G/T - k(\text{dBW}) \tag{12.9}$$

$$L_{fs} = 32.44 + 20 \log d + 20 \log f \tag{12.10}$$

● Key Terms

apogee   point in a satellite orbit that is farthest from the earth

bent-pipe configuration   a design of satellite transponder in which signals are amplified and shifted in frequency but do not undergo any other processing

constellation   in satellite telephony, a group of satellites coordinated in such a way as to provide continuous communication

crosslink   a radio or optical connection directly between satellites, without going through an earth station

equatorial orbit   a satellite orbit that is entirely above the equator

footprint   depiction of the signal strength contours from a satellite transmitter on the earth

geostationary orbit   satellite orbit in which the satellite appears to remain stationary at a point above the equator

geosynchronous orbit   satellite orbit in which the satellite's period of revolution is equal to the period of rotation of the earth

hemispheric beam   antenna beam on a geostationary satellite that is adjusted to cover the whole earth

low-earth-orbit (LEO) satellite   an artificial satellite orbiting the earth at an altitude less than about 1500 kilometers

medium-earth-orbit (MEO) satellite   a satellite in orbit at a distance above the earth's surface of approximately 8,000 to 20,000 km

perigee   the point in a satellite orbit that is closest to the earth

polar orbit   a satellite orbit passing over the north and south poles

spot beam   in a satellite system, a focused beam of energy that covers a relatively small area on the earth, produced by a high-gain antenna on the satellite

store-and-forward technique   in digital communication, the use of a device (repeater) to receive one or more data packets, store them, and retransmit them at a later time

tracking   in satellite communication, continuously adjusting the position of a directional antenna on the ground, so that it always points at the satellite

transponder   in satellite communication, a repeater located on a satellite

● Questions

1. Compare the following in terms of cost and practical communication distance:
   (a) repeaters on towers
   (b) satellites in low earth orbit
   (c) geostationary satellites

2. How does the orbital period of a satellite change as it moves farther from the earth?

3. Sketch the earth and the orbit of a geostationary satellite, approximately to scale.

4. Why do all geostationary satellites orbit the earth at the same distance and above the equator?

5. Why are geostationary satellites unusable from earth stations in the polar regions?

6. What are the Van Allen belts and what effect do they have on the placement of satellites?

7. Explain the advantages of the use of elliptical orbits for satellite communication.

8. Why are spot beams used with geostationary satellites?

9. How many geostationary satellites are necessary for a system to have worldwide coverage (except in the polar regions)?

10. What is Doppler shift? Why does the effect increase as the height of a satellite's orbit decreases?

11. Why is it necessary for mobile systems to use antenna tracking, even with geostationary satellites?

12. Why is it necessary to use multiple satellites for real-time coverage with LEO and MEO systems?

13. How can the use of multiple satellites be avoided for data communication with MEO and LEO satellites?

14. What advantages and disadvantages do satellites in low earth orbit have compared with geostationary satellites for mobile communication systems?

15. What is meant by a "bent-pipe" satellite transponder?

16. What is meant by the term *backoff* with respect to a transponder power amplifier, and when is it necessary to use it?

17. What are crosslinks and how are they useful in a satellite communication system?

18. What satellite communication system is especially designed for use by ships at sea?

19. Compare the Inmarsat mini-M and MSAT services in terms of coverage and convenience.

20. Compare Iridium with MSAT in terms of coverage and convenience.

21. Compare Iridium with Globalstar in terms of the way the networks are organized.

22. How does the Teledesic system differ from the other LEO systems described in this chapter?

23. What are the differences between "big LEO" and "little LEO" systems?

24. Why are store-and-forward techniques unsuitable for voice communication?

25. Explain how the Ellipso system can achieve worldwide coverage with fewer satellites than the other systems discussed in this chapter.

26. Compare the LEO and MEO concepts for voice communication. What advantages does each have?

● Problems

1. Find the orbital velocity and period for a satellite that is 1000 km above the earth's surface.

2. The moon orbits the earth with a period of approximately 28 days. How far away is it?

3. What velocity would a satellite need to have to orbit just above the surface of the earth? Why would such an orbit be impossible in practice?

4. A receiving installation has a *G/T* of 30 dB. If its antenna gain is 50 dBi and the antenna noise temperature is 25 K, what is the equivalent noise temperature of the receiver?

5. How could the *G/T* of the system described in the preceding problem be improved to 35 dB? Give two ways, and perform calculations for each.

6. Find the noise temperature of a receiver with a noise figure of 1 dB.

7. Find the noise temperature of an antenna on a satellite if it looks at the earth (giving it a "sky" noise temperature of 290 K) and is coupled to the reference plane by a waveguide with a loss of 0.3 dB.

8. Compare the time delay for a signal transmitted by geostationary satellite to that for a signal transmitted via terrestrial microwave radio link over a distance of 2000 km. Assume the distance from ground to satellite is 37,000 km at both ends of the path. Ignore any time delay in the electronics for both systems.

9. Compare the signal strengths from two satellites as received on the ground. One is geostationary, with a path length of 40,000 km; the other is in low earth orbit, with a path length of 500 km. Assume all other factors are equal. Which signal is stronger, and by how many decibels?

10. It is possible to communicate between points on earth by using the moon as a passive reflector. In fact, it is done routinely by radio amateurs. Calculate the time delay and round-trip path loss at 1 GHz using the moon as a reflector. (The distance to the moon is calculated in problem 2.) Note that your path loss calculation ignores losses in the reflection at the moon's surface. What can you conclude about the likely commercial possibilities of this idea?

11. (a) Calculate the path loss for a satellite in the Iridium system on the link from satellite to mobile, assuming that the satellite is directly overhead.

    (b) Calculate the round-trip time delay for the Iridium system under the same conditions as in part (a).

12. Repeat the previous problem, but use the ICO MEO system. Compare the results and draw some conclusions about the differences between LEO and MEO systems.

13. Calculate the velocity and orbital period for each of the following types of satellite:

    (a) Iridium

    (b) Globalstar

    (c) Teledesic

    (d) ORBCOMM

13

Paging and Wireless Data Networking

Objectives

After studying this chapter, you should be able to:

• Describe and explain the operation of several systems used for one- and two-way paging.

• Compare paging systems with respect to capabilities and complexity.

• Describe the operation of voice paging systems.

• Describe the operation of wired Ethernet LANs.

• Explain the need for wireless LAN equipment.

• Discuss the IEEE 802.11 and Bluetooth standards and suggest which would be preferred for given applications.

• Explain the need for and the operation of wireless Ethernet bridges and modems.

• Describe the operation of infrared LANs and compare them to wired LANs and wireless LANs using radio.

• Describe and compare public packet-data networks and compare them with other kinds of wireless data communication.

13.1 Introduction

In this chapter we look at some forms of wireless communication that are perhaps less glamorous than PCS and satellite telephony but are nonetheless important. Pagers have been with us for many years, and many people predicted that cellular and PCS phones would make them obsolete. That hasn't happened; in fact, paging features have been built into personal communication systems. Meanwhile, paging systems have become more capable.

Wireless modems and LANs have also been around for a good many years but have not been very popular outside niche markets until recently. The current vogue for notebook and subnotebook computers and for connected organizers like the Palm Pilot™ has spurred interest in these technologies as well. In this chapter we will look at some of the emerging standards in this field.

13.2 Paging and Messaging Systems

Paging systems have undergone changes not unlike those that have occurred with wireless voice communication. The traditional paging system uses widely spaced transmitters, each covering a considerable geographic area. In addition, all transmitters in a given system operate on the same frequency, either in the VHF range at about 152 or 158 MHz or in the UHF range from 454.025 to 454.650 MHz, and all pages are sent by all the transmitters in the system. The system, wasteful of spectrum though it is, works because short paging messages require little time or bandwidth.

The traditional pager is simply a fixed-tuned receiver that uses a transmitted code to identify messages meant for it. The earliest pagers simply beeped when they received a message. That signaled the pager owner to call a central office to find out the details of the page. Most modern systems transmit a short numeric or alphanumeric message to the pager. The numeric message is usually a phone number for the owner to call, but some inventive pager owners, particularly teenagers, have invented numerical codes for standard messages.

Simple paging systems like these remain very popular in spite of the increased popularity of cellular and PCS telephones. The advantages of pagers are their small size, low cost, and long battery life (typically several weeks, compared to hours for cell phones and days for PCS phones). The extended battery life of one-way pagers results partly from the lack of any need for a transmitter or for power-consuming vocoders and audio amplifiers in the receiver, and partly from the ability of pagers to go into a "sleep" mode, emerging at intervals to check for messages.

The ultimate in paging is for the user to be able to send a reply. Such systems exist and we'll look at them too. Some systems even allow a voice message to be received and answered.

One-Way Paging Systems

The traditional way to handle paging is to have a network of relatively powerful transmitters (on the order of hundreds of watts), all of which transmit all pages on the same frequency. Figure 13.1 shows how this works. Superficially the system may resemble a cellular network, but in reality the technology is much simpler. Frequencies are reused by using the same frequencies for every transmitter. With a voice system this would give a very low capacity, since one call would tie up a pair of frequencies over the whole network. Paging messages, however, typically require only a few seconds to transmit and are relatively infrequent, so many pagers can share a frequency using TDMA. Each pager has a unique address called a **capcode**, which is sent every time it is paged. Transmissions addressed to other pagers are simply ignored.

The fact that all pages are sent from all transmitters means that there is no need for the system to know the location of any pager. In fact, since the pagers are receive-only devices with no transmission capability, there is no way for the system to tell where an individual pager is or even whether it is turned on. The use of multiple transmitters has the added advantages of reducing the number of dead spots for the signal and completely avoiding any need for handoffs.

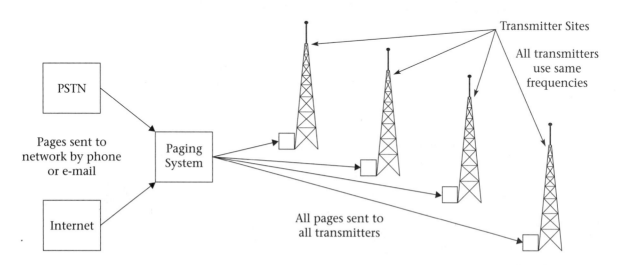

FIGURE 13.1 Basic paging network

When a paging system covers a very wide area, such as the United States or Europe, roaming systems may be used in which the pager owner informs the system when he or she moves away from the local network. This removes the need to send all pages over a very large area. Satellites are often used to transmit pages great distances to local systems, which relay them via terrestrial transmitters. In addition, as we discussed in the previous chapter, there are LEO satellite systems that send paging messages directly to individual pagers. These systems are more expensive but are useful in remote areas without terrestrial paging systems. See Figure 13.2 for an illustration of the difference between these two types of satellite-based paging.

FIGURE 13.2
Satellite-based
paging systems

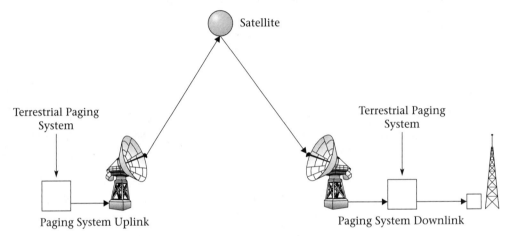

(a) Terrestrial system using satellite for national coverage

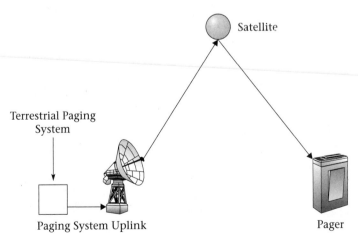

(b) Direct satellite transmission to pagers

The most common protocol for one-way paging systems is called *POCSAG,* which is an acronym for *Post Office Code Standardization Advisory Group;* it is so named because it was originated by the British Post Office. It can be transmitted at three different data rates: 512, 1200, and 2400 b/s. Currently the 1200 b/s variation is the most common; the 512 b/s system is gradually being phased out, and the 2400 b/s system is used mainly for nationwide systems. The modulation scheme is FSK with ±4-kHz deviation, in an RF channel that is 25 kHz wide. The protocol can handle about two million pager addresses. This seems like a lot but remember, even in a nationwide system, each pager must have a unique capcode.

POCSAG messages are sent in batches. Several batches may be sent together in one transmission. See Figure 13.3 for an illustration of a POCSAG signal. Each transmission begins with a preamble in the form of 576 bits of the sequence 1010101010. Each batch begins with a 32-bit synchronizing pattern followed by eight frames with two codewords per frame. Each codeword can be defined as an address or a message codeword. These are easily distinguished: the address codeword begins with a 0 and the message codeword begins with a 1. The address codeword has an 18-bit address field. Following are two bits that allow for four different paging sources, and ten error-correction bits.

The number of possible addresses that can be transmitted in one frame is $2^{18} = 262{,}144$ addresses. This is not enough for a large system. To increase the number of possible addresses, each pager is programmed with a 3-bit code that tells it in which of the eight frames in the batch it should look for its address. This gives a total possible number of addresses equal to $2^{21} = 2{,}097{,}152$ addresses.

FIGURE 13.3
POCSAG paging protocol

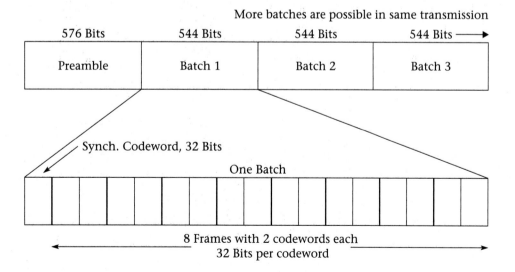

More batches are possible in same transmission

576 Bits | 544 Bits | 544 Bits | 544 Bits ⟶

Preamble | Batch 1 | Batch 2 | Batch 3

Synch. Codeword, 32 Bits

One Batch

8 Frames with 2 codewords each
32 Bits per codeword

Codewords in a batch that are not used for addresses are used for messages. (If there are not enough messages to fill a batch, then an idle message is sent.) There are 20 message bits per codeword, and as many codewords may be used as necessary to transmit the message. Simple tone pagers do not need any message frames; an address is sufficient. Alphanumeric pagers require more message frames than do the simpler numeric pagers.

EXAMPLE 13.1 ▼

Suppose the POCSAG system is used with simple tone pagers, which require only an address field. If all the frames are used for addresses, how many pages could be transmitted by this system in one minute if it operates at the slowest POCSAG rate of 512 b/s? Assume that only one preamble is needed.

SOLUTION

The preamble uses 576 bits. To find how many bits remain in one minute we can find the total number of bits and subtract 576. Each frame can contain two address codewords. Let us assume that things are balanced and all codewords can be used. That is, one-eighth of the pagers being paged have addresses in each of the eight possible codewords. This is obviously a best-case assumption.

$$\text{Total number of usable bits/minute} = 512 \times 60 - 576 = 30{,}144 \text{ bits}$$

Each batch has one 32-bit synchronizing codeword and sixteen 32-bit address codewords for a total of $17 \times 32 = 544$ bits. Therefore,

$$\text{batches/min} = \frac{\text{bits/min}}{\text{bits/batch}}$$

$$= \frac{30{,}144}{544}$$

$$= 55.412 \text{ batches/min}$$

Each batch has 16 addresses so the number of pages that can be transmitted in one minute is

$$\text{addresses/min} = \text{batches/min} \times \text{addresses/batch}$$

$$= 55.412 \times 16$$

$$= 886$$

A newer family of paging protocols, introduced by Motorola, is the *FLEX™* group, consisting of *FLEX* for one-way paging, *ReFLEX™* for two-way alphanumeric paging, and *InFLEXion™* for voice paging.

The FLEX system uses four-level FSK to allow operation at 1600, 3200, and 6400 b/s. There are four possible transmit frequencies rather than two as for conventional FSK, which is used by POCSAG. The possible frequency deviations from the carrier frequency are ±800 Hz and ±2400 Hz. This provides four possibilities so that two information bits can be transmitted per symbol, allowing a data rate of 6400 b/s with a symbol rate of 3200 baud in the same 25-kHz channel as is used for a POCSAG system.

EXAMPLE 13.2

Calculate the efficiency, in terms of bits per second per hertz of RF bandwidth, for the FLEX system at its maximum data rate.

SOLUTION

The channel width is 25 kHz and the maximum bit rate is 6400 b/s. This represents

$$\frac{6400 \text{ b/s}}{25 \text{ kHz}} = 0.256 \text{ b/s/Hz}$$

This is a rather low figure, indicating a robust system. In fact, this system can operate in much less bandwidth. The two-way ReFLEX system, described next, can operate with three paging channels in one 50-kHz RF channel.

Two-Way Paging Systems

The Motorola ReFLEX™ system is the de facto standard for two-way alphanumeric pagers. Two-way paging is much more complex and expensive than the more common one-way operation. Every pager needs a transmitter, with a power output on the order of one watt; this adds greatly to the cost of these devices.

On the other hand, there are advantages in addition to the obvious one of being able to reply to pages without having to find a phone. The presence of a transmitter allows pagers to acknowledge the receipt of messages, making for more reliable service. It is also possible for a two-way paging system to employ frequency reuse, much like a cellular or PCS telephone system. Since the pagers have transmitters, they can respond to queries from the system and can be located within a particular cell.

ReFLEX paging systems operate in the frequency ranges of 928–932 and 940–941 MHz for the **outbound channel** (base to pager) and 896–902 MHz for the **inbound channel** (pager to base). One outbound carrier can be carried in a 25-kHz channel, or three carriers in a 50-kHz channel. The available outbound data rates are 1600, 3200, and 6400 b/s. Inbound communication uses a 12.5-kHz channel and can support up to a 9600 b/s data rate.

Voice Paging

Voice paging has been available for many years but was not widely used until recently. It allows a pager to function in a manner similar to a telephone answering machine.

The Motorola InFLEXion™ system is the most popular voice paging protocol. It uses analog compression and SSB AM to transmit voice messages from the base station to the pagers. Both upper and lower sidebands are used, but each sideband constitutes a separate voice channel. This system allows two voice messages to be transmitted simultaneously on a channel 6.25 kHz wide. A pilot carrier is also transmitted for ease in demodulation. See Figure 13.4. Time-division multiplexing is also used so that many pagers can share a single voice channel.

FIGURE 13.4
InFLEXion voice paging signal

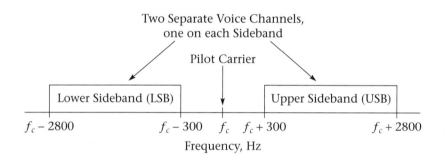

InFLEXion voice pagers normally allow a text (not voice) reply in a manner similar to two-way alphanumeric pagers. In that respect they are less flexible than cellular or PCS telephones. Since voice paging is fairly expensive and the price of cellular and PCS phones is dropping, it remains to be seen how popular voice paging will become.

13.3   Wireless Local-Area Networks

Many offices and even some homes use local-area networks to connect microcomputers. Traditionally, these networks have used coaxial cable, twisted-pair wire, or in some cases fiber optics, to connect the computers to each other, usually through a central hub. This is quite suitable for installations where the computers are fixed in place and where wiring can be run without too much difficulty.

Some offices with staff who bring notebook computers to meetings or who move about regularly find wireless network connections useful. There is also an emerging market in home networking. The problem in the home is not so much mobility as the inability or reluctance of homeowners and tenants to run wiring through or along the walls. Wireless networks seem to be much easier to install—and in some cases that may even be true.

LAN Topologies   There are three basic ways to organize a local-area network. As shown in Figure 13.5, they are the star, the ring, and the bus. The star is the traditional way to connect a central mainframe to dumb terminals. The ring topology is used with some LANs, notably those using the IBM token-ring system. By far

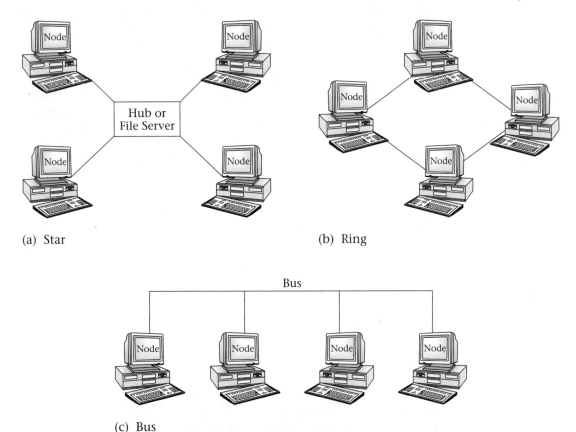

(a) Star

(b) Ring

(c) Bus

FIGURE 13.5   LAN topologies

the most common logical topology for LANs, however, is the bus. The majority of local-area networks use the **Ethernet** standard, a logical bus topology usually running at either 10 or 100 Mb/s (now up to 1 Gb/s). Ethernet can use either coaxial cable or twisted-pair wiring, but the latter is cheaper and more commonly seen nowadays. In this chapter we shall focus on networks using Ethernet, because practically all wireless networking equipment is designed to connect to Ethernet systems.

With a basic Ethernet system, all stations are connected to the bus at all times, and any station can transmit data packets at any time. Each packet has an address and so does each station; the station to which a packet is addressed will accept it and read the data, while all other stations simply ignore it.

Any bus network has the possibility of **collisions**; an occurrence where two stations transmit at the same time, thereby rendering the data from both unusable. Ethernet systems try to minimize collisions with a set of simple rules. Each station "listens" before transmitting to make sure the bus is not being used. If it is, the station waits until all is quiet. In the event that two stations do transmit at the same time, each will detect this, stop transmitting immediately, and wait a random time before beginning the transmission again. This is a type of protocol known as **carrier-sense multiple-access with collision detection (CSMA/CD)**.

The physical layout of an Ethernet LAN using twisted-pair wiring usually looks more like a star than a bus. All the computers are connected to a central hub, as shown in Figure 13.6, using two pairs of wire in a single cable. Appearances are deceiving, however. Logically the network is still a bus, and the central hub simply makes wiring easier and less expensive. Hubs can usually do some network management, such as disconnecting a defective station. It is also possible to use a switch instead of a hub at the central location, which can make the network more efficient by avoiding collisions. The address of each packet is examined as it arrives at the switch, and the packet

FIGURE 13.6
Ethernet network using twisted-pair wiring

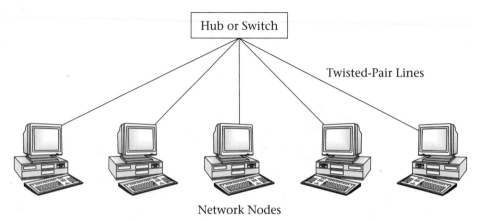

Hub or Switch

Twisted-Pair Lines

Network Nodes

is sent only to the node to which it is addressed, eliminating the possibility of a collision.

The alternative, for 10 Mb/s networks only, is to use a true bus using 50 Ω RG-58/U coaxial cable. This is sometimes done, but it has disadvantages. Coaxial cable is more expensive than twisted-pair wiring, and connecting a station to the network requires breaking the cable and inserting a tee connector. See Figure 13.7.

FIGURE 13.7
Ethernet using coaxial cable

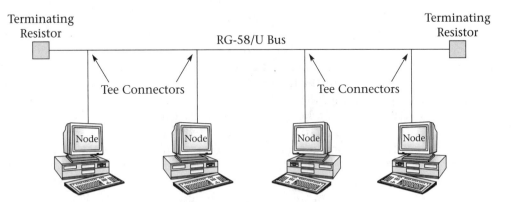

There is a third type of Ethernet, which uses thicker RG-8/U coaxial cable and transceivers, but this is seen so seldom that it will not be covered here.

Ethernet systems using twisted-pair wiring are called 10BaseT (for 10 Mb/s, baseband, twisted pair) or 100BaseT for the 100 Mb/s version. Practically all wireless LAN equipment is designed to connect to this type of network.

Ethernet networks have restrictions on the distances between terminals. The reason for this limitation can be found in the CSMA/CD protocol. Before transmitting, a node "listens" for a period of 9.6 microseconds. If another station is transmitting at the beginning of this period, it will be detected by the end of the listening period, provided that its distance from the receiving station is less than the distance a signal can travel during the listening period. Recall that the speed of transmission along a line can be found by multiplying a number called the velocity factor, which is a characteristic of the cable, by the speed of light.

The velocity factor for typical coaxial cable is about 0.66, so the distance traveled in 9.6 μs is given by:

$$d = vt$$
$$= v_f ct$$
$$= 0.66 \times 3 \times 10^8 \text{ m/s} \times 9.6 \text{ μs}$$
$$= 1900 \text{ m}$$

where

d = distance traveled
v = velocity of propagation along the cable
t = propagation time
v_f = velocity factor
c = speed of light

It would not be wise to have an Ethernet segment this long, however. If a station begins transmitting during the listening period for another station, it might not be detected. The Ethernet specifications call for maximum cable runs of 500 m on thick coax, 200 m on thin coax, and 100 m on twisted-pair.

Of course, it is still possible that another station will begin transmitting while the first station is listening. If the station is too far away, the transmission will not be detected and a collision may result. In that case, the collision should be detected relatively early in the transmission, so that the process can be stopped and restarted with as little wasted time as possible. The system is most efficient when the packets of data are relatively long.

To determine how long it can take to detect a collision, consider the scenario shown in Figure 13.8 on page 499. Nodes A and B are at opposite ends of the bus. Suppose node A transmits. The signal from node A moves down the line toward node B, but it takes some time to get there. Meanwhile, node B listens and, hearing nothing, begins to transmit just as the signal from A reaches it. Node B stops transmitting, but both transmissions now have errors and have to be discarded. Meanwhile, node A knows nothing about the collision until the signal from node B reaches it. Then it stops transmitting and the network is clear for one or the other to try again.

EXAMPLE 13.3

Two nodes on an Ethernet LAN using coaxial cable are 200 m apart. How long will it take the stations to detect a collision?

SOLUTION

The total time used by this process is the time it takes for a signal to travel down the line and back. That is, it is two times the propagation delay along the line. Recall that for any line,

$$v = v_f c$$

The time for a signal to travel a given distance along the cable is given by

$$t = \frac{d}{v}$$

FIGURE 13.8
Ethernet distance
restrictions

(a) Physical situation

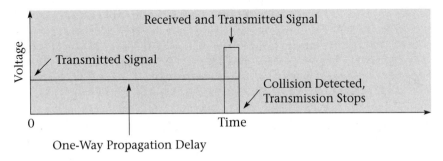

(b) Signal at node A

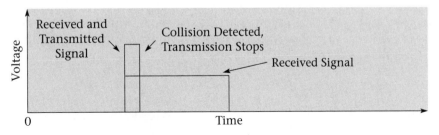

(c) Signal at node B

For an Ethernet system with a length of 200 m on coaxial cable with a velocity factor of 0.66, the propagation delay is:

$$t = \frac{d}{v}$$

$$= \frac{200 \text{ m}}{0.66 \times 300 \times 10^6 \text{ m/s}}$$

$$= 1 \text{ μs}$$

The maximum time it takes to detect a collision is twice this, or 2 μs. At 10 Mb/s, this corresponds to the time taken by 2 μs × 10 Mb/s = 20 bits.

Ethernet packets are generally much longer than this. Long packets can extend the distances possible on the net, but they reduce the network efficiency when there are collisions, since the whole packet must be retransmitted.

Wireless connections to LANs have timing constraints similar to those seen with wired LANs. Of course, the velocity factor is equal to one for a wireless connection. On the other hand, since wireless LAN transceivers cannot receive while they are transmitting, a collision cannot be detected until the end of the packet. This makes shorter packets more desirable with wireless systems.

Ethernet Bridges

When a network becomes too large to satisfy the distance constraints just described, it can be broken into two sections connected by a **bridge**, as shown in Figure 13.9. A bridge connects to both network segments. It looks at each packet of data on each segment of the network, and from its address, determines whether or not its destination is on the same segment as its source. If the source and destination are on the same segment, it does nothing. If they are on different segments, it places the packet on the segment holding the destination. This is called *selective forwarding*.

FIGURE 13.9
Ethernet bridge

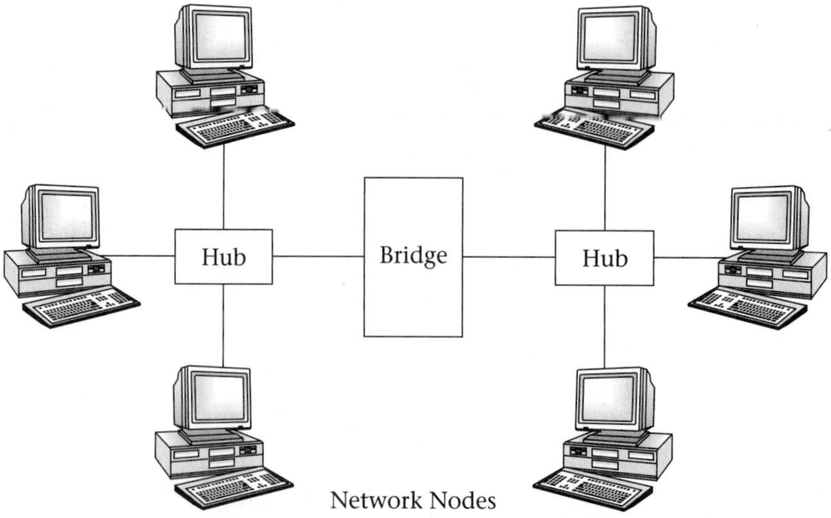

Network Nodes

Wireless LANs   In general, wireless local-area networks are slower and more expensive than wired LANs. Therefore, the use of wireless LANs is indicated only where there are compelling advantages to being free of cabling. Offices where people frequently change work stations, work locations like warehouses or hospitals where people have a need to communicate while moving about, and places where it is inconvenient or expensive to run wires, are examples. Ordinary wired Ethernet networks commonly run at either 10 or 100 Mb/s, and gigabit Ethernet is now available. By contrast, wireless LAN speeds tend to be in the 1- to 2-Mb/s range. This is fast enough for many applications, however, particularly if the application program runs locally and the network is used only to transfer data. For instance, a spreadsheet program running on a notebook computer can usually share files with others on the network without requiring a great deal of bandwidth.

Radio LANs   There has been a proliferation of proprietary radio standards for LANs. Most have used unlicensed frequency bands in the 900-MHz range, and more recently the 2.4-GHz range. However, there was no standardization until very recently. This meant that network components from different manufacturers would not work together. Wired Ethernet, by contrast, has quite rigid standards: any network adapter designed for a 10BaseT system will work with any other adapter made for the same system.

Recently, however, there has been some improvement in this chaotic situation. Two relatively new standards are particularly interesting. The first that we will consider, called IEEE 802.11, is a general-purpose standard for LANs that cover a typical office environment. The second, with the interesting name of Bluetooth, is intended for much shorter range, up to 10 m, and is intended to provide a low-cost, built-in capability for devices such as cellular phones and personal digital assistants (PDAs). Let us look at each in turn.

IEEE 802.11   The Institute of Electrical and Electronic Engineers worked for many years to bring some order to the chaotic wireless LAN environment, finally approving the 802.11 standard in 1998. It envisions spread-spectrum operation on the unlicensed *ISM (industrial, scientific, and medical)* frequency band from 2.4 to 2.484 GHz. This is the same part of the spectrum that is used for microwave ovens; the use of spread-spectrum transmission allows these networks to operate in the presence of interference, both from microwave ovens and from other nearby LANs. The 2.4-GHz band was chosen for the 802.11 standard primarily because, unlike other unlicensed frequency ranges such as 902–928 MHz, it is available in most of the world and until recently has been largely unused because of the interference problem from microwave ovens. The standard also allows for infrared operation, but this provision is seldom used.

Both the direct-sequence and frequency-hopping variations of spread-spectrum operation can be used. The following statements establish a few basics on which the standard is built:

- A set of wireless nodes is called a *Basic Service Set* (*BSS*), and a network consisting of only a BSS with no access points is clled an *ad-hoc network*.

- A network can consist only of wireless nodes communicating with each other—an *Independent Basic Service Set* (*IBSS*)—but more usually a number of wireless nodes in a BSS communicate with an *access point* that is also part of a wired Ethernet network.

- There can be multiple access points for extended coverage.

- A network with multiple access points is called an *Extended Service Set* (*ESS*).

- Wireless units can roam within the ESS. See Figure 13.10 on page 503.

The 802.11 system specifies a **CSMA/CA (Carrier-Sense Multiple-Access with Collision Avoidance)** protocol. The main difference between this and the CSMA/CD system used with wired Ethernet is that the radios are only half-duplex, so they cannot detect collisions once they have begun to transmit a packet of data. This can reduce the network capacity by increasing the time taken to detect collisions.

The choice between frequency-hopping and direct-sequence spread-spectrum versions of the specification is not clear-cut, and both types of products are available. In current practical implementations, the direct-sequence version offers greater range, but the frequency-hopping system often seems to perform better in cluttered offices with many reflections and sources of interference.

The frequency-hopping version of 802.11 uses 79 separate channels, 1 MHz apart and visited at a rate of approximately ten channels per second. The maximum transmitter power allowed is 1 W, and devices have to be able to reduce this to not more than 100 mW when the full power level is not needed. When systems are interference-limited, as is often the case in LANs, there is no benefit in increased power: it only shortens battery life in portable equipment.

The basic frequency-hopping standard uses frequency-shift keying with two transmitted frequencies. The bit rate is 1 Mb/s. There is an optional standard, using four-level frequency-shift keying (two different shifts, much like the FLEX paging system that we looked at earlier in this chapter). With the same symbol rate of 1 Mbaud, this system achieves a transmitted bit rate of 2 Mb/s. The idea behind the two speed standards is to allow an intermix of lower-cost lower-speed and higher-cost higher-speed devices.

The standard does not envisage any central control over hopping sequences to prevent two transmitters from transmitting at once. The protocol will prevent this from happening at the start of a transmission, but it can happen later. There are 78 predefined hopping patterns organized into three

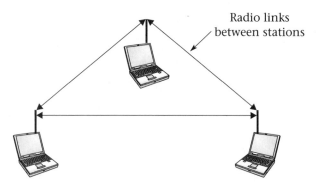

Radio links between stations

(a) "Ad Hoc" network: 1 BSS, no access points

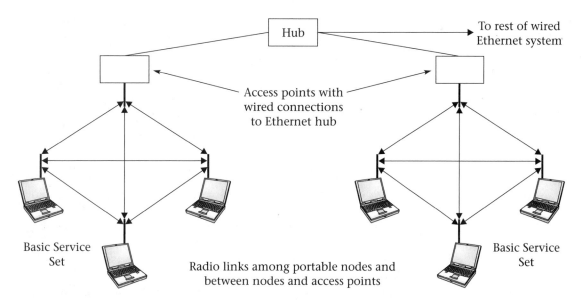

Hub

To rest of wired Ethernet system

Access points with wired connections to Ethernet hub

Basic Service Set

Basic Service Set

Radio links among portable nodes and between nodes and access points

(b) Extended service set

FIGURE 13.10 IEEE 802.11 networks

sets of 26 patterns each. In spite of collisions, the data throughout the whole system keeps increasing until there are about 15 networks operating in the same area.

The direct-sequence version of the specification is quite different, as you would expect. The bit rates allowed are the same: 1 and 2 Mb/s. The way this is achieved is different, however. Either binary (for 1 Mb/s) or quadrature delta phase-shift keying (DBPSK or DQPSK) can be used. In either case, the symbol rate is 1 Mbaud. See Figure 13.11.

FIGURE 13.11
DBPSK and DQPSK

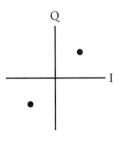

Bit	Phase change from previous symbol
0	0°
1	180°

(a) DBPSK

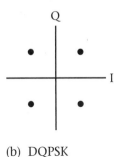

Bits	Phase change from previous symbol
00	0°
01	90°
10	−90°
11	180°

(b) DQPSK

The symbol rate is increased to 11 Mbaud by using ten extra bits, called chips, for each data bit. The result is an 11:1 spreading gain. That is, the bandwidth is increased by a factor of 11. The strength of an interfering signal will also be reduced by a factor of 11 (10.4 dB) during the de-spreading process.

Even with the spreading due to the chipping bits, the signal from a single direct-sequence wireless node occupies only about 5 MHz. Therefore, several frequency channels can be assigned; this system combines FDMA with CDMA. The exact number of channels varies with the country, since the width of the band allocated varies somewhat. In North America there are 11 available channels.

Many companies have introduced wireless LAN equipment using the 802.11 standard. Most of the wireless nodes consist of network cards that are intended to be installed in laptop and notebook computers. Access points are more likely to be stand-alone boxes. In both cases there is a small antenna that should be oriented for best results. Examples are shown in Figure 13.12.

FIGURE 13.12
Wireless Ethernet devices
(Courtesy of Lucent Technologies)

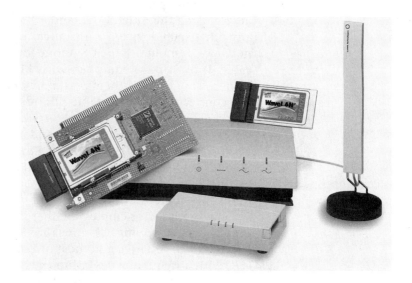

The distance obtainable with a wireless Ethernet connection varies greatly with the environment. As we saw when we discussed propagation, the square-law rule for attenuation does not apply once we leave a free-space environment. The typical office is anything but free space, being full of objects of varying conductivity as well as quite a few sources of electrical noise. Usable ranges are therefore very difficult to predict. Typical systems work over a distance of about 100 to 200 m in an office environment.

IEEE 802.11—a and b Versions

- The IEEE 802.11b standard was introduced in September 1999. It allows for a maximum bit rate of 11 Mb/s using direct-sequence spread-spectrum operation in the 2.4-GHz band. Equipment designed to this standard can interoperate with the 1-Mb/s and 2-Mb/s equipment previously described.

- The IEEE 802.11a project is designed as a future standard for data rates of up to 54 Mb/s in the 5-GHz band. It is still being developed.

Bluetooth   The Bluetooth specification is a joint venture involving several companies: Ericsson, IBM, Intel, Nokia, and Toshiba. It is designed to be an open standard for short-range systems; the usable operating range is specified as 10 cm to 10 m and can be extended to 100 m using RF amplifiers for the transmitters. The intention is to make Bluetooth devices small enough and inexpensive enough that they can be built into many types of equipment: cellular and PCS phones; notebook computers and personal digital assistants (PDAs); and computer peripherals such as printers, modems, and loudspeakers. Any device using the Bluetooth standard should be able to communicate with any other such device. For instance, at present it is necessary to connect a notebook computer to a PCS phone for wireless data by using a proprietary cable. If both were equipped with Bluetooth transceivers it would only be necessary to have them reasonably close together to make a truly wireless connection. The portable phone could even be left, powered on, in a briefcase.

Technically, the Bluetooth standard resembles the frequency hopping version of the IEEE 802.11 wireless Ethernet standard. Like the LAN system, Bluetooth operates in the 2.4-GHz ISM band. The reason is the same as cited by the IEEE: this band is available worldwide, though the exact frequency limits vary somewhat from place to place.

Bluetooth radios employ a rather simple spread-spectrum technique using frequency hopping. Channels are 1 MHz apart, giving room for 79 channels in North America and most of Europe. The transmitter uses two-level FSK, with a frequency deviation (measured each way from the carrier frequency) of between 140 kHz and 175 kHz. The channel bit rate is 1 Mb/s, and since this is a two-level system, the bit rate and baud rate are equal.

Bluetooth systems divide data into packets and send one packet per hop. The length of the packet (and the corresponding hop) can be from one to five *slots;* the length of a slot is 625 μs.

EXAMPLE 13.4

Calculate the maximum and minimum hopping rate for the Bluetooth system.

SOLUTION

The maximum hopping rate is

$$f_h(\text{max}) = \frac{1}{625\,\mu s}$$

$$= 1600\,\text{Hz}$$

and the minimum is

$$f_h(\text{min}) = \frac{1}{5 \times 625\,\mu s}$$

$$= 320\,\text{Hz}$$

Bluetooth differs from the wireless Ethernet systems in that it envisions both audio and data transmission. Generally the audio is synchronous and the data asynchronous. Audio is coded at 64 kb/s per stream, using either linear CVSD (continuously variable slope delta modulation) or companded PCM. The companding can follow either the North American µ-law or the European A-law standard, in order to be compatible with wireline telephony. The Bluetooth system can carry up to three of these 64 kb/s streams in each direction.

When used for data, the system can support 432 kb/s in each direction simultaneously, or alternatively, 721 kb/s in one direction and 57.6 kb/s in the other. The latter alternative is useful for asymmetrical applications like file downloads, where most of the data flows one way with the other direction used only for acknowledgements.

Full-duplex operation is accomplished by switching from transmit to receive between slots. This is called **TDD (time-division duplex)** and is similar to the procedure used in the TDMA and GSM cellular systems and PCS.

The simplest Bluetooth networks, called *piconets,* can have from two to eight nodes. All Bluetooth nodes have the same capabilities, but whichever node initiates the communication acts as the master for timing and frequency-hopping information, with the others acting as slaves.

One Bluetooth device can operate simultaneously on two piconets, acting as a bridge between the two. A conglomeration of two or more piconets is called a *scatternet.* See Figure 13.13 on page 508 for a depiction of Bluetooth piconets and scatternets.

When a Bluetooth device is in standby mode and not connected to a network, it listens every 1.28 seconds on a set of 32 hop frequencies, going through each in turn. When a station wishes to connect, it sends a *Page* message (if it already knows the network address of the intended recipient from previous communication) or an *Inquiry* message (if it does not know what other devices are on the network). Page and Inquiry messages are first sent to 16 channels, then to the other 16 wake-up channels if there is no response. Thus, it can take up to $2 \times 1.28 = 2.56$ seconds for a connection to be set up.

Bluetooth can be used to set up a wireless LAN or to connect to a wired LAN, just as with the IEEE 802.11 standard. The main differences are that the

FIGURE 13.13
Bluetooth
networks

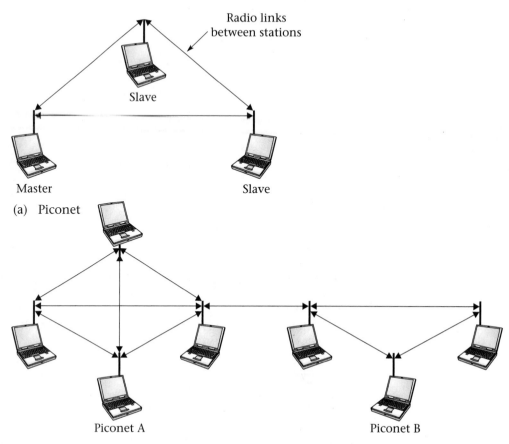

(a) Piconet

(b) Scatternet

data rate and range are less with Bluetooth. On the other hand, once companies tool up for production, the cost of adding Bluetooth capability to a piece of equipment is expected to be only a few dollars. This compares with typical wireless Ethernet costs of several hundred dollars per node.

Wireless Bridges   Wireless bridges connect LAN segments. Often these are in different buildings, so it is necessary to use a system with more range than those described above. Some bridges use high-speed dedicated microwave links. These can be fast but they are expensive and require licensing. Lower-cost bridges using the 902-MHz or 2.4-GHz unlicensed bands are available from several vendors. A typical example is shown in Figure 13.14. It can operate over a distance of about 10 km, under good line-of-sight conditions, at a data rate

FIGURE 13.14
Wireless bridge
(Courtesy of Wi-LAN Inc.)

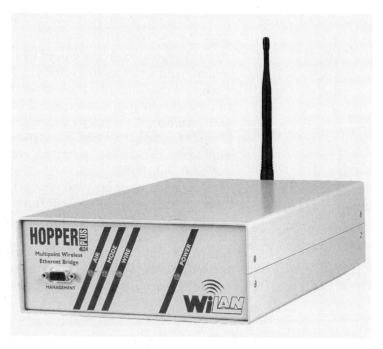

of about 2 Mb/s. Line-of-sight propagation conditions are more likely between buildings than within them. The built-in antenna shown can be replaced with a directional rooftop antenna for greater range.

Connections Using Infrared

Though the great majority of the wireless systems described in this book use radio communication, we should note that light waves can also be employed. You are probably familiar with fiber optics; it is not really a wireless technology because the fiber cable, though it may not contain copper, certainly has the effect of restricting mobility. Obviously it is possible to transmit light directly through the air, however, without benefit of fiber. Generally invisible infrared light is used. The common television remote control is an example of a simple infrared communication system.

Anyone who has used an infrared remote control is familiar with some of the limitations of optical systems. The range tends to be short, the signal will not pass through walls or other opaque objects, and usually the transmitter must be carefully pointed in the direction of the receiver. All of these issues tend to limit the usefulness of infrared communication systems. We should note in passing, though, that the inability of the light signal to pass through walls can be an advantage in some cases, as it reduces the danger of deliberate eavesdropping and of accidental interference between systems.

A short-range infrared system called *Infrared Data Association* (*IRDA*) has been used for some time to allow two devices to communicate with each other. A typical example would be the synchronization of data between a desktop and a notebook computer. The system is deliberately restricted to a range of one meter to allow several independent infrared links to operate simultaneously in the same room. LEDs operating at a wavelength of about 860 nm are used. Several data rates are specified, up to a maximum of 4 Mb/s, but most systems operate at a significantly lower data rate, since it is common to connect an infrared transceiver to the serial port on computers without built-in infrared ports.

At least one commercial system does take advantage of the 4-Mb/s data rate possible with built-in infrared ports. Clarinet Systems makes the EthIRLAN™, which uses a wired hub with infrared transceivers connected to it by wires. The hub is connected to a portable computer by placing one of its transceivers near the infrared port on the computer. What results is similar to a wired LAN except that portable computers can be added to and removed from the LAN at will without disturbing any electrical connections. See Figure 13.15.

FIGURE 13.15
Infrared connections to LAN

(Courtesy of Clarinet Systems "EthIRLAN" System, © 1999 Clarinet Systems. All rights reserved.)

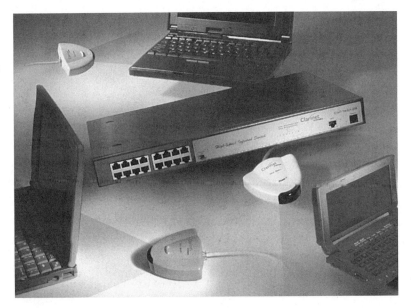

It is also possible to build a complete wireless LAN with relatively widely separated computers. Each computer needs an infrared transceiver, which typically connects to a standard Ethernet card. A central transceiver connects to the network. In this case it is better to use diffused infrared light than a narrow beam. Diffused infrared does not need a direct line of sight: reflections from walls and ceiling can be used. This allows the computers to be moved around, but they must, of course, remain in the same room.

Figure 13.16 shows one of these systems. It has a range of about 10 m. Several computers with infrared transceivers can form a simple peer-to-peer network. If one transceiver is connected to an existing Ethernet LAN, all of the computers with transceivers become part of that network.

FIGURE 13.16
Infrared LAN
(Courtesy of iRLan Ltd.)

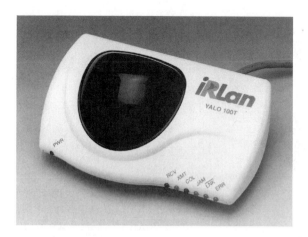

Systems like this, while they work, have failed to become very popular. Perhaps this is because they are rather cumbersome for use with portable devices, and they are much more expensive than wired systems for desktop computers. If a network is confined to one room, in most cases it is not difficult to install the wire for a conventional Ethernet system.

Wireless Modems

In a sense all wireless data communication devices are modems, since the data must be modulated onto a carrier for transmission. This section deals with point-to-point, rather than networked, data transfer at distances of up to a few kilometers: larger distances than are found in a typical LAN, but within line of sight. For distances greater than line of sight, a wireless modem can be used as a repeater, or the user can resort to commercial packet-radio, paging, cellular, or PCS networks.

A typical wireless modem is shown in Figure 13.17. It can work with one other modem, or with several by polling each in turn. The data rate is fairly low, with a maximum of 19.2 kb/s, but this is sufficient for many applications. Such devices are used for traffic control, process monitoring (at oil wells, for instance), and similar applications.

FIGURE 13.17
Wireless modem
(Courtesy of Wi-LAN Inc.)

Many wireless modems use the 2.4-GHz band; most of the others use another ISM unlicensed band from 902 to 928 MHz. Most use spread-spectrum operation to reduce the possibility of interference. Since the frequency ranges where they operate are unlicensed, there is no coordination of frequencies.

Most wireless modems operate at transmitter power levels similar to the wireless Ethernet systems previously described, but they use much lower speeds. These systems can therefore work over greater distances, up to several kilometers in open terrain.

Wireless modems are similar in operation to conventional telephone modems and quite different from wireless LAN systems. Typically they attach to a computer's serial port and operate with any ordinary modem software package.

13.4   Wireless Packet-Data Services

So far we have looked at a number of wireless data communication schemes, each with its own niche. Wireless LANs give high-speed short-distance connectivity; wireless bridges and modems can extend proprietary systems for a few kilometers. Paging systems give local, national, and even international (using satellite systems) access to short text messages.

Note that there is some overlap. PCS have short-messaging features that are very similar to those offered by paging systems, for instance. Personal communication systems are also introducing packet-radio services, as we noted earlier. These duplicate the functions of CDPD (which we studied earlier) and also of dedicated wireless packet-data networks (which we shall look at now). The competition between services is vigorous, and it is impossible to tell at this writing which services will prosper, which will find niche markets, and which will disappear.

Public packet-data networks predate PCS and even cellular telephony. They are intended strictly for data transfer, which can include such jobs as vehicle tracking, remote order entry, and even internet access.

There are two major public wireless packet-data networks in North America, Mobitex™ and ARDIS™. There are also a number of private networks. Keep in mind that both these services compete with CDPD, analog cell phones using modems, and PCS messaging.

Mobitex The Mobitex standard for wireless packet-switched data was created by the Swedish telephone company for use by its field personnel. It was introduced as a commercial system in 1986 and is now developed by Ericsson, a Swedish manufacturer of wireless equipment, which also manufacturers most of the equipment for this system. The specifications are public, though, so there are other suppliers as well. In the United States, RAM Mobile Data, a subsidiary of BellSouth, is the main user of Mobitex; in Canada, Cantel AT&T operates a Mobitex network. Between them these systems cover most North American cities.

Mobitex networks use a cellular structure in the 900-MHz band (896–902 MHz and 935–941 MHz) in North America (frequencies differ in other countries), but they are separate from the AMPS cellular system—except that the two systems sometimes use the same towers. Narrow channels only 12.5 kHz wide are used with GMSK (Gaussian minimum shift keying) modulation to support a data rate of 8 kb/s. Since this is a packet-switched system, there is no need for connections or multiplexing. Data is sent by landline to a Mobitex transmitter, which transmits packets as received. Each mobile terminal receives all packets, but ignores all except those addressed to it. Similarly, the mobile transmits packets to the base station when it is available using a contention protocol similar to that used with Ethernet.

Being a packet-switched system, Mobitex can at best be called "near real-time." This is not good enough for audio or video, but is quite satisfactory for e-mail, database access, dispatching, vehicle tracking, web surfing, and many other activities.

ARDIS   The *Advanced Radio Data Information Services (ARDIS)* system was created by
IBM as a joint venture with Motorola for its own use in working with its out-
side sales and service personnel. It is now owned by the American Mobile
Satellite Corporation. It is available in major metropolitan areas in the
United States, and also in Canada, where the Bell Mobility data network uses
the ARDIS system.

ARDIS differs from Mobitex in that all its cells use the same frequencies,
much like a paging system. This is wasteful of bandwidth but it does allow
for diversity in that a mobile unit can receive signals from, and transmit to,
more than one base station.

The normal data rate with ARDIS is 4.8 kb/s in a 25-kHz channel; there is
a high-speed version with a data rate of 19.2 kb/s but having limited availability. ARDIS systems compete directly with Mobitex for the portable fax,
e-mail, and tracking market.

Summary

The main points to remember from this chapter are:

- There are many options for wireless data communication, including paging systems, cellular and PCS radio, wireless LANs and modems, and
  packet-data networks.

- There are many different types of paging systems, including simple
  one-way beepers, one-way numeric, one-way alphanumeric, two-way
  alphanumeric, and voice pagers.

- One-way paging systems transmit all pages throughout the whole system
  to avoid having to locate the recipient.

- Two-way paging systems can locate pagers in much the same way as a
  cellular radio system.

- The most common type of voice pager, the Motorola InFLEXion system,
  uses analog compressed voice in an SSB AM system.

- Most wireless LAN systems work as extensions to wired Ethernet systems.

- Most current wireless LAN equipment follows the IEEE 802.11 standard,
  and uses either frequency-hopping or direct-sequence spread-spectrum
  radio in the 2.4-GHz unlicensed frequency band.

- The Bluetooth standard resembles IEEE 802.11, but is modified for
  low-cost short-range operation. It uses frequency hopping at 2.4 GHz.

- Many wireless LAN bridges and wireless modems use proprietary systems
  at 902 MHz or 2.4 GHz in order to achieve ranges of several kilometers.
  Some long-range bridges use microwave links that require licenses.

- Infrared light is common for very short-range connections and is occasionally used for wireless LANs.
- Public packet-data networks are used for e-mail, keeping contact with employees in the field, limited worldwide web browsing, and similar low-data-rate applications.

Key Terms

bridge   a device to connect two segments of a network

capcode   unique address for a pager

carrier-sense multiple-access with collision avoidance (CSMA/CA)   method of reducing contention in a network, involving each station checking for interference before transmitting

carrier-sense multiple-access with collision detection (CSMA/CD)   method of reducing contention in a network, involving each station checking continuously for interference before and during transmissions

collision   attempt by two transmitters to use the same channel simultaneously

Ethernet   form of local-area network using CSMA/CD and a logical bus structure

inbound channel   communication channel from mobile to base station

outbound channel   a radio channel used for communication from a base station to mobile stations

time-division duplexing (TDD)   transmission of data in two directions on a channel by using different time slots for each direction

Questions

1. Why is it necessary for one-way paging systems to transmit all messages from all transmitters?

2. What factors account for the long battery life of pagers compared with cell phones?

3. What modulation schemes and data rates are used with the POCSAG protocol?

4. What limits the number of pagers in a POCSAG system, and approximately what is the maximum number of pagers that can be used in such a system?

5. How does the POCSAG system accommodate the differing message-length requirements of tone-only, numeric, and alphanumeric pagers?

6. Compare POCSAG with FLEX in terms of data rate and modulation scheme.

7. How does the ReFLEX system differ from FLEX?

8. What advantages does a two-way paging system have over a one-way system?

9. How does the InFLEXion system transmit voice?

10. What are the three topologies that are used with local-area networks?

11. What logical topology is used with Ethernet?

12. What physical topology is used in a typical Ethernet network? Is there any other possible physical topology?

13. What transmission speeds are commonly used with wired Ethernet?

14. Does a practical wired Ethernet network always attain the data rates specified in your answer to the previous question? Explain.

15. What limits the distance between nodes in a wired Ethernet?

16. Describe the function of an Ethernet bridge.

17. How can using a switch instead of a hub improve the performance of an Ethernet LAN?

18. What frequency range is used for the IEEE 802.11 and Bluetooth wireless network standards, and why?

19. Why is it necessary to use spread-spectrum technology in the frequency range used by IEEE 802.11 and Bluetooth?

20. What data rates are possible with IEEE 802.11(b) wireless LANs?

21. Explain the difference between a Basic Service Set, an Independent Service Set, and an Extended Service Set in the IEEE 802.11 specification.

22. What two types of modulation are used for the frequency-hopping version of 802.11? Explain how they differ.

23. Define the two terms CSMA/CD and CSMA/CA, describe how they differ, and state which is used in wireless LANs, and why.

24. What modulation schemes are used for the direct-sequence version of 802.11?

25. How is the spreading of the spectrum accomplished in each of the two RF variants of 802.11?

26. What is the spreading gain in the direct-sequence variant of 802.11?

27. What is the anticipated range for devices using the Bluetooth system?

28. What modulation scheme and spread-spectrum technique are used for Bluetooth?

29. What is the maximum bit rate for the Bluetooth system?

30. How is audio transmitted using Bluetooth?

31. What is a piconet? How does it differ from a scatternet?

32. What are the advantages and disadvantages of infrared wireless LAN systems, compared with radio-frequency systems?

33. What is the maximum range for the IRDA system? Why is it designed to have such a short range?

34. Why is diffused infrared preferred for an optical LAN in a situation where computers may be moved?

35. How does a wireless bridge differ from a wireless Ethernet card for a computer?

36. How does a wireless modem differ from a wireless bridge?

37. Compare the Mobitex and ARDIS systems in terms of data rate.

● Problems

1. Calculate the efficiency of the POCSAG paging system in terms of bits/second per hertz of RF bandwidth at each of its three specified data rates.

2. Suppose a POCSAG system is set up so that each frame contains one address codeword and one message codeword. A certain type of page requires 500 message bits. Allowing for all the other types of bits involved, how many frames and how much time will it take to transmit this message at 1200 b/s? Assume that all the required frames can be transmitted in one transmission with one preamble.

3. Suppose a POCSAG system at 1200 b/s is replaced with a FLEX system at 6400 b/s. Give an estimate of the increase in capacity. (The answer will not be precise because the two protocols differ in the amount of redundancy they include.)

4. Suppose that instead of four-level FSK the FLEX system used eight-level FSK. What would be the maximum bit rate, assuming the baud rate remained unchanged? Would there be any disadvantage to doing this?

5. Calculate the efficiency in bits per second per hertz of bandwidth of the ReFLEX system operating at its maximum rate with three paging channels in one 50-kHz RF channel.

6. Suppose two stations on a wireless LAN are 150 m apart. How long does it take one station to detect that another is transmitting?

7. What is the spreading gain for a wireless LAN system using the frequency-hopping version of the 802.11 system?

8. Suppose the Bluetooth system is used to connect stereo headphones to a computer. What data rate could be supported for each of the two channels?

9. Prepare a table comparing the data rates delivered by the Mobitex and ARDIS systems described in this chapter with the CDPD, cellular modem, and PCS data transmission systems previously discussed.

10. Which of the systems described in the previous problem uses frequency spectrum most efficiently?

The Future of Wireless

Objectives

After studying this chapter, you should be able to:

- Describe and compare wireless alternatives to the copper local loop.

- Explain the process by which standards for the third generation of wireless communication are being developed.

- Compare third-generation wireless proposals.

- Explain the third-generation requirements and compare them with systems currently available.

- Explain the difference between MMDS and LMDS and discuss their contribution to the total wireless picture.

- Explain what is meant by convergence and discuss the place of wireless communication in the future.

- Discuss safety and esthetic factors that could affect the progress of wireless communication.

- Make some predictions of your own about the future of wireless communication.

14.1 Introduction

There is no doubt that we are still near the beginning of the wireless revolution. The cellular concept started the revolution by allowing frequency reuse at relatively short distances. As higher frequencies become economical for everyday use, there is almost no limit to the amount of data that can be sent by wireless means.

At the same time, the more traditional wired media have not remained static. The development of high-speed fiber-optic networks, as well as new technologies that increase the capacity of existing twisted-pair and coaxial cables, have implications for wireless systems as well. In this chapter we look at some emerging wireless technologies, then consider the interactions among wireless and wired technologies. Finally, the author will go out on a very thin limb and make some rash predictions about the future. New developments in the wireless field are announced almost daily, so this chapter cannot hope to be exhaustive. Rather, we will try to find some pattern to current and possible future developments, so that the reader can make sense of a very confusing area of technology.

14.2 Wireless Local Loop

The "last mile" of any telephone system is the most costly. Running twisted-pair wire to every house is expensive even in urban settings, but in a remote area the costs and difficulties are much greater.

Wireless local loops using radio transmission in the VHF and UHF bands have been employed in isolated areas for some time, because they can transmit signals for several kilometers without the need for stringing cable. Until recently, however, this was definitely a niche market. These systems use the traditional method of building one tall tower and using it to get as much range as possible without frequency reuse. One or sometimes two subscribers use a dedicated RF channel. The amount of spectrum allocated to this type of service is not sufficient to allow wireless local-loop service in any but the most sparsely populated areas.

Figure 14.1 shows a traditional wireless local-loop setup. Modulation is analog FM and transmitter power is 1 to 10 W, depending on the distance to be covered.

Recent developments in wireless technology, especially the cellular concept and spread-spectrum communication, have caused the idea of wireless local loops to be revisited. Several companies offer fixed wireless systems based on this modern technology. See Figure 14.2(a) for a block diagram of this type

FIGURE 14.1
FM wireless
local loop

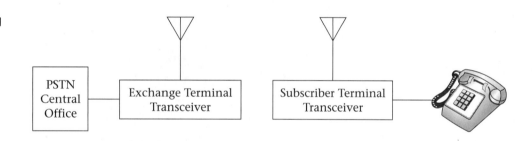

of system, with examples. Figure 14.2(b) shows a commercial system using CDMA in a dedicated frequency band. The subscriber terminal provides the proper signals, including analog voice, dial tones, and ringing voltage, to work with standard phones and fax machines. The system uses QPSK on dedicated RF channels in the range from 1.35 to 3.6 GHz. Each RF channel can carry 30 voice channels at the full landline-standard 64 kb/s. This standard has the advantage of allowing the lines to work with 56-kb/s modems, but it is also possible to operate the system at a reduced bit rate of 32 or 16 kb/s per voice channel to achieve greater capacity.

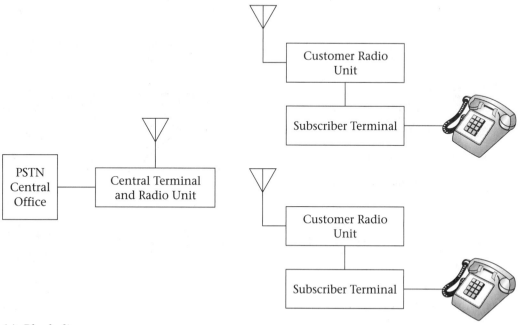

(a) Block diagram

FIGURE 14.2(a)   Digital wireless local loop

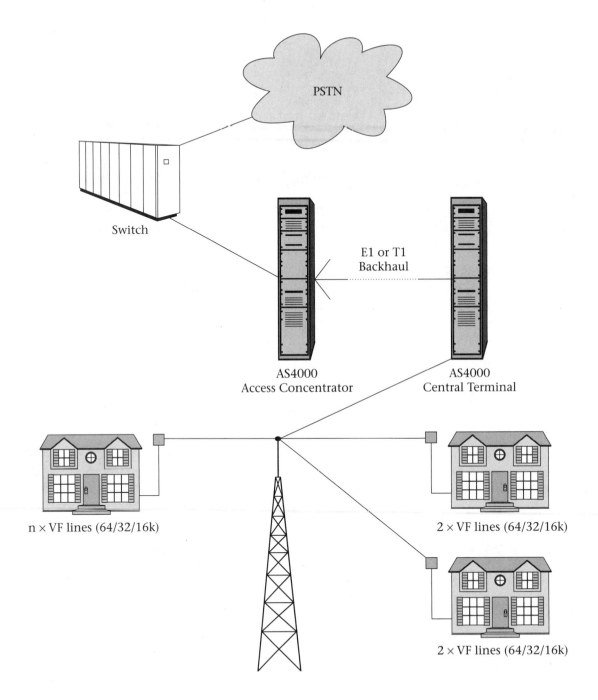

FIGURE 14.2(b)   Commercial system using CDMA

There has also been considerable interest in using established PCS networks to provide fixed service. The most obvious way to do this is simply to use a PCS phone at home, instead of a wired phone, or perhaps instead of getting an additional phone line. However, several companies provide interface units that will allow ordinary telephones to connect through existing house wiring to a PCS transceiver.

Wireless local loops are still not very popular in areas which already have telephone wiring in place, but they are attractive for new installations (particularly in developing countries that lack extensive copper infrastructure). One of the advantages of a wireless system in this kind of environment is that it is possible to begin slowly, with a small number of base stations, and increase the number of cells as the number of customers grows. A wired system, on the other hand, must be fully installed before any revenue comes in. This can cause severe cash-flow problems in a relatively undeveloped area where only a few people are likely to become customers at first.

In developed countries, far-sighted thinkers expect that eventually the wired local loop will be replaced for telephony, not by a fixed wireless system, but by a variation of PCS that is flexible enough, and cheap enough, to be used everywhere for both voice and data. However, such a system might well involve tiny cells (*picocells*) in users' homes. These would have to be connected to the PCS network somehow, so the copper local loop might well survive in a new guise—as part of a PCS network rather than part of the wired PSTN. The next section deals with developments in PCS that could lead to this situation.

14.3   Third Generation PCS

Cellular radio and PCS have been very successful, but they are not the ultimate in wireless personal communication. Manufacturers, service providers and standards bodies were already working on improvements when the first systems went into operation. At this time, third-generation (3G) wireless systems are about to be deployed.

Requirements for the Third Generation   There are several areas in which improvements to the present systems are needed. The following paragraphs provide a brief description of some of them.

Improved Data Communication   All the digital personal communication systems we have looked at so far were designed mainly for voice communication. The emphasis was on keeping bandwidth requirements low, along with minimizing power requirements for the mobile unit. We have seen that these systems also work well for e-mail and other short messages, but the available bit rate for data communication is quite low, even compared with ordinary telephone modems. The reasons for this are easily understood, especially once we realize that these systems transmit digitized voice at much lower bit rates than does the PSTN. The recent development of PCS phones with internet browsing capabilities, and of notebook computers and personal digital assistants that can connect to a portable phone or can have a cellular or PCS radio built in for web browsing, only underscores the need for a higher data rate for serious data communication. A great deal of work has gone into specialized browsers and web sites, all with the aim of minimizing the amount of data that must be transmitted over the air interface. Nevertheless, users who are used to connecting to the internet over office LANs or from home using cable modems or ADSL telephone-line connections still find wireless access painfully slow. Other high-speed data applications, such as streaming multimedia or video telephony, are simply impossible with current systems.

Greater Capacity For some time it has seemed obvious that the next generation of wireless communication must incorporate the possibility of using much higher data rates when needed. On the other hand, there is no need to use higher rates for ordinary telephone-quality voice; indeed, using higher rates than necessary merely wastes spectrum. It also wastes battery power in the mobile, since higher data rates need greater bandwidth and therefore greater transmitter power for an equivalent signal-to-noise ratio at the receiver. In fact, the trend in vocoder design is toward lower bit rates for telephone-quality voice. Therefore, the third generation should have a variable data rate, using whatever bit rate is necessary for the application.

One of the reasons for the creation of second-generation (PCS) wireless systems was that the first-generation cell phone systems were becoming overloaded in major metropolitan areas. At current growth rates, it will not be long before the same thing happens to the second-generation networks (it is already happening in Japan). Therefore, a third-generation wireless network should have greater capacity for ordinary voice calls, as well as allowing higher data rates for digital communication.

Adaptability to Mobile, Pedestrian, and Fixed Operation For some time there has been considerable interest in making it possible to use the same wireless phone in a variety of situations. We have already noted, that even though the AMPS cellular system was really designed with mobile use (from vehicles) in mind, at the present time there are more portable (hand-carried) than mobile AMPS phones in use. The second-generation systems were designed for portable phones and incorporate more extensive use of micro-cells and picocells (very small cells in office buildings, shopping malls, and so on.) For the third generation it is expected that in addition to vehicle and pedestrian use some people will use their wireless phones from fixed locations, such as a home or office. The requirements and the difficulties in meeting these requirements differ in each situation. There is probably less need for high-speed data in a mobile than in a fixed environment, for instance. This is actually rather fortunate, since the frequent handoffs required when a vehicle moves at high speed make high-speed data connections difficult.

The idea is to have a *Universal Personal Telecommunications* (*UPT*) terminal that would work anywhere, using a variety of RF standards for fixed, mobile and perhaps even satellite operation. The whole system would be transparent to the user, who would not need to know how the connection was made any more than a present-day wireline telephone user needs to know the details of call switching and routing.

Greater Standardization   As we noted earlier, there is a lack of standardization in the second generation of wireless systems, especially in North America, where three incompatible systems compete and where backward compatibility is still required with analog AMPS if a phone is to be used outside cities. In addition to the various terrestrial systems, each existing and proposed satellite system uses its own protocols and modulation scheme. It would certainly be useful to have one worldwide standard for the third generation. Failing that, a small group of standards that allowed manufacturers of portable terminals to produce equipment that would work anywhere in the world would be an improvement on the current situation.

Any of the second-generation systems could be extended by increasing its data rate but leaving the basic system unchanged. Obviously, adopting a single third-generation specification using any of the existing standards would give an advantage to those manufacturers and service providers who already have experience with that system. Consequently, there has been a good deal of argument about the merits of various proposals for a 3G system. Not all of the arguments have had a strictly technical basis.

The IMT-2000 Specification   By autumn of 1999, general agreement was reached on the outline of a third-generation system, designated by the International Telecommunications Union (ITU) as IMT-2000 (*IMT* stands for *International Mobile Telecommunications,* and 2000 refers both to the approximate implementation date and to the fact that the proposed system will operate at about 2000 MHz), which meets most of the above requirements. Maximum data rates will be

144 kb/s for mobile users in high-speed vehicles, about 384 kb/s for pedestrians and perhaps for slow-moving vehicles, and up to 2 Mb/s for stationary users. There are two major reasons for the differences in data rates. The first is that pedestrians, because of their lower speed, encounter handoffs less frequently than users in moving vehicles. Stationary users, of course, are not subject to handoffs at all. Secondly, stationary users are much less affected by multipath fading.

While no one standard for the air interface has been approved, the number of different standards to be included has, after much negotiation, been reduced from about fifteen to five. This is a decided improvement, though still far from ideal. The specification includes one TDMA standard, one FDMA standard, and one CDMA standard with three variations. There will probably be multimode phones that can cope with all these standards. Figure 14.3 illustrates how the standards come together.

The need for a capacity increase necessitates a greater spectrum allocation for the new system. The World Administrative Radio Conference in 1992 assigned the frequency bands from 1885–2025 MHz and 2110–2200 MHz for this service. In North America, as we have seen, much of this band is already in use for PCS; nonetheless, there will be some increase in the spectrum allocation. Part of this range, from 1980–2010 and 2170–2200 is also used for satellite telephony, but this sharing of bands is deliberate as satellites are incorporated in the IMT-2000 standard.

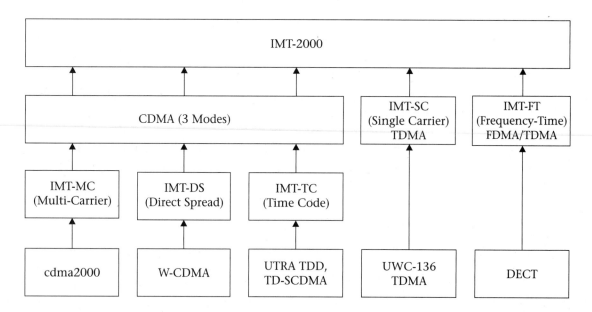

FIGURE 14.3 Evolution of third-generation PCS standards

It is questionable whether the 1992 spectrum allocation will be sufficient. If there is a great deal of demand for high-speed data services, it will probably be necessary to revisit the question of spectrum allocation for third-generation PCS.

Wideband CDMA Systems

The IMT-2000 system will incorporate three variations of CDMA. See Table 14.1. The modes differ in how duplexing is accomplished and how many carriers are used. All variations operate in a 5-MHz channel, as compared to 1.25 MHz for CDMAOne (also called IS-95), the current CDMA PCS.

The new CDMA standard is an amalgam of several proposals that eventually were combined into three. One variant, called cdma2000 and proposed by users of the North American IS-95 CDMA system, can use three separate carriers within a single 5-MHz channel, each modulated with a chipping rate of 1.2288 Mb/s. This is the same rate as used with IS-95, so this variation essentially uses three IS-95 signals in one larger channel, combining the data from the three to give some combination of greater numbers of voice calls and higher-speed data transmission. In fact, the three carriers do not have to be carried in contiguous spectrum; three 1.25-MHz channels at different points in the spectrum can be used. This is to allow for backward compatibility with the existing IS-95 system. The cdma2000 system can also use a single carrier operating at a higher chipping rate for new systems where backward compatibility is not required.

The other variation, proposed by a consortium of European and Japanese interests and called *W-CDMA* (for *Wideband CDMA*), uses a chipping rate of 4.096 Mb/s with only one carrier per 5-MHz channel and is therefore

TABLE 14.1   CDMA Comparison

Specification	IS-95 (Current system)	cdma2000 (U.S.A.)	W-CDMA (Japan/Europe)
Channel bandwidth	1.25 MHz	5 MHz	5 MHz
Number of carriers	1	3	1
Chip rate	1.2288 Mb/s	1.2288 Mb/s per carrier	4.096 Mb/s
Data rate	9.6 kb/s	Variable to 2 Mb/s	Variable to 2 Mb/s

not backward compatible with IS-95. It also does not rely on the synchronization of base stations by means of the American GPS satellites. This provides advantages in the case of indoor base stations where the reception of satellite signals is difficult.

See Table 14.1 for a comparison of the two major CDMA systems.

The harmonized standard will allow two CDMA modulation schemes: a multi-carrier system resembling cdma2000, and a single-carrier scheme as in W-CDMA.

In addition to two ways of dividing the channel, there are two ways of providing full-duplex operation. You will recall that the IS-95 system, like all first- and second-generation wireless systems, provides full-duplex operation by using separate carrier frequencies for the forward and reverse channel. With a digital system there is another possibility called **time-division duplexing (TDD)**, in which different time slots are used for each direction, but the same RF channel is used for both. Digital PCS already use different time slots for the two directions, but they also use different RF channels. The new third-generation specification will allow for either TDD or the more conventional frequency-division duplexing. The IMT-TC specification allows for either spread-spectrum or TDMA operation in each direction.

Wideband TDMA   The other variation that seems likely to be incorporated in the third-generation PCS standard is a wideband version of the North American TDMA system, with some elements incorporated from the European GSM system. This proposal is known as UWC-136 and was proposed by the *Universal Wireless Communications (UWC)* consortium, representing operators of and equipment providers for North American TDMA cellular systems and PCS using the IS-136 standard.

The new TDMA standard envisions a wide variety of channel widths, data rates, and modulation schemes, so as to allow a gradual migration from current TDMA and GSM systems to the third-generation standard. North American TDMA operators can begin by increasing the data rate in their 30-kHz channels, while GSM systems can do the same with their 200-kHz channels. The plan is eventually to move to channels 1.6 MHz wide. The wider channel will support a channel bit rate of up to 5.2 Mb/s using 8-DPSK, that is, phase-shift keying using eight different phase angles, which allows three bits to be transmitted per symbol. See Figure 14.4. Once the final standard is in place, all the IMT-2000 requirements should be supported. As mentioned above, TDMA systems can use either frequency- or time-division duplexing.

In general, it appears that the likely result of all the complicated negotiations is that the third generation of wireless communication standards will be only slightly less chaotic than the second. A fourth generation, with data

FIGURE 14.4
8-DPSK

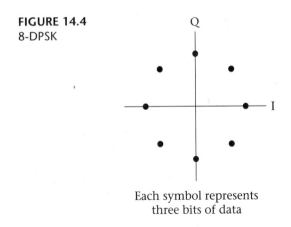

Each symbol represents
three bits of data

rates of 150 Mb/s and more, is already being considered. Perhaps by then it will be possible to arrive at one worldwide standard. Don't count on it though.

14.4 Residential Microwave Communication Systems

Local multipoint distribution systems (LMDS) use terrestrial microwave transmission to provide a number of services to homes and businesses. These services can include broadcast television, high-speed internet access, and telephony. In Canada the same technology is called *LMCS (local multipoint communication system)*. Most systems use frequencies in the 28-GHz range. There is also a variant known as **MMDS (multichannel multipoint distribution system)** which operates in the 2-GHz range. We'll look at both, beginning with MMDS because it is simpler and was first historically.

MMDS   Multichannel multipoint distribution systems have been operating for several years in the 2-GHz range. The first systems were analog but recent versions are digital. MMDS began as an alternative to cable television using coaxial cable and is sometimes called by the rather contradictory name *wireless cable* for that reason.

Figure 14.5 shows the layout of a typical MMDS. Local television stations are received off the air at the headend, which also picks up cable-network stations by satellite. The signals are rebroadcast in a digital format called *MPEG* (for *Motion Picture Experts Group*) by microwave from one or more tall towers to small wall-mounted antennas on houses and apartment buildings.

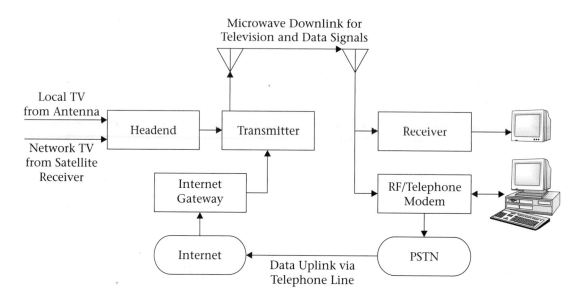

FIGURE 14.5 MMDS system

The system works over line-of-sight distances of up to about 50 km, depending on the antenna heights. Receiving antennas must have a direct line of sight to the transmitter, so the system is uneconomical in hilly areas or dense metropolitan areas with many tall buildings that may block the signal. There is sufficient spectrum for about 100 channels.

MMDSs are also used for internet access. Typically, all data requests from all subscribers are transmitted on one RF channel, and each home terminal accepts only those packets that are addressed to it. In this way the system resembles a LAN.

Multichannel multipoint distribution systems are one-way: a bidirectional system would require dividing the coverage area into much smaller cells, which would be expensive. With an existing MMDS, any return link that may be required for uploading data to the internet or for ordering pay-per-view television programs must use some other technology. Usually upstream communication is accomplished using the PSTN, but subscribers find this inconvenient, especially for internet access, as it ties up a telephone line. Recently some MMDS operators have begun to investigate the possibility of using a low-speed wireless link, such as PCS, for upstream communication. The only problem with this is that it adds considerably to the cost of the system, which must remain competitive with cable television in order to survive.

Most observers expect that MMDS services will eventually be replaced by LMDS, which we look at next. Another possibility, since many MMDS services are owned by telephone companies, is that these services will be replaced by video services delivered by telephone line, when these services become available.

LMDS The concept behind an LMDS is similar to that behind the earlier MMDS, but with some major changes. First, moving the frequency to 28 GHz allows much freer use of spectrum; about 1.3 GHz is allocated to this service in the United States, and 1 GHz in Canada. Second, the high frequency causes severe attenuation problems in the presence of rain, and even foliage from trees can block the signal. This, coupled with the difficulty of generating substantial amounts of power at 28 GHz, forces the system to use much smaller cells with a maximum diameter on the order of 5 km. Finally, the LMDS is designed to be bidirectional. This allows it to be used for internet access without using the PSTN and, in fact, allows the LMDS to be used for telephone communication instead of the PSTN. See Figure 14.6.

However, early versions of LMDS, such as SPEEDUS™ in New York City, still rely on a wireline modem and a PSTN connection for upstream traffic.

Because of the very large amount of spectrum available, LMDS can offer very high data rates. SPEEDUS, for example, advertises rates up to 48 Mb/s, which is much greater than its two nearest competitors (cable modems using

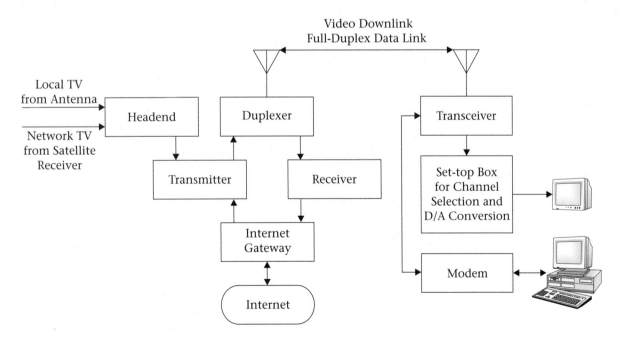

FIGURE 14.6 LMDS system

cable-television coaxial cable and ADSL using twisted-pair telephone wiring). Both of these offer data rates on the order of 1 Mb/s.

As with MMDS, LMDS is essentially a line-of-sight technology. However, some research shows, that due to the high frequency and short wavelength, LMDS may be able to take advantage of reflections from buildings to achieve coverage in some areas that are not in the direct line of sight from a transmitter. See Figure 14.7.

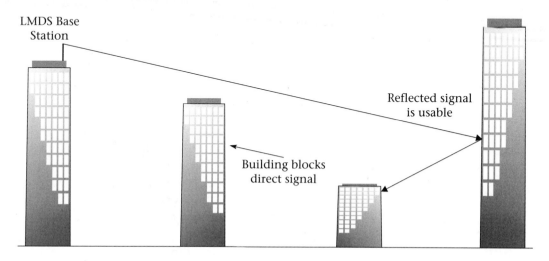

FIGURE 14.7 Multipath propagation and LMDS systems

 ## 14.5 Convergence

There has been a good deal of talk, most of it rather vague, about the convergence of various forms of communication. To some extent this is indeed happening, but it seems from what we have seen so far that an equally good case could be made for divergence.

Insofar as the idea of convergence has a technical, rather than merely a marketing, component, it could loosely be summed up in the statement: as all forms of communication become digital they begin to resemble each other more than previously. A few years ago such systems as television broadcasting, radio broadcasting, voice communication, and data transfer were considered separately. Each had its own technology, modulation schemes, technical and corporate structures, history, and indeed culture. All these and more can now be viewed as data streams, differing from one another in bit rate, need for error correction, need for real-time delivery and so on, but definitely all aspects of the same technology. Therefore, any medium capable of carrying a sufficient bit rate could be used for any of the above services, as well as others yet to be developed.

This kind of thinking leads companies that are already involved in one or more of the above areas to speculate about moving into others. Thus we have internet access and, shortly, telephony on cable-television systems; internet access and, shortly, television on telephone wire (in addition to the basic telephone service, which can be used simultaneously with the new high-speed services); and, more germane to the focus of this book, internet access and, eventually, video on portable wireless communication devices.

The dream of many people involved in the industry is of one medium bringing all communication services to the consumer. Cable television people think of this medium as the coaxial cable they already have installed to most places. Telephone people are stuck with lower-bandwidth twisted-pair wiring, but they point out that their switched network gives them a big advantage. Both telephone and cable-television people dream of running fiber optics to every residence in order to provide virtually unlimited bandwidth, but that is so expensive it is unlikely to happen for some time. Look for fiber to the home in new housing developments, though.

From our point of view as students of wireless technology, convergence means that we can take another look at wireless as a means of carrying services and traffic that were formally the domain of wired systems or broadcast radio and television. LMDS proponents claim that their system can carry video, audio, internet access, and telephony, and do it more economically than any of the other systems. Even third-generation PCS, with a maximum data rate of 2 Mb/s, can be used for reasonably high-speed data and can make a reasonable attempt at video, as well as providing telephony services reliable enough to make a wired telephone unnecessary for most people. These systems have the additional advantage of portability.

14.6 Divergence

In the previous section we looked at convergence. Now let us look again, this time from the point of view that the multiplicity of media offering very similar services can really be seen more as divergent than otherwise.

A few years ago the communication situation was much simpler than now, with fewer choices. For telephone, there was the Bell System with twisted-pair wire. For television there were three choices: over-the-air, cable TV, and satellite systems. None of these were interactive. When internet access came along, it was provided via telephone lines. Later, download was available by satellite at quite a large cost premium, but uploads still needed a telephone line.

Portable telephony within the house used a cordless phone. Outside, AMPS cell phones were the rule for those who could afford them. Airtime was too expensive for anyone to use a cell phone when at home, instead of a wired or cordless phone. Video on demand consisted of renting a VHS tape at the local video rental outlet. The user's communication experience was fragmented but clear: one medium was used for each type of communication.

Now most consumers still have cable television and wireline telephone, but they can use either for internet access and will shortly be able to use the phone line for television and the cable TV line for telephony. Their PCS wireless phone is reliable enough and almost cheap enough to use as their main phone, but then they would not be in the phone book (and if everyone did this, current PCS would be overloaded and would quickly become unreliable). In addition, the consumer can choose direct-broadcast satellite or LMDS for television, perhaps internet access, and, in the case of LMDS, for telephone service as well. This is not even to mention the bewildering choice of local and long-distance telephone companies and internet service providers that can face the consumer once he or she has decided on the medium to use. To some this sounds like convergence, but not to this writer.

Part of the convergence idea is that equipment will also converge. People will watch television on their computer, and send e-mail with their television. Indeed this is already possible. Whether it will become popular is another question, however. If more entertainment becomes interactive, the idea of a combined big-screen television and computer will make sense. Otherwise people are more likely to watch movies in the living room and communicate by e-mail in the study, as now. This implies two separate pieces of equipment, so it matters little how much technology they share.

14.7 Safety and Esthetic Concerns

The proliferation of wireless systems has caused some concerns about safety and esthetics. At this time it does not look like either type of concern will have too much effect on the proliferation of these systems, but we should look at them. Safety, in particular, should be kept in mind by anyone working on wireless systems.

Safety Safety concerns surrounding wireless equipment are of three kinds: distraction of drivers by mobile-phone use, causing accidents; health effects due to radiation; and possible adverse effects due either to radiation or possible sparks, in dangerous environments.

The first problem is easily dealt with. Portable phones are a distraction to drivers, of course, and must be dealt with like any other distraction. Some jurisdictions have made it illegal to use a portable phone when driving; in others, users should exercise caution. Hands-free setups and speed dialing can obviously help.

The second problem is more complex. Certainly wireless equipment is not harmful in the same way that X rays and gamma rays are. It is well known that ionizing radiation, including X rays and gamma rays, can cause cancer by breaking molecular bonds and altering cell structure. Television receivers and computer monitors emit low levels of X rays, and household smoke detectors emit gamma rays. There is no safe level for this radiation, but traditionally the danger has been assumed to be negligible if the user's total radiation dose is not increased significantly above the background level that is present all the time.

Ionization depends on the energy per photon of radiation, which increases with frequency. Radiation from radio equipment is at far too low a frequency to cause ionization, regardless of its power level.

On the other hand, there are other physiological effects due to RF radiation. Localized heating is the best documented. Such heating effects are particularly dangerous to the eyes, which have insufficient circulation to remove heat. Radar technicians have become blind from working near operating radar equipment. However, the power density levels from wireless equipment are much lower than that, and as long as reasonable precautions are taken when working on base-station transmitting antennas, there should be no danger on that score.

Whether there is any danger at all from exposures at levels below those which cause damage from heating is not clear. There are many anecdotal reports of people who use cell phones getting brain cancer for instance, but this kind of juxtaposition hardly constitutes proof. After all, people who do not use cell phones also get cancer.

Studies of the effects of radiation, or of any other environmental factor, are of two basic types. There are epidemiological studies, which try to link rates of various diseases to the exposure of the people to radiation. These use statistical methods in an attempt to determine whether people exposed to the environmental factor have more illness than those whose lives are similar except for this factor. Epidemiological studies are not concerned with the way in which the damage is done.

There have also been laboratory studies made on mice that have been exposed to radiation. To date no serious epidemiological studies have shown any correlation between RF fields at the levels encountered by wireless equipment users and illness. There has been some evidence that quite high levels of radiation (well above current safety standards) may contribute to cancer in mice. Whether the effect on humans would be similar is not clear.

The second major way to study radiation, or other environmental factors, is to look for physiological changes in the presence of radiation. Some studies have looked at the cellular level and claim to have found some effect on cellular biology; again, quite high levels of radiation were used.

The problem with research into radiation safety is that it is impossible, or nearly so, to prove that a particular radiation level is safe at all times and for all people. A consensus has to be developed, subject to change as more data is received. There are many variables, including power level, frequency and duration of exposure, distance of the source from the person, frequency of the radiation, and even the type of modulation in use (for instance, pulsed radiation may have more damaging effects than continuous radiation.)

Most countries have standards that limit exposure to RF radiation. In the United States and Canada, the limit at cellular frequencies is 0.57 mW/cm^2; at PCS frequencies it is 1.2 mW/cm^2. These standards are set at 2% of the level at which it appears there may be some effects.

Certainly studies must be continued to see whether there are any problems with radiation that meets current standards. Obviously the question of danger from RF radiation at low levels is of great concern to the whole wireless industry, but at present the attitude of the industry seems to be to keep quiet and hope the problem goes away. There may be some logic to this: as the industry gradually moves to systems that use smaller cells and lower transmitted power levels, the danger due to RF radiation, is there is any, is likely to diminish.

Sometimes cellular providers have trouble establishing base stations at good locations because of neighbors' concerns about radiation. However, the effects of radiation, if any, will undoubtedly be greater from using a phone in close proximity to one's body than from a tower-mounted antenna hundreds of meters away. Even though base-station transmitters are more powerful than those in portable phones, the field strength on the ground from such transmitters is very much lower than that from a phone held at one's ear, because of the square-law attenuation of free space. Most such concerns are probably more esthetic than safety-oriented.

Interaction of radio transmitters with the nonhuman environment is much easier to understand. For many years highway construction crews have posted notices for motorists to turn off radio transmitters in the neighborhood of blasting operations. This is because of the possibility that the wiring from the detonator to the blasting cap may pick up radiation, creating an electrical current in the wire that could prematurely set off an explosive charge. This is much less likely with cell phones and PCS than with the higher-powered transmitters found in taxicabs, police cars and utility vehicles, but it would still be wise for the cell phone user to comply.

Other situations pose interference dangers. Cell phone use is not allowed on airplanes for two reasons: the signals could possibly affect aircraft navigation or operation, but more likely a cell phone signal at an altitude of several kilometers could easily be received by many cell sites, placing a greater than normal load on the system.

Cell phones can also cause problems with sensitive electronic equipment in hospitals. The best solution to that problem would be better shielding of the affected equipment, but in the meantime it would be prudent for wireless phone users to obey admonitions to turn off their phones in hospitals and similar locations.

On the other hand, there are some alleged dangers that are more urban myth than reality. For instance, some gasoline stations demand that people refrain from using cell phones while their cars are being refilled. This is supposed to be due to the danger of a spark setting off an explosion. The RF radiation from cell phones is certainly not sufficient to cause arcing in external devices and the chances of an internal spark setting off an explosion are less from portable phones than from simpler devices, such as flashlights, since once the phone is turned on, all the switching in cell phones is done electronically. Still, it could do no harm to avoid using *anything* electrical while refilling a vehicle.

In all of these situations, it should be remembered that cell phones, PCS phones and two-way pagers may transmit even while no one is making a call. The only way to make sure they do not is to turn them off completely.

Esthetics Esthetic considerations are not subject to objective measurement and evaluation. The proliferation of wireless devices using many different standards has caused an equivalent proliferation of towers and antennas. This situation will become worse as cell sites become smaller. PCS base stations must be located in residential areas, for instance, if people are to have coverage in their homes. This causes problems for people who are used to seeing antenna towers in industrial and commercial areas but not in their own neighborhood.

Various techniques are available to minimize the impact of antenna installations. Antennas can be installed on, and sometimes even inside, buildings (church steeples are a favorite location, though sometimes the parishioners complain about an imagined danger to their health). Antenna masts have even been disguised as trees. See Figure 14.8 for some disguised antennas.

As cellular and PCS antennas become more ubiquitous and as more people use them, the chances are good that people will stop noticing them, in much the same way they tend to ignore utility poles or street lights (in fact, antennas for microcells are often fastened to street light poles). It certainly appears that wireless communication is here to stay, and people who want to get its benefits will have to put up with some minor nuisance.

FIGURE 14.8
Camouflaged
antennas

(Courtesy of Valmont
Microflect)

14.8 Some Rash Predictions

Let us conclude with a few predictions. First, the future looks pretty chaotic, at least in the near term. As we have seen, the third generation of personal wireless communication will have almost as many conflicting standards as the second. In addition, there will be competition between different methods of delivering almost identical services; this is bound to become confusing.

Perhaps the future will eventually look something like this. For voice communication, wireless makes sense. Bandwidth demands for voice are small and shrinking, and voice communication devices are easy to miniaturize. People want to be able to call and be called wherever they are (provided they have the option to send calls to voice mail instead of answering them when they need privacy). At some point in the fairly near future people will become used to carrying a miniature phone with them and using it wherever they are. This is possible even today and will shortly become ubiquitous.

Data is more problematic. Quite a few people need mobile e-mail, which can be accomplished quite easily even with current PCS, but high-speed connectivity for internet access and streaming video is a different story. For easy browsing or viewing of video, the portable device needs to be larger than pocket size, which means most people will not want to carry it around with them. Also, outside some niche markets, few people have a real need for high-speed data everywhere they go. (They may find it a fun idea, but let's define "need" as being willing to pay for the extra convenience.) Even third-generation wireless data will be slow and expensive compared with wired solutions using cable modems or ADSL on telephone lines. Therefore, for the foreseeable future, it seems likely that most people will continue to rely on wired services (with perhaps a wireless network confined to the home or office) for high-speed data. The possible exception is that fixed wireless services like LMDS may be competitive with wired services in some markets.

It also seems likely that some of the more radical developments will take place outside the major cities on which most of our attention tends to be focused. Such cities already have extensive wired infrastructure, most of which is already paid for. That lowers the cost for wired systems and makes it harder for wireless to compete. The use of cellular and PCS phones instead of wired phones is already common in developing countries that have little installed copper, and this trend seems likely to continue.

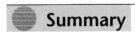 **Summary**

The main points to remember from this chapter are:

- Until recently wireless local loops have been used only where the cost or difficulty of installing wire is prohibitive.
- Modern CDMA equipment makes wireless local loops practical in developing countries, in rural areas, and sometimes even for extra lines where wired service is already provided.
- The third generation of personal wireless systems will feature higher maximum data rates, greater capacity for voice calls, and the ability to work with a wide range of cell sizes and types. It may also be somewhat more standardized than the second generation.
- Both CDMA and TDMA systems appear likely to be part of the third-generation specifications.
- Terrestrial microwave systems at 28 GHz are beginning to be used to deliver television, internet, and telephone services to individual residences. They will probably supplant an older one-way microwave system.
- Many people are expecting that the digitization of practically all communication systems will lead to a gradual convergence of systems, but it is doubtful whether this will actually happen in the near future.

- Safety and esthetic concerns could slow the development of wireless technology.
- The future of wireless seems assured, particularly in the areas of voice and low-speed data. Truly high-speed data may have to wait some time for the fourth generation of wireless.

● Key Terms

local multipoint distribution system (LMDS)   network using microwaves for two-way transmission of telephony, television, and high-speed data

multichannel multipoint distribution system (MMDS)   terrestrial microwave system for the distribution of television, internet, and telephone services to businesses and residences

time-division duplexing (TDD)   transmission of data in two directions on a channel by using different time slots for each direction

● Questions

1. Under what circumstances are wireless local loops more economical than ordinary twisted-pair wiring?

2. What is the difference between simply using a PCS phone at home and using a modern wireless local loop?

3. What advantages do wireless local loops have in areas where phone services are just being introduced?

4. What are the major improvements to be expected from a third-generation PCS standard?

5. Why do third-generation PCS standards specify three different maximum data rates?

6. What are the variations in mobility specified for the third PCS generation? Why does the speed at which a user is moving make a difference?

7. What are the data rates envisioned by the IMT-2000 specifications, and how do they compare with current systems?

8. The second generation of PCS uses three incompatible standards in North America. Does the third generation improve on this situation?

9. Explain time-division duplexing. How does it compare with traditional duplexing methods?

10. How does the cdma2000 system maintain backward compatibility with IS-95? Under what circumstances would this be a useful feature?

11. Explain the difference between MMDS and LMDS.

12. How does an individual user communicate with the headend in an MMDS? Are there any obvious problems with this method?

13. Why do maximum propagation distances at 28 GHz tend to be less than at lower frequencies?

14. Why is a direct line of sight not always required for LMDS?

15. Why can LMDS, but not MMDS, be used for local telephony?

16. Compare the radiation from cell phones with that from X-ray machines and smoke detectors, in terms of known hazards.

17. Why is there no "safe" level for gamma radiation?

18. What is the major known danger from high-power microwave sources like radar transmitters and microwave ovens?

19. Compare the reasons for restricting the operation of wireless equipment in aircraft, hospitals, gas stations, and movie theatres.

● Problems

1. Compare the use of wireless and wired technologies for each of the following services:

   (a) Local telephone service

   (b) Internet access

   (c) Broadcast television

   (d) Interactive video

2. List all of the ways by which the following can be accomplished in your area, and discuss each method, including advantages, disadvantages, and price.

   (a) Local telephone service

   (b) Reception of broadcast television stations

   (c) Reception of television signals that are not broadcast over the air (for instance, news and sports stations, movie stations)

   (d) Music service (radio broadcasting or equivalent)

   (e) Internet access (Be sure to get approximate data rates, and find out, in the case of non-telephone services, whether a telephone modem is still required for uploads.)

(f) E-mail (Remember that this is not necessarily the same as full internet access.)

(g) Paging services

3. Suppose a wireless system operates at 200 kb/s using BPSK. What data rate would be possible in the same bandwidth using

(a) QPSK?

(b) 8-DPSK?

4. Find out what internet service providers are available in your area, and their approximate cost, for

(a) Telephone modems

(b) Cable modems, if available

(c) ADSL on telephone lines, if available

(d) LMDS, if available

(e) Satellite service (Don't forget to include the cost of the uplink using a telephone line and internet service provider.)

5. Suppose that a line-of-sight radio path has a path loss of 50 dB at 1 GHz.

(a) What will be the free-space path loss, in dB at 28 GHz?

(b) What accounts for the increase in loss?

(c) Are there any factors that are likely to make the path loss at 28 GHz higher than that calculated in part (b)? If so, what are they?

6. What zoning restrictions, if any, exist in your area to restrict either:

(a) towers for base stations?

(b) antennas attached to houses, for direct-broadcast satellites or LMDS?

7. Compare the data rate achievable with 3G PCS with that available from telephone modems, cable modems, and ADSL using telephone lines.

8. Make some rash predictions of your own concerning the future of wireless (and wired) communication.

Appendix A   Impedance Matching

In Chapter 6 we noted the deleterious effects of impedance mismatches on transmission lines. Mismatches result in power being reflected back to the source and in higher-than-normal voltages and currents that can stress the line and connected equipment. In general, best results are obtained when the load is matched to the characteristic impedance of the line. When this is not the case, it is often useful to connect some sort of matching network to correct the mismatch. Similarly, when either the load or the line is unbalanced, and the other is balanced, it is highly desirable to install a device called a balun network to convert one to the other. Matching networks can be constructed using lumped constants (inductors, capacitors, and transformers), transmission line sections, or waveguide components. Matching networks are sometimes broadband, so as to pass a wide range of frequencies, but often they are made narrow in bandwidth to allow them to act as bandpass filters as well.

We noted earlier that the impedance looking into a transmission line varies with its length. For a lossless line the impedance repeats every one-half wavelength. If it is acceptable to have a mismatch on a short section of line near the load, it is often possible to simplify a matching problem by backing off a short distance (less than one-half wavelength) from the load and installing the matching device at a point on the line where the impedance is easier to work with. It should be obvious that this is a narrowband technique, since the electrical length (in degrees) of a particular physical length of line (in meters) varies with frequency.

The Smith Chart

For many years (since 1944 in fact), a special circular graph called a Smith® chart has been used to indicate complex impedances and admittances and the way in which they vary along a line. It is more common nowadays to do transmission line calculations with the aid of a computer than to use pencil and compass, but many of the computer programs use the Smith chart to display their output. Test instruments, called vector network analyzers, that measure complex functions, also tend to use this display format. Therefore it is important to understand the layout of the Smith chart and to be able to interpret it—the actual calculations can be safely left to a computer.

Figure A.1 shows a conventional paper Smith chart. A distance of one-half wavelength on the line corresponds to one revolution around the chart. Clockwise rotation represents movement toward the generator, and counterclockwise rotation represents progress toward the load, as shown on the

IMPEDANCE OR ADMITTANCE COORDINATES

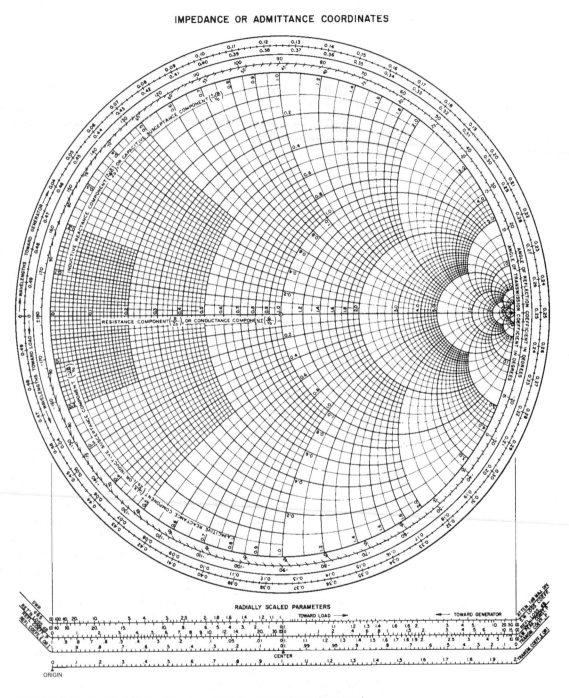

FIGURE A.1 Smith Chart (Courtesy of Analog Instruments Company)

chart itself. For convenience there are two scales (in decimal fractions of a wavelength) around the outside of the chart, one in each direction. Each scale runs from zero to 0.5 wavelength.

The body of the chart is made up of families of orthogonal circles; that is, they intersect at right angles. The impedance or admittance at any point on the line can be plotted by finding the intersection of the real component (resistance or conductance), which is indicated along the horizontal axis, with the imaginary component (reactance or susceptance), shown above the axis for positive values and below for negative.

Because of the wide variation of transmission line impedances, most paper Smith charts use normalized impedance and admittance to reduce the range of values that have to be shown. The value of 1 in the center of the chart represents an impedance equal to Z_0. Paper charts are also available with an impedance of 50 Ω in the center, and computerized displays can have any value for Z_0, with no normalization required.

To normalize an impedance, simply divide it by the characteristic impedance of the line.

$$z = \frac{Z}{Z_0} \qquad\qquad\qquad (A.1)$$

where

z = normalized impedance at a point on the line
Z = actual impedance at the same point
Z_0 = characteristic impedance of the line

Since z is actually the ratio of two impedances, it is dimensionless.

EXAMPLE A.1

Normalize and plot a load impedance of $100 + j25$ Ω on a 50-Ω line.

SOLUTION

From Equation (A.1),

$$z = \frac{Z}{Z_0}$$

$$= \frac{100 + j25\,\Omega}{50\,\Omega}$$

$$= 2 + j0.5$$

Figure A.2 shows how this impedance would be plotted on a normalized chart.

IMPEDANCE OR ADMITTANCE COORDINATES

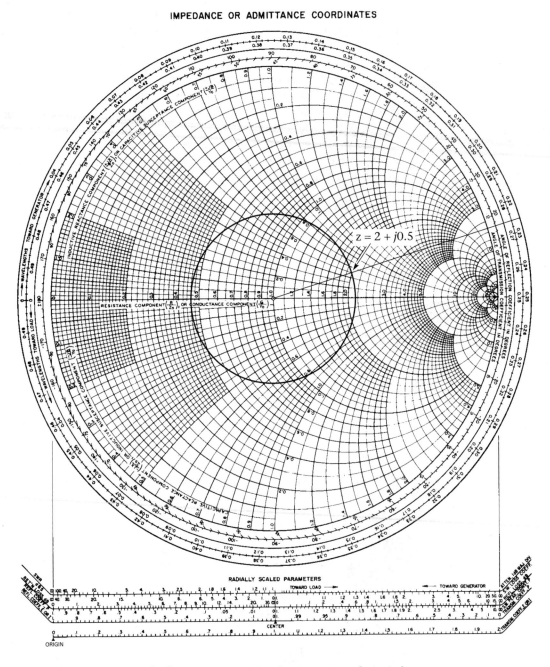

FIGURE A.2 Use of Smith Chart (Courtesy of Analog Instruments Company)

Once the normalized impedance at one point on the line has been plotted, the impedance at any other point can be found very easily. Draw a circle with its center in the center of the chart, which is at the point on the horizontal axis where the resistive component is equal to one. Set the radius so that the circle passes through the point just plotted. Then draw a radius through that point, right out to the outside of the chart, as shown in Figure A.2. Move around the outside in the appropriate direction, using the wavelength scale as a guide. Just follow the arrows. If the first point plotted is the load impedance, then move in the direction of the generator. Once the new location on the line has been found, draw another radius. The normalized impedance at the new position is the intersection of the radius with the circle. In other words, the circle is the locus of the impedance of the line. Every point on the circle represents the impedance at some point on the line.

The radius of the circle represents the *SWR* on the line. In fact, it is usually referred to as the *SWR circle*. The SWR can easily be found by reading the normalized resistance value where the circle crosses the horizontal axis to the right of the center of the chart. Another way is to mark the radius of the circle on the "standing wave voltage ratio" scale at the bottom of the chart. If required, the SWR can also be found in decibels by using the adjoining scale. For this example, the SWR is read as 2.16.

Figure A.3 on page 548 is a computer printout from a program called WinSmith®, which is one of many that perform Smith chart calculations. The output is provided on a Smith chart, as well as in tabular form.

We noted before that the SWR depends on the magnitude of the reflection coefficient. There is another scale on the chart that allows this magnitude to be found directly. See the *reflection coefficient voltage* scale at the lower right of the chart. Marking off the radius of the SWR circle on this scale will give the magnitude of Γ, which can be read as 0.37, on both the paper and computerized charts. (On the computer chart, Γ is represented by G and *VSWR* by V.)

It is possible to use the Smith chart to find the impedance at any point along a line. The next example illustrates this.

EXAMPLE A.2

A 50-Ω line operating at 100 MHz has a velocity factor of 0.7. It is 6 m long and is terminated with a load impedance of $50 + j50\ \Omega$. Find the input impedance for the line.

SOLUTION

Now that we have seen how much the computer display resembles the paper chart, let us use the computer. See Figure A.4.

FIGURE A.3
Printout from
*win*Smith software
(Courtesy of Noble
Publishing)

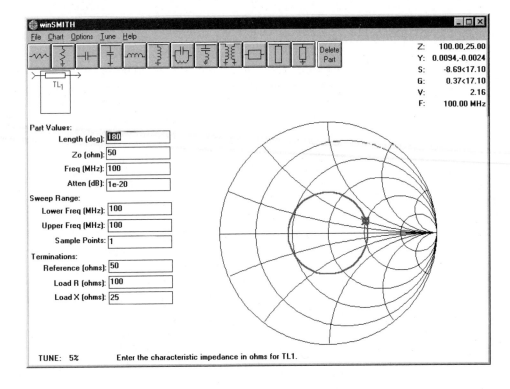

We need to know the length of the line in degrees. First find the wavelength using Equation (6.1).

$$\lambda = \frac{v_p}{f}$$

$$= \frac{v_f c}{f}$$

$$= \frac{0.7 \times 300 \times 10^6}{100 \times 10^6}$$

$$= 2.1 \text{ m}$$

Now we can find the length in degrees with Equation (6.7).

$$\phi = \frac{360L}{\lambda}$$

$$= \frac{360 \times 6}{2.1}$$

$$= 1029°$$

From the plot in Figure A.4, we can see that the input impedance (represented by the small circle) is $19.36 + j5.44$ Ω.

FIGURE A.4
Printout from
*win*Smith software
(Courtesy of Noble
Publishing)

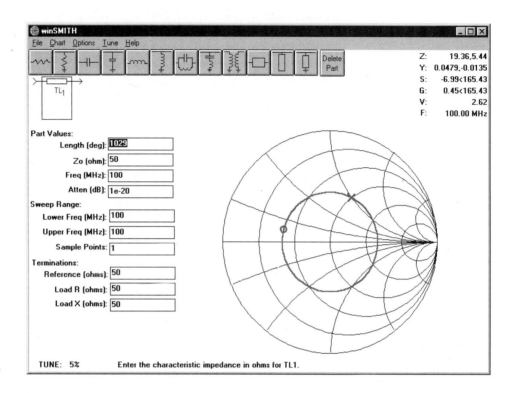

Now that we understand how impedances and lines are plotted on the Smith chart, it should be possible to use the chart as an aid for impedance matching. Since the center of the chart always represents the characteristic impedance of the system, matching a line involves moving its input impedance into the center of the chart. The progress of the solution to a matching problem can be monitored by observing the input impedance: the closer it is to the center of the chart, the better the match.

Probably the best way to see how this works is to try a few examples using various matching techniques. Each technique will be described, then followed with an example. Smith charts will be used to illustrate the technique.

Matching Using a Transformer

A transformer can be used to match impedances, provided the load impedance is real at the point where the transformer is inserted. RF transformers are usually toroidal, using ferrite or powdered iron cores as illustrated in Figure A.5, although air-core transformers are also common. Recall from basic electrical theory that the impedance ratio is the square of the turns ratio, that is:

$$\frac{Z_1}{Z_2} = \left(\frac{N_1}{N_2}\right)^2 \tag{A.2}$$

FIGURE A.5
Toroidal
transformer

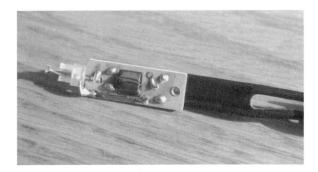

Transformers are also useful for connecting balanced lines to unbalanced lines, since there is no electrical connection between windings. A transformer used in this way is called a *balun transformer* or just a *balun*. A good example of a balun transformer can be found on the back of many television sets, where it adapts 75-Ω coaxial (unbalanced) cable-television cable to an antenna input designed for 300-Ω twin-lead (balanced). Figure A.6 is a photo of a typical TV balun. Since the required impedance ratio is 4:1, the turns ratio must be 2:1.

FIGURE A.6
TV balun

A transformer can be used to match loads with complex impedances by installing it a little distance from the load in the direction of the source. The next example illustrates this.

EXAMPLE A.3 ──▼

Find the correct turns ratio and location for a transformer that is required to match a 50-Ω line to a load impedance of 75 + j25 Ω.

SOLUTION

We solve the problem by moving along the line in the direction of the generator until the impedance looking into the line is resistive. The arc representing the line crosses the horizontal axis of the chart at a point 16.8° from the load. See Figure A.7(a).

FIGURE A.7(a)
Printout from *win*Smith software
(Courtesy of Noble Publishing)

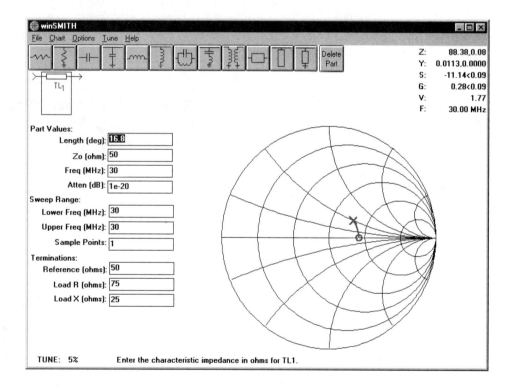

Now the requirement is simply to match the 50-Ω line to the impedance at this point on the line, which is 88.38 Ω, resistive. The required turns ratio can be found from Equation (A.2):

$$\frac{Z_1}{Z_2} = \left(\frac{N_1}{N_2}\right)^2$$

$$\frac{N_1}{N_2} = \sqrt{\frac{Z_1}{Z_2}}$$

$$= \sqrt{\frac{50}{88.38}}$$

$$= 0.752$$

Figure A.7(b) shows what happens when we add a transformer with this ratio at the point found above. The line is matched. Note, however, that a change in frequency will change the electrical length of the line and destroy the match. This can be seen by changing the frequency from 30 MHz to 40 MHz while leaving everything else alone. See the result in Figure A.7(c) on page 553. There is no longer a perfect match; however, a closer look shows that the SWR is still a very reasonable 1.12, so the setup is still usable.

FIGURE A.7(b)
Printout from
*win*Smith software
(Courtesy of Noble
Publishing)

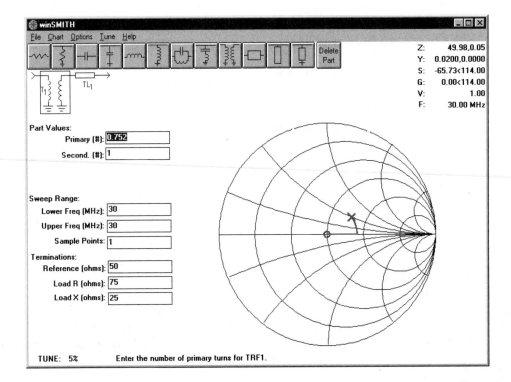

FIGURE A.7(c)
Printout from
*win*Smith software

(Courtesy of Noble
Publishing)

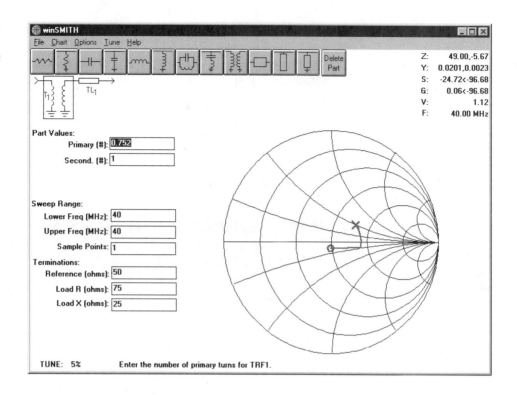

A one-quarter wavelength of transmission line can also be used as a transformer. A quarter wavelength along a line represents one-half a rotation around the chart, so a one-quarter wavelength of line of the right impedance can transform one real impedance into another. The characteristic impedance Z_0' of the line for the transformer can be found from the following equation:

$$Z_0' = \sqrt{Z_0 Z_L} \tag{A.3}$$

EXAMPLE A.4

Solve the problem of the previous example using a quarter-wave transformer.

SOLUTION

Since a quarter-wave transformer, like a conventional transformer, can only match real impedances, it is necessary to place the transformer at the same distance from the load as in the previous example. From the previous problem we see that the load impedance to be matched at this point is 88.38 Ω. This can be matched with a quarter-wave section of line having a characteristic impedance, from Equation (A.3), of

$$Z_0' = \sqrt{Z_0 Z_L}$$

$$= \sqrt{50 \times 88.38}$$

$$= 66.48 \ \Omega$$

Figure A.8 shows that this does indeed result in a match.

FIGURE A.8
Printout from
*win*Smith software
(Courtesy of Noble
Publishing)

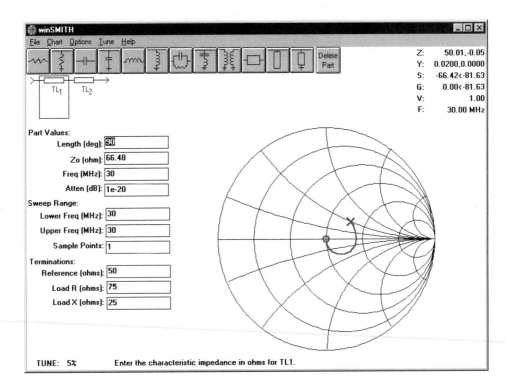

Series Capacitance and Inductance

Where the resistive part of the load impedance is correct, the reactive part can be canceled by adding a series reactance of the opposite type. If the resistive part is not correct, the reactance can be installed at the correct distance from the load to bring the resistive part of the impedance to the correct value.

EXAMPLE A.5

Use a series capacitor or inductor to match a 50-Ω line to each of the following loads at a frequency of 100 MHz:

(a) $50 + j75 \; \Omega$

(b) $150 + j75 \; \Omega$

SOLUTION

(a) Since the real part of the impedance is correct, we need only add a capacitive impedance of $-j75 \; \Omega$, that is, a capacitor with 75-Ω reactance. Recall from electrical fundamentals that

$$X_c = \frac{1}{2\pi f C}$$

where

$$X_c = \text{capacitive reactance in ohms}$$
$$f = \text{frequency in hertz}$$
$$C = \text{capacitance in farads}$$

In this case, we know f and X_c, so we rearrange the equation:

$$C = \frac{1}{2\pi f X_c}$$

$$= \frac{1}{2\pi \times 100 \times 10^6 \times 75}$$

$$= 21.2 \text{ pF}$$

It was not necessary to use the Smith chart, whether on paper or on a computer, to solve this problem, but the computerized version of the chart verifies the match. See Figure A.9 on page 556, and notice that the arc on the chart follows one of the circles of constant resistance to the center of the chart.

FIGURE A.9
Printout from
*win*Smith software
(Courtesy of Noble
Publishing)

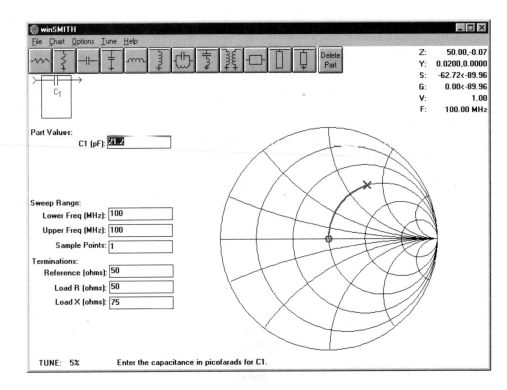

(b) Since the real part of the load impedance is not equal to 50 Ω, it is necessary to add enough line to reach a point on the circle representing a resistance of 50 Ω, that is, the circle that passes through the center of the chart. Figure A.10(a) on page 557 shows that this takes a length of 35 electrical degrees. At this point the resistive component of the impedance is correct but there is a capacitive reactance of 72.6 Ω. Figure A.10(b) shows that the system can be matched by using a series inductance of 115 nH.

FIGURE A.10(a)
Printout from
*win*Smith software
(Courtesy of Noble
Publishing)

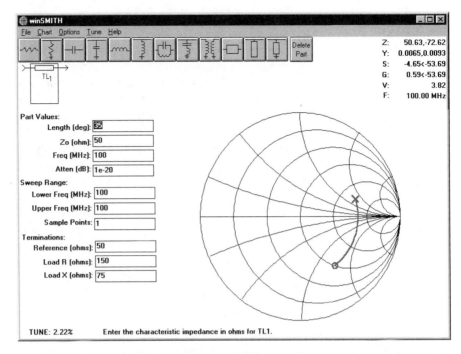

FIGURE A.10(b)
Printout from
*win*Smith software
(Courtesy of Noble
Publishing)

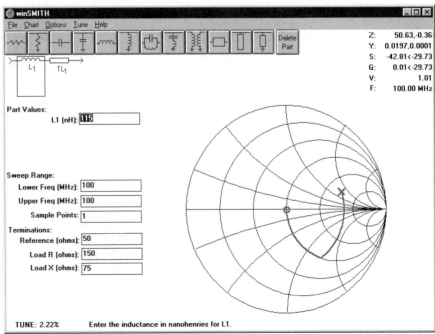

Stub Matching

As noted earlier, shorted transmission line stubs are often used instead of capacitors or inductors at VHF and above. Usually these are placed in parallel with the main line, rather than in series. In this case it is easier to handle the problem from an admittance rather than an impedance point of view, since parallel admittances add. Shorted transmission lines can compensate only for the imaginary component of the load admittance, so if the load admittance is complex it is once again necessary to install the matching component at some distance from the load. In this case we back off until the real component of the admittance is equal to the characteristic admittance of the line.

Smith charts are available with admittance as well as impedance coordinates. It is also possible to use a conventional chart and convert from impedance to admittance. Computer programs invariably offer a choice of coordinates. The WinSmith program introduced earlier allows for either or both to be displayed. An example will demonstrate transmission line matching with a single shorted stub.

EXAMPLE A.6

Match a line with a characteristic impedance of 72 Ω to a load impedance of 120 − j100 Ω using a single shorted stub.

SOLUTION

The stub must be inserted at a point on the line where the real part of the load admittance is correct. This value is

$$\frac{1}{72\,\Omega} = 0.0139\text{ S}$$

Actually, we can simply find the conductance circle that passes through the center of the chart. Figure A.11(a) on page 559 shows a chart with both impedance and admittance circles. This is a little confusing, so in Figure A.11(b) the impedance circles have been removed. It can be seen from either figure that a distance of 42° is required between the load and the stub.

Next we add a shorted stub and adjust its length to bring the input impedance to the center of the chart. It can be seen from Figure A.11(c) on page 560 that this requires a stub length of 40°.

FIGURE A.11(a)
Printout from
*win*Smith software
(Courtesy of Noble
Publishing)

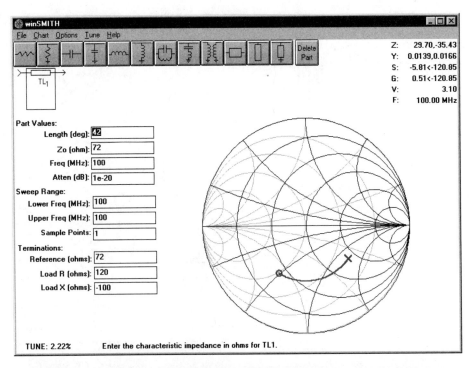

FIGURE A.11(b)
Printout from
*win*Smith software
(Courtesy of Noble
Publishing)

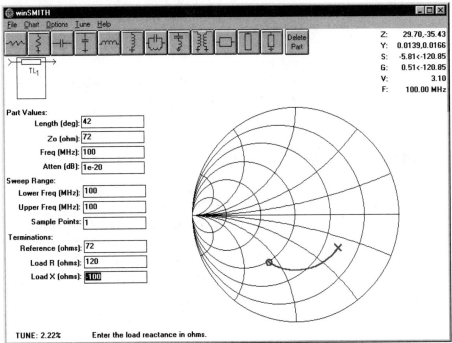

FIGURE A.11(c)
Printout from
*win*Smith software
(Courtesy of Noble
Publishing)

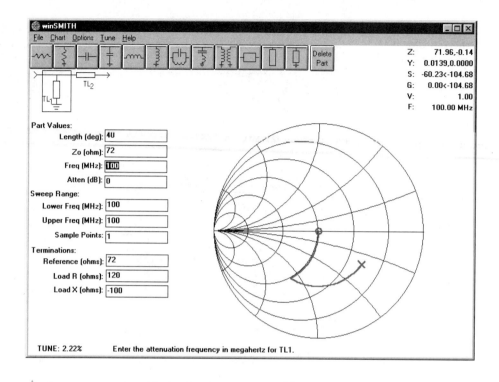

The foregoing is only a sample of the ways in which lines can be matched to loads.

The problems that follow can be solved using a paper Smith chart or a computer program, if available.

● Problems

1. A 75-Ω transmission line is terminated with a load having an impedance of 45 − j30 Ω. Find:

    (a) the distance (in wavelengths) from the load to the closest place at which a quarter-wave transformer could be used to match the line

    (b) the characteristic impedance that should be used for the quarter-wave transformer

2. A 50-Ω coaxial transmission line with a velocity factor of 0.8 is connected to a load of 89 – j47 Ω. The system operates at 300 MHz. Find the proper position and component value to match this line using:

   (a) a series capacitor

   (b) a series inductor

   (c) a transformer

3. Find the SWR for each setup of problem 2, at a frequency of 350 MHz.

4. A transmitter supplies 100 W to a 50-Ω lossless line that is 5.65 wavelengths long. The other end of the line is connected to an antenna with a characteristic impedance of 150 + j25 Ω.

   (a) Find the SWR and the magnitude of the reflection coefficient.

   (b) How much of the transmitter power reaches the antenna?

   (c) Find the best place at which to insert a shorted matching stub on the line. Give the answer in wavelengths from the load.

   (d) Find the proper length for the stub (in wavelengths).

Appendix B   Frequencies for Common Wireless Systems

CITIZENS' BAND RADIO: 27-MHz Band

Channel	Frequency (MHz). (Same channel used for both directions.)	Channel	Frequency (MHz). (Same channel used for both directions.)
1	26.965	21	27.215
2	26.975	22	27.225
3	26.985	23	27.255
4	27.005	24	27.235
5	27.015	25	27.245
6	27.025	26	27.265
7	27.035	27	27.275
8	27.055	28	27.285
9	27.065	29	27.295
10	27.075	30	27.305
11	27.085	31	27.315
12	27.105	32	27.325
13	27.115	33	27.335
14	27.125	34	27.345
15	27.135	35	27.355
16	27.155	36	27.365
17	27.165	37	27.375
18	27.175	38	27.385
19	27.185	39	27.395
20	27.205	40	27.405

CORDLESS TELEPHONE FREQUENCIES: 43–49 MHz Band

Channel	Base	Handset
1	43.720	48.760
2	43.740	48.840
3	43.820	48.860
4	43.840	48.920
5	43.920	49.000
6	43.960	49.080
7	44.120	49.100
8	44.160	49.160
9	44.180	49.200
10	44.200	49.240
11	44.320	49.280
12	44.360	49.360
13	44.400	49.400
14	44.460	49.480
15	44.480	49.500
16	46.610	49.670
17 (B)	46.630	49.845
18 (C)	46.670	49.860
19	46.710	49.770
20 (D)	46.730	49.875
21 (A)	46.770	49.830
22 (F)	46.830	49.890
23	46.870	49.930
24	46.930	49.990
25	46.970	49.970

The letters A through F denote channels that are also used for baby monitors.

NORTH AMERICAN CELLULAR RADIO FREQUENCIES

Base Frequencies (forward channels)	Mobile Frequencies (reverse channels)	Type of Channel	Carrier
869.040–879.360	824.040–834.360	Voice	A
879.390–879.990	834.390–834.990	Control	A
880.020–880.620	835.020–835.620	Control	B
880.650–889.980	835.650–844.980	Voice	B
890.010–891.480	845.010–846.480	Voice	A*
891.510–893.970	846.510–848.970	Voice	B*

Table denotes transmit carrier frequencies. Mobile transmits 45 MHz below base.

A = non-wireline carrier (RCC)   B = wireline carrier (Telco)
* = frequencies added in 1986

BROADBAND PCS BAND PLAN

Allocation	Base transmit (Forward Channel or Downlink)	Mobile transmit (Reverse Channel or Uplink)
A	1850–1865	1930–1945
B	1870–1885	1950–1965
C	1895–1910	1975–1990
D	1865–1870	1945–1950
E	1885–1890	1965–1970
F	1890–1895	1970–1975

Frequencies are in MHz.

Answers to Odd-Numbered Problems

Chapter 1

1. 62.8×10^3 rad/s

3. 16 dB

5. See Figure 1.

FIGURE 1

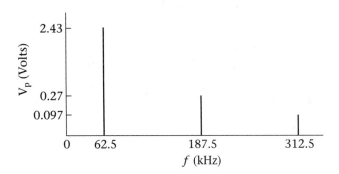

7. See Figure 2.

FIGURE 2

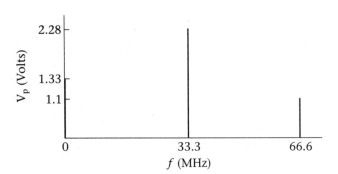

9. 750 THz, 429 THz

Chapter 2

1. (a) 7.4 MHz
   (b) 7.0 kHz
   (c) 0.267
   (d) 15 V

(e) See Figure 3.

FIGURE 3

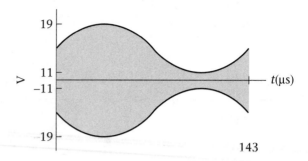

143

3. (a) 8 V

(b) 50%

(c) See Figure 4.

FIGURE 4

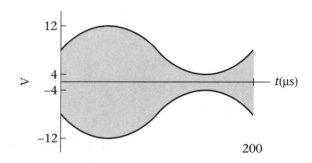

200

(d) $v(t) = (8 + 4 \sin(31.4 \times 10^3 t))\sin(62.8 \times 10^6 t)$ V

5. (a) 0.1

(b) 10 V

(c) 1 V

7. 30 kHz

9. See Figure 5.

FIGURE 5

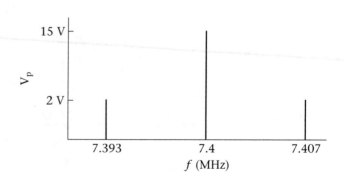

11. (a) 15.8 mW

(b) 0.796

13. See Figure 6.   **FIGURE 6**

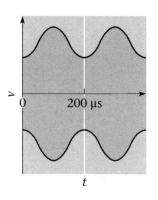

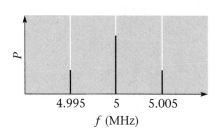

(a)

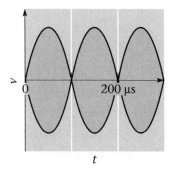

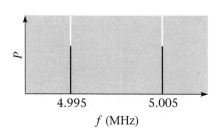

(b)

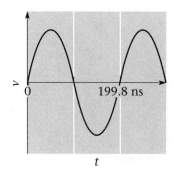

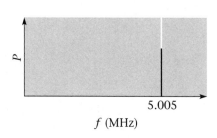

(c)

15. (a) 16 W
    (b) 8 dB

17. 10 kHz

19. (a) No effect

    (b) Doubling V_m doubles the modulation index.

21. (a) 2.5 kHz

    (b) 484 mW

    (c) 96.6%

23. 154 W

25. (a) 316 μV

    (b) 0.0316 rad

    (c) 50 dB

27. For full-carrier AM, sideband power is in addition to carrier power, which remains constant. For SSBSC AM, the carrier power is zero.

Chapter 3

1. (a) 7 kHz

   (b) Practical filters need some distance between passband and stopband.

3. See Figure 3.4. Make the time intervals 250 μs and the peak voltage 1 V in part (a).

5. 75.2%

7. 11101111

9. 11110011

Chapter 4

1. (a) 60 kb/s

   (b) 120 kb/s

3. 0.083 b/s/Hz; No

5. (a) The number of states (angles, in this case) is $2^3 = 8$

   (b) See Figure 7.

   (c) Advantage: higher data rate for a given bandwidth. Disadvantage: less robust in the presence of noise or interference.

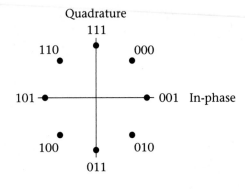

FIGURE 7

7. (a) 8
   (b) 640 Mb/s
9. 30
11. (a) 12 Mb/s
    (b) 187
    (c) 8
13. 500
15. (a) 1 Mb/s
    (b) 667 kHz
    (c) 8 dB

17. (a) 320 kHz
    (b) 40 kHz
    (c) 60 kHz
    (d) 30 kHz
    (e) 60 kHz
    (f) 300 kHz

Chapter 5

1. (a) 50 mA
   (b) 760 Ω
3. 852 Hz, 1336 Hz
5. (a) 16.128 Mb/s
   (b) 5.88%

7. 33.3%, 0.52% (assuming no bit robbing)
9. (a) upstream 10, downstream 23.4
   (b) upstream 5, downstream 11.7

Chapter 6

1. (a) 1.875 m
   (b) 37.5 cm
   (c) 15 cm

3. 429 THz to 750 THz
5. 47.6 Ω
7. 240 m

9. See Figure 8.

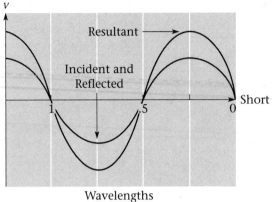

(a)

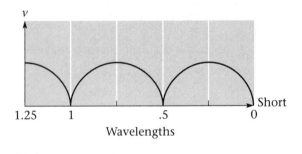

(b)   (c)

FIGURE 8

11. (a) 1.5

    (b) −0.2

    (c) 400 μW

13. Series resonant

15. 0.594 μV

17. 6.55 GHz, 6.55–13.1 GHz

19. 6.6 ns

21. (a) See Figure 9(a).
    (b) See Figure 9(b).
    (c) The circulator does not require a load.

FIGURE 9

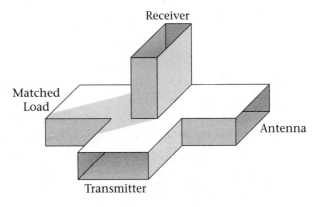

(a)

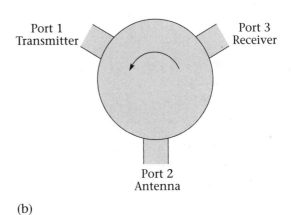

(b)

Chapter 7

1. 107×10^6 m/s

3. 35.4 nW/m², 3.65 mV/m

5. 135 Ω

7. (a) 49 dBW
   (b) 178 pW

9. (a) 12.5 km
   (b) 25.6 km
   (c) 51.2 km

11. (a) 150 dB
    (b) 160 dB

13. 3.6 min.

15. 167 μs, 3.33 ms, 240 ms

17. 581 fW/m²

Chapter 8

1. 950 mm

3. 2.71 mV/m

9. (a) 750 MHz
   (b) 6.76 dBi
   (c) 33°

11. $\theta_H = 24°$ $\theta_E = 19.3°$

13. (a) parasitic, end-fire, to the right
    (b) phased, end-fire, to the left
    (c) phased, broadside, all directions except left and right
    (d) phased, broadside, into and out of the page
    (e) phased, end-fire, to the left

15. (a) 3 dB increase in gain
    (b) cancellation

17. (a) 39 dBi   (b) 39 dBi

19. (a) 262 μW   (b) 8.28 μW

21. 7.68 dBi

5. 15.7 dBW

7. 71.25 cm

Chapter 9

1. (a) 2000
   (b) 2000 to 2400

3. See Figure 10.

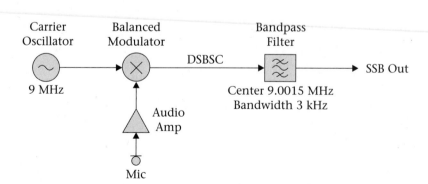

FIGURE 10

5. (a) 220 MHz, 20 MHz

    (b) 20 MHz, 100 MHz, 120 MHz, 200 MHz, 220 MHz, 240 MHz

7. 2.5 μW

9. (a) 1.65 MHz

    (b) High-side

11. 108 dB

13. (a) 17.5 MHz

    (b) 14 MHz

15. 15 kHz

17. (a) High-side

    (b) 750 kHz

    (c) $f_{LO1} = 25.5$ MHz, $f_{LO2} = 9.25$ MHz

19. (a) See Figure 11.

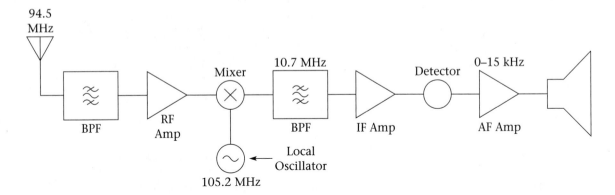

FIGURE 11

    (b) 115.9 MHz

    (c) Improve bandpass filter before mixer; raise IF

21. (a) 156.7 MHz, 139.8 MHz

    (b) Direct-FM using PLL

    (c) 16 kHz (from Carson's rule)

Chapter 10

1. (a) 427 nW

    (b) 380 µW; effective area of receiving antenna is smaller

3. 2 W; suitable for mobile but not portable cell phone

5. (a) 70.7 m

    (b) (i) Less interference

         (ii) No change

    (c) (i) Improvement

         (ii) No change

7. 112.5 m

9. 40 000

11. (a) 34 cells with a radius of 920 m

    (b) better for digital phones, much worse for analog

    (c) leave more channels analog

13.

User Action	Channel	Data Sent
Key in no.; Press send	reverse control	MIN, ESN, number called
	forward control	no. of voice channel, CNAC, DCC
	forward voice	confirmation, SAT
	reverse voice	confirmation, SAT
Talk and listen	forward and reverse voice	conversation, SAT

15. (a) 12.5 km

    (b) 9.3 km

    (c) Shadowing by hills or buildings, multipath propagation, attenuation due to operating portable inside a car or building, sensitivity of cell phone receiver, etc.

17. 192 voice channels

19.

System		P (mW)	P (dBm)
Class IV digital cell phone	(Minimum)	0.4	−4
49-MHz cordless phone		0.03	−15.2
900-MHz cordless phone		16	+12

Chapter 11

1. (a) 82.8 mm, 37.3 mm

   (b) same

   (c) 800 MHz, by 6.7 dB

3. (a) 2.53 kb/s

   (b) 506 b/s

5. (a) TDMA 1.62 b/s/Hz; GSM 1.35 b/s/Hz; CDMA 0.98 b/s/Hz

   (b) TDMA 0.81 symbol/s/Hz; GSM 1.35 symbol/s/Hz; CDMA 0.49 symbol/s/Hz

7. (a) 52.1 µs

   (b) 7.3 years

9. 6.67 µs, 3.33 µs, 0

Chapter 12

1. 7.35 km/s; 1.76 hours

3. 7.91 km/s; air resistance

5. Increase antenna gain to 55 dBi or reduce noise temperature to 6.6 K.

7. 290 K

9. LEO by 38 dB

11. (a) 154 dB

    (b) 5.2 ms

13. (a) 7464 m/s, 1.67 hours

    (b) 7155 m/s, 1.91 hours

    (c) 7173 m/s, 1.89 hours

    (d) 7461 m/s, 1.68 hours

Chapter 13

1. 0.02 b/s/Hz, 0.048 b/s/Hz, 0.096 b/s/Hz

3. approximately 5.3 times

5. 0.38 b/s/Hz

7. 19 dB

9.

Format	Data Rate	Network Type
Mobitex	8 kb/s	Packet-switched
ARDIS	4.8 kb/s standard; 19.2 kb/s in some areas	Packet-switched
CDPD	19.2 kb/s	Packet-switched
Analog Cellular with Modem	9.6 kb/s approx., depending on connection	Circuit-switched
TDMA PCS	9.6 kb/s	Circuit-switched
GSM PCS	9.6 kb/s	Circuit-switched
CDMA PCS	14.4 kb/s	Circuit-switched

Chapter 14

1. (a) Wireless is more expensive in built-up areas with high telephone density, often less expensive in rural areas or areas with few customers. Wireless has the advantage of portability and the promise of a single portable phone number. Wired technology has arguably better security, though new digital wireless technologies are better than older ones.

   (b) Wired technologies are currently faster and cheaper, especially cable modems and ADSL. Wireless service offers portability. Fixed wireless using LMDS/LMCS is competitive with cable modems and ADSL; fixed wireless is not portable.

   (c) Wired service (CATV) is economical in high-density areas; in rural areas LMCS/LMDS and satellite service are competitive. Conventional terrestrial broadcasting is cost-effective but offers a limited number of channels.

   (d) This can be supported on CATV, LMDS/LMCS, and in the near future, on telephone lines. The three systems appear to be competitive, with CATV having the advantage of a large installed base. Low-quality interactive video is possible on the internet, though at present wireless internet speeds are barely adequate even for this.

3. (a) 400 kb/s

   (b) 600 kb/s

5. (a) 78.9 dB

   (b) decrease in effective area for receiving antenna of equivalent gain

   (c) attenuation by trees, rain, etc.

7. All are variable. Let us assume telephone modems operate at 64 kb/s and both ADSL and cable modems at about 1 Mb/s; for 3G wireless, maximum data rates will be 144 kb/s for mobile users in high-speed vehicles, about 384 kb/s for pedestrians and perhaps for slow-moving vehicles, and up to 2 Mb/s for stationary users. All types of 3G are faster than telephone modems, while only stationary 3G PCS users will have data rates comparable with and perhaps exceeding cable modems and ADSL.

Appendix A

1. (a) 49°

   (b) 44.6 Ω

3. (a) 2.07

   (b) 1.27

   (c) 1.52

Advanced Mobile Phone Service (AMPS) North American first-generation cellular radio standard using analog FM

air interface in wireless communication, the radio equipment and the propagation path

aliasing distortion created by using too low a sampling rate when coding an analog signal for digital transmission

amplitude-shift keying (ASK) data transmission by varying the amplitude of the transmitted signal

angle modulation term that applies to both frequency modulation (FM) and phase modulation (PM) of a transmitted signal

angle of elevation angle measured upward from the horizon. Used to describe antenna patterns and directions

apogee point in a satellite orbit that is farthest from the earth

array combination of several antenna elements

asymmetrical digital subscriber line (ADSL) method of providing high-speed data transmission on twisted-pair telephone loops by using high-frequency carriers

audio frequency-shift keying (AFSK) use of an audio tone of two or more different frequencies to modulate a conventional analog transmitter for data transmission

B (bearer) channels in ISDN, channels that carry subscriber communication

Bandwidth portion of frequency spectrum occupied by a signal

base station controller (BSC) in cellular and PCS systems, the electronics that controls base station transmitters and receivers

baseband information signal

baud rate speed at which symbols are transmitted in a digital communication system

beamwidth angle between points in an antenna pattern at which radiation is 3 dB down from its maximum

bent-pipe configuration a design of satellite transponder in which signals are amplified and shifted in frequency but do not undergo any other processing

bit rate speed at which data is transmitted in a digital communication system

bit robbing use of bits that normally carry payload information for other purposes, such as controlling the communication system

bit stuffing addition of bits to a bitstream to compensate for timing variations

blank-and-burst signaling in cellular communication, interrupting the voice channel to send control information

bridge a device to connect two segments of a network

call blocking failure to connect a telephone call because of lack of system capacity

capcode unique address for a pager

capture effect tendency of an FM receiver to receive the strongest signal and reject others

carrier high-frequency signal which is modulated by the baseband signal in a communication system

carrier-sense multiple-access with collision avoidance (CSMA/CA) same as CSMA/CD except used in situations where a station cannot receive and transmit at the same time. With CSMA/CA, a collision cannot be detected once a transmission has started.

carrier-sense multiple-access with collision detection (CSMA/CD) method of reducing contention in a network, involving each station checking continuously for interference before and during transmissions

cellular digital packet data (CDPD) method of transmitting data on AMPS cellular telephone voice channels that are temporarily unused

central office switch in a telephone system that connects to local subscriber lines

characteristic impedance ratio between voltage and current on a transmission line, on a waveguide, or in a medium

chip extra bit used to spread the signal in a direct-sequence spread-spectrum system

circuit-switched network communication system in which a dedicated channel is set up between parties for the duration of the communication

citizens' band (CB) radio short-distance unlicensed radio communication system

codec (coder-decoder) device that converts sampled analog signal to and from its PCM or delta modulation equivalent

code-division multiple access (CDMA) system to allow multiple users to use the same frequency using separate PN codes and a spread-spectrum modulation scheme

coding conversion of a sampled analog signal into a PCM or delta modulation bitstream

collision attempt by two transmitters to use the same channel simultaneously

common-channel signaling use of a separate signaling channel in a telephone system, so that voice channels do not have to carry signaling information

companding combination of compression at the transmitter and expansion at the receiver of a communication system

compression deliberate reduction of the dynamic range of a signal by applying more gain to lower-amplitude than to higher-amplitude components

constellation in satellite telephony, a group of satellites coordinated in such a way as to provide continuous communication

constellation diagram in digital communication, a pattern showing all the possible combinations of amplitude and phase for a signal

control mobile attenuation code (CMAC) information sent by the base station in a cellular radio system to set the power level of the mobile transmitter

crosslink a radio or optical connection directly between satellites, without going through an earth station

crosstalk interference between two signals multiplexed into the same channel

D (data) channel in ISDN, a communication channel used for setting up calls and not for user communication

decoding conversion of a PCM or delta modulation bitstream to analog samples

delta modulation a coding scheme that records the change in signal level since the previous sample

delta phase-shift keying (DPSK) digital modulation scheme that represents a bit pattern by a change in phase from the previous state

demodulation recovery of a baseband signal from a modulated signal

deviation in FM, the peak amount by which the instantaneous signal frequency differs from the carrier frequency in each direction

dibit system any digital modulation scheme that codes two bits of information per transmitted symbol

digital color code (DCC) signal transmitted by a cell site to identify that site to the mobile user

digital signal processing (DSP) filtering of signals by converting them to digital form, performing arithmetic operations on the data bits, then converting back to analog form

dipole any antenna with two sections

direct-sequence spread spectrum technique for increasing the bandwidth of a transmitted signal by combining it with a pseudo-random noise signal with a higher bit rate

directional coupler device that allows signals to travel between ports in one direction only

directivity gain of an antenna with losses ignored

dispersion pulse spreading caused by variation of propagation velocity with frequency

downlink signal path from a base station or a satellite to a mobile station or a ground station

driver amplifier immediately preceding the power amplifier stage in a transmitter

dropped call a telephone connection that is unintentionally terminated while in progress

dual-tone multi-frequency (DTMF) dialing signaling using combinations of two audio tones transmitted on the voice channel

duplexer combination of filters to separate transmit and receive signals when both use the same antenna simultaneously

dynamic range ratio, usually expressed in decibels, between the strongest and weakest signals that can be present in a system

effective isotropic radiated power (EIRP) transmitter power that would, if used with an isotropic radiator, produce the same power density in a given direction as a given transmitting installation

effective radiated power (ERP) transmitter power that would, if used with a lossless dipole oriented for maximum gain, produce the same power density in a given direction as a given transmitting installation

electronic serial number (ESN) number assigned to a cell phone by the manufacturer as a security feature. It is transmitted with the phone's telephone number to authorize a call.

element an antenna used as part of an array

end office see **central office**

envelope imaginary pattern formed by connecting the peaks of individual RF waveforms in an amplitude-modulated signal

equatorial orbit a satellite orbit that is entirely above the equator

Ethernet form of local-area network using CSMA/CD and a logical bus structure

expansion restoration of the original dynamic range to a previously-compressed signal by applying more gain to higher-amplitude components than to lower-amplitude components

fading reduction in radio signal strength, usually caused by reflection or absorption of the signal

far-field region distance from an antenna great enough to avoid local magnetic or electrical coupling, and great enough for the antenna to resemble a point source

fast associated control channel (FACCH) in a digital cellular system or PCS, control information that is transmitted by "stealing" bits that are normally used for voice information

fiber-in-the-loop (FITL) use of optical fiber for telephone connections to individual customers

flat-topped sampling   sampling of an analog signal using a sample-and-hold circuit, such that the sample has the same amplitude for its whole duration

foldover distortion   see **aliasing**

footprint   depiction of the signal strength contours from a satellite transmitter on the earth

forward channel   communication from a cell site or repeater to the mobile unit

Fourier series   expression showing the structure of a signal in the frequency domain

framing bits   bits added to a digital signal to help the receiver to detect the beginning and end of data frames

frequency diversity   use of two or more RF channels to transmit the same information, usually done to avoid fading

frequency domain   method of analyzing signals by observing them on a power-frequency plane

frequency hopping   form of spread-spectrum communication in which the RF carrier continually moves from one frequency to another according to a prearranged pseudo-random pattern

frequency modulation   modulation scheme in which the transmitted frequency varies in accordance with the instantaneous amplitude of the information signal

frequency modulation index   peak phase shift in frequency-modulated signal, in radians

frequency multiplier   circuit whose output frequency is an integer multiple of its input frequency

frequency synthesizer   device to produce programmable output frequencies that are accurate and stable

frequency-division multiple access (FDMA)   sharing of a communication channel among multiple users by assigning each a different carrier frequency

frequency-division multiplexing (FDM)   combining of several signals into one communication channel by assigning each a different carrier frequency

frequency-shift keying (FSK)   digital modulation scheme using two or more different output frequencies

full-duplex communication   two-way communication in which both terminals can transmit simultaneously

Gaussian minimum-shift keying (GMSK)   variant of FSK which uses the minimum possible frequency shift for a given bit rate

geostationary orbit   satellite orbit in which the satellite appears to remain stationary at a point above the equator

geosynchronous orbit   satellite orbit in which the satellite's period of revolution is equal to the period of rotation of the earth

group velocity   speed at which signals move down a transmission line or waveguide

half-duplex communication   two-way communication system in which only one station can transmit at a time

handoff   transfer of a call in progress from one cell site to another

hemispheric beam   antenna beam on a geostationary satellite that is adjusted to cover the whole earth

hybrid coil   a specialized transformer (or its electronic equivalent) that allows telephone voice signals to travel in both directions simultaneously on a single twisted-pair loop

image frequency reception of a spurious frequency by a superheterodyne receiver, resulting from mixing of the unwanted signal with the local oscillator signal to give the intermediate frequency

Improved Mobile Telephone Service (IMTS) a mobile telephone service, now obsolescent, using trunked channels but not cellular in nature

in-band signals control signals sent in a voice channel at voice frequencies

inbound channel communication channel from mobile to base station

in-channel signals control signals using the same channel as a voice signal

integrated services digital network (ISDN) telephone system using digital local loops for both voice and data, with the codec in the telephone equipment

intelligence information to be communicated

interleaving changing the time order of digital information before transmission to reduce the effect of burst errors in the channel

International Mobile Subscriber Identification (IMSI) in the GSM system, a telephone number that is unique to a given user, worldwide

isotropic radiator an antenna that radiates all power applied to it, equally in all directions. It is a theoretical construct and not a practical possibility.

justification addition of bits to a digital signal to compensate for differences in clock rates; informally known as *bit stuffing*.

local access and transport area (LATA) in a telephone system, the area controlled by one central office switch

local area network a small data network, usually confined to a building or cluster of buildings

local loop in a telephone system, the wiring from the central office to an individual customer

local multipoint distribution system (LMDS) network using microwaves for two-way transmission of telephony, television, and high-speed data

low-earth-orbit (LEO) satellite an artificial satellite orbiting the earth at an altitude less than about 1500 kilometers

mark in digital communication, a logic one

medium earth orbit (MEO) satellite a satellite in orbit at a distance above the earth's surface of between approximately 8,000 and 20,000 km

microcell in cellular radio, a small cell designed to cover a high-traffic area

microstrip transmission line consisting of a circuit-board trace on one side of a dielectric substrate (the circuit board) and a ground plane on the other side

mixer nonlinear device designed to produce sum and difference frequencies when provided with two input signals

mobile identification number (MIN) number that identifies a mobile phone in a cellular system; the mobile telephone number

mobile switching center (MSC) switching facility connecting cellular telephone base stations to each other and to the public switched telephone network

mobile telephone switching office (MTSO) see **mobile switching center (MSC)**

modem acronym for modulator-demodulator; device to enable data to be transmitted via an analog channel

modulating signal the information signal that is used to modulate a carrier for transmission

modulation process of using an information signal to vary some aspect of a higher-frequency signal for transmission

modulation index number indicating the degree to which a signal is modulated

monopole an antenna with only one conductor, generally using ground or a ground plane to represent a second conductor

multi-channel multipoint distribution system (MMDS) terrestrial microwave system for the distribution of television, internet, and telephone services to businesses and residences

multipath distortion distortion of the information signal resulting from the difference in arrival time in signals arriving via multiple paths of different lengths

multiple access use of a single channel by more than one transmitter

multiplexing use of a single channel by more than one signal

natural sampling sampling of an analog signal so that the sample amplitude follows that of the original signal for the duration of the sample

near/far effect in a spread-spectrum system, the tendency for stronger signals to interfere with the reception of weaker signals

near-field region the region of space close to an antenna, where the radiation pattern is disturbed by induced, as well as radiated, electric and magnetic fields

network an organized system for communicating among terminals

noise an unwanted random signal that extends over a considerable frequency spectrum

noise power density the power in a one-hertz bandwidth due to a noise source

number assignment module (NAM) in a cellular phone, a memory location that stores the telephone number(s) to be used on the system

orbital satellite any satellite that is not in a geostationary orbit

outbound channel a radio channel used for communication from a base station to mobile stations

out-of-band in telephone signaling, a control signal that is outside the voice frequency range

overmodulation modulation to an extent greater than that allowed for either technical or regulatory reasons

packet-switched network a communication system that works using data divided into relatively short transmissions called packets; these are routed through the system without requiring a long-term connection between sender and receiver

perigee the point in a satellite orbit that is closest to the earth

personal communication system (PCS) a cellular telephone system designed mainly for use with portable (hand-carried) telephones

phase modulation communication system in which the phase of a high-frequency carrier is varied according to the amplitude of the baseband (information) signal

phase velocity the apparent velocity of waves along the wall of a waveguide

phase-shift keying (PSK) digital modulation scheme in which the phase of the transmitted signal is varied in accordance with the baseband data signal

picocells very small cells in a cellular radio system

point of presence (POP) place where one telephone network connects to another

polar orbit a satellite orbit passing over the north and south poles

private branch exchange (PBX) small telephone switch located on customer premises

processing gain improvement in interference rejection due to spreading in a spread-spectrum system

pseudo-random noise (PN) sequence a transmitted series of ones and zeros that repeats after a set time, and which appears random if the sequence is not known to the receiver

public switched telephone network (PSTN) the ordinary public wireline phone system

pulse-amplitude modulation (PAM) a series of pulses in which the amplitude of each pulse represents the amplitude of the information signal at a given time

pulse-code modulation (PCM) a series of pulses in which the amplitude of the information signal at a given time is coded as a binary number

pulse-duration modulation (PDM) a series of pulses in which the duration of each pulse represents the amplitude of the information signal at a given time

pulse-position modulation (PPM) a series of pulses in which the timing of each pulse represents the amplitude of the information signal at a given time

pulse-width modulation (PWM) see **pulse-duration modulation (PDM)**

quadrature AM (QAM) modulation scheme in which both the amplitude and phase of the transmitted signal are varied by the baseband signal

quadrature phase-shift keying (QPSK) digital modulation scheme using four different transmitted phase angles

quantizing representation of a continuously varying quantity as one of a number of discrete values

quantizing errors inaccuracies caused by the representation of a continuously varying quantity as one of a number of discrete values

quantizing noise see **quantizing errors**

radiation resistance representation of energy lost from an antenna by radiation as if it were dissipated in a resistance

radio common carrier (RCC) a company that acts as a carrier of radiotelephone signals

rake receiver a radio receiver that is capable of combining several received signals with different time delays into one composite signal

Rayleigh fading variation in received signal strength due to multipath propagation

repeater a transmitter-receiver combination used to receive and retransmit a signal

reverse channel communication channel from mobile station to base station

ringback signal in telephony, a signal generated at the central office and sent to the originating telephone to indicate that the destination telephone is ringing

roamer a cellular customer using a network other than the subscriber's local cellular network

run-length encoding method of data compression by encoding the length of a string of ones or zeros instead of transmitting all the one or zero bits individually

sectorization division of a cell into several sections all radiating outward from a cell site

selectivity ability of a receiver to discriminate against unwanted signals and noise

sensitivity ability of a receiver to detect weak signals with a satisfactory signal-to-noise ratio

short messaging service (SMS) transmission of brief text messages, such as pages or e-mail, by a cellular radio or PCS system

side frequencies frequency components produced above and below the carrier frequency by the process of modulation

sideband a group of side frequencies above or below the carrier frequency

sidetone in telephony, the presence in the receiver of sounds picked up by the transmitter of the same telephone

signaling system seven (SS7) system used in telephony which transmits all call setup information on a packet-data network that is separate from the voice channels used for telephone conversations

signal-to-noise ratio ratio between the signal power and noise power at some point in a communication system

simplex a unidirectional communication system, for example, broadcasting

slope overload in delta modulation, an error condition that occurs when the analog signal to be digitized varies too quickly for the system to follow

slow associated control channel (SACCH) in a digital cellular system or PCS, control information that is transmitted along with the voice

smart card a card with an embedded integrated circuit, that can be used for functions such as storing subscriber information for a PCS system

soft handoff in a cellular or PCS system, connecting a mobile to two or more base stations simultaneously

space binary zero

spatial diversity use of two or more physically separated antennas at one end of a communication link

spectrum analyzer test instrument that typically displays signal power as a function of frequency

splatter frequency components produced by a transmitter that fall outside its assigned channel

spot beam in a satellite system, a focused beam of energy that covers a relatively small area on the earth, produced by a high-gain antenna on the satellite

spreading gain improvement in interference rejection due to spreading in a spread-spectrum system

spurious responses in a receiver, reception of frequencies other than that to which it is tuned

spurious signals unwanted signals accidentally produced by a transmitter

star network a computer network topology in which each terminal is connected to a central mainframe or server

station class mark (SCM) code which describes the maximum power output of a cellular phone

store-and-forward repeater in digital communication, a device which receives one or more data packets, stores them, and retransmits them at a later time

stripline transmission line consisting of a circuit board having a ground plane on each side of the board, with a conducting trace in the center

subscriber ID module (SIM) in the GSM system, a "smart card" containing all user information, which is inserted into the phone before use

superheterodyne principle use of a mixer and local-oscillator combination to change the frequency of a signal

supervisory audio tone (SAT) in the AMPS system, a sine wave above the voice frequency range, transmitted on the voice channel along with the voice, used by the base station to detect loss of signal

symbol in digital communication, the state of the signal at a sampling time

system identification number (SID) in the AMPS system, a number transmitted by the base station to identify the system operator

tandem office telephone switch that connects only to other switches, and not to individual customers

telepoint a very small cell used with some cordless phones to allow their use in public areas

time-division duplexing (TDD) transmission of data in two directions on a channel by using different time slots for each direction

time-division multiple access (TDMA) system to allow several transmissions to use a single channel by assigning time slots to each

time-division multiplexing (TDM) system to combine several data streams onto a single channel by assigning time slots to each

time domain representation of a signal as a function of time and some other parameter, such as voltage

tracking in satellite communication, continuously adjusting the position of a directional antenna on the ground, so that it always points at the satellite

transceiver a combination transmitter and receiver

transponder in satellite communication, a repeater located on a satellite

trunk lines transmission line carrying many signals, either on multiple pairs or multiplexed together on a single twisted-pair, coaxial cable, or optical fiber

uplink transmission channel to a satellite or base station

variable-frequency oscillator (VFO) an oscillator whose frequency can be changed easily, usually by means of a variable capacitor or inductor

vocoder circuit for digitizing voice at a low data rate by using knowledge of the way in which voice sounds are produced

voltage standing-wave ratio (VSWR) ratio between maximum and minimum peak or rms voltage on a transmission line

voltage-controlled oscillator (VCO) oscillator whose frequency can be changed by adjusting a control voltage

Walsh code class of orthogonal spreading codes used in CDMA communication

waveguide metallic tube down which waves propagate

wavelength distance a wave travels in one period

white noise noise containing all frequencies with equal power in every hertz of bandwidth

Index